METHODS IN MOLECULAR BIOLOGY™

For further volumes:
http://www.springer.com/series/7651

Cyclic Nucleotide Signaling in Plants

Methods and Protocols

Edited by

Chris Gehring

Division of Chemical and Life Sciences and Engineering,
King Abdullah University of Science and Technology, Thuwal, Saudi Arabia

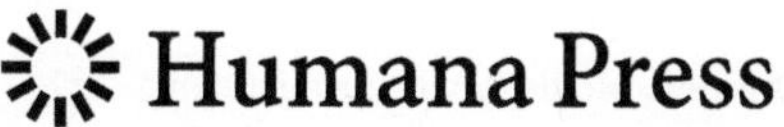

Editor
Chris Gehring
Division of Chemical and Life Sciences and Engineering
King Abdullah University of Science and Technology
Thuwal, Saudi Arabia

ISSN 1064-3745 ISSN 1940-6029 (electronic)
ISBN 978-1-62703-440-1 ISBN 978-1-62703-441-8 (eBook)
DOI 10.1007/978-1-62703-441-8
Springer New York Heidelberg Dordrecht London

Library of Congress Control Number: 2013937953

Printed on acid-free paper

Humana Press is a brand of Springer
Springer is part of Springer Science+Business Media (www.springer.com)

Preface

Understanding how whole organisms, tissues, and cells perceive and process signals has given rise to the biological discipline of "Signal Transduction." Signal transduction research in turn is rapidly evolving not least due to novel and improved analytical methods that have led to an increase in our understanding of the molecular mechanisms underlying cellular signaling. Progress has been made both at the level of single-component analysis and in vivo imaging that can reveal rapid changes at the cellular level as well as at the systems level where transcriptomics and phosphoproteomics, in particular, afford a window into complex biological responses including long-term adaptive responses.

The last two decades have seen a growing interest in cyclic nucleotide research in plants with an emphasis on the elucidation of the roles of cGMP and, perhaps to a lesser extent, cAMP. Here we detail both established and novel techniques and approaches to better understand the biological role of this important signaling system. Chapter 1 summarizes major trends in plant signal transduction and cyclic nucleotide research with an emphasis on molecular methods. The subsequent chapters cover two major themes. The first is centered around the detection and quantification of cyclic nucleotides and the discovery and characterization of novel nucleotide cyclases as well as experimental procedures to elucidate cyclic nucleotide-dependent cellular processes (Chapters 2–12). The second main theme covers bioinformatic methods to identify candidate nucleotide cyclases and cyclic nucleotide-gated channels. In addition, we also detail a computational method to infer biological functions of candidate nucleotide cyclases (Chapters 13–15).

Further to the above-mentioned themes, one chapter is dedicated to methods for identifying and characterizing cyclic nucleotide phosphodiesterases that obviously play an important part in cyclic nucleotide signaling and cyclic nucleotide homeostasis (Chapter 16). Additionally, two chapters on the measurement of reactive oxygen species and nitric oxide in plant tissues have been included since these compounds are critical components of biotic and abiotic plant stress responses and are associated with cyclic nucleotide transients as well as downstream responses (Chapters 17, 18). The final chapter (Chapter 19) details a method that allows the quantification of photosynthetic responses to cyclic nucleotides.

Thuwal, Saudi Arabia *Chris Gehring*

Preface

Contents

Contributors

RASHID ALI • *Center for Vascular Biology, University of Connecticut Health Center, Farmington, CT, USA*

MAY ALQURASHI • *Division of Chemical and Life Sciences and Engineering, King Abdullah University of Science and Technology, Thuwal, Saudi Arabia*

VAGNER BENNEDITO • *Genetics and Developmental Biology Program, Division of Plant and Soil Sciences, West Virginia University, Morgantown, WV, USA*

GERALD A. BERKOWITZ • *Agricultural Biotechnology Laboratory, Department of Plant Science, University of Connecticut, Storrs, CT, USA*

HEIKE BURHENNE • *Research Core Unit for Mass Spectrometry - Metabolomics, Institute of Pharmacology, Hannover Medical School, Hannover, Germany*

PATIENCE CHATUKUTA • *Department of Biological Sciences, School of Environmental and Health Sciences, North-West University, Mmabatho, South Africa*

ADAM DAWE • *Division of Computer, Electrical and Mathematical Sciences and Engineering, King Abdullah University of Science and Technology, Thuwal, Saudi Arabia*

BRIDGET T. DIKOBE • *Department of Biological Sciences, School of Environmental and Health Sciences, North-West University, Mmabatho, South Africa*

LARA DONALDSON • *Division of Chemical and Life Sciences and Engineering, King Abdullah University of Science and Technology, Thuwal, Saudi Arabia*

LYUDMILA V. DUBOVSKAYA • *Institute of Biophysics and Cell Engineering, National Academy of Sciences of Belarus, Minsk, Belarus*

CHRIS GEHRING • *Division of Chemical and Life Sciences and Engineering, King Abdullah University of Science and Technology, Thuwal, Saudi Arabia*

RICHARD M. GRAEFF • *Department of Physiology, The University of Hong Kong, Hong Kong, China*

ARNOUD GROEN • *Department of Biochemistry, Cambridge Centre for Proteomics, Cambridge Systems Biology Centre, University of Cambridge, Cambridge, UK*

HELEN R. IRVING • *Monash Institute of Pharmaceutical Sciences, Monash University, Parkville, VIC, Australia*

JEAN-CHARLES ISNER • *Guard Cell Group, University of Bristol, Bristol, UK*

VOLKHARD KAEVER • *Research Core Unit for Mass Spectrometry - Metabolomics, Institute of Pharmacology, Hannover Medical School, Hannover, Germany*

YUN KANG • *Plant Biology Division, Samuel Roberts Noble Foundation, Ardmore, OK, USA*

DAVID T. KAWADZA • *Department of Biological Sciences, School of Environmental and Health Sciences, North-West University, Mmabatho, South Africa*

ALEYSIA KLEINERT • *Botany and Zoology Department, Faculty of Science, University of Stellenbosch, Matieland, South Africa*

LUSISIZWE KWEZI • *Department of Biological Sciences, School of Environmental and Health Sciences, North-West University, Mmabatho, South Africa*
HON CHEUNG LEE • *Department of Physiology, The University of Hong Kong, Hong Kong, China*
FOUAD LEMTIRI-CHLIEH • *Division of Chemical and Life Sciences and Engineering, King Abdullah University of Science and Technology, Thuwal, Saudi Arabia*
KATHRYN LILLEY • *Department of Biochemistry, Cambridge Centre for Proteomics, Cambridge Systems Biology Centre, University of Cambridge, Cambridge, UK*
NDIKO LUDIDI • *Department of Biotechnology, University of the Western Cape, Belville, South Africa*
FRANS J.M. MAATHUIS • *Biology Department, University of York, York, UK*
GRACE H. MABADAHANYE • *Department of Biological Sciences, School of Environmental and Health Sciences, North-West University, Mmabatho, South Africa*
MARTIN R. MCAINSH • *Lancaster Environment Centre, Lancaster University, Lancaster, UK*
CLAUDIUS MARONDEDZE • *Division of Chemical and Life Sciences and Engineering, King Abdullah University of Science and Technology, Thuwal, Saudi Arabia*
STUART MEIER • *Division of Chemical and Life Sciences and Engineering, King Abdullah University of Science and Technology, Thuwal, Saudi Arabia*
VICTOR MULEYA • *Monash Institute of Pharmaceutical Sciences, Monash University, Parkville, VIC, Australia*
NATALIA MARIA ORDOÑEZ • *Division of Chemical and Life Sciences and Engineering, King Abdullah University of Science and Technology, Thuwal, Saudi Arabia*
ERIN B. PURCELL • *Department of Microbiology and Immunology, University of North Carolina at Chapel Hill, Chapel Hill, NC, USA*
STEPHEN K. ROBERTS • *Biomedical and Life Sciences Division, Lancaster University, Lancaster, UK*
OZINIEL RUZVIDZO • *Department of Biological Sciences, School of Environmental and Health Sciences, North-West University, Mmabatho, South Africa*
LANA SHABALA • *School of Agricultural Science, University of Tasmania, Hobart, TAS, Australia*
SERGEY SHABALA • *School of Agricultural Science, University of Tasmania, Hobart, TAS, Australia*
RITA TAMAYO • *Department of Microbiology and Immunology, University of North Carolina, Chapel Hill, NC, USA*
LUDIVINE THOMAS • *Division of Chemical and Life Sciences and Engineering, King Abdullah University of Science and Technology, Thuwal, Saudi Arabia*
ALEX VALENTINE • *Botany and Zoology Department, Faculty of Science, University of Stellenbosch, Matieland, South Africa*
ROBIN K. WALKER • *Department of Physiology and Biophysics, College of Medicine, Howard University, Washington, DC, USA*
JANET I. WHEELER • *Monash Institute of Pharmaceutical Sciences, Monash University, Parkville, VIC, Australia*
ALOYSIUS WONG • *Division of Chemical and Life Sciences and Engineering, King Abdullah University of Science and Technology, Thuwal, Saudi Arabia*
ALICE K. ZELMAN • *Agricultural Biotechnology Laboratory, Department of Plant Science, University of Connecticut, Storrs, CT, USA*

Chapter 1

Molecular Methods for the Study of Signal Transduction in Plants

Helen R. Irving and Chris Gehring

Abstract

Novel and improved analytical methods have led to a rapid increase in our understanding of the molecular mechanism underlying plant signal transduction. Progress has been made both at the level of single-component analysis and in vivo imaging as well as at the systems level where transcriptomics and particularly phosphoproteomics afford a window into complex biological responses. Here we review the role of the cyclic nucleotides cAMP and cGMP in plant signal transduction as well as the discovery and biochemical and biological characterization of an increasing number of complex multi-domain nucleotide cyclases that catalyze the synthesis of cAMP and cGMP from ATP and GTP, respectively.

Key words Plant signal transduction, Receptors, Second messengers, In vivo imaging, Systems analysis, cAMP, cGMP, Nucleotide cyclases

1 Signal Transduction: From Single-Component Imaging to Systems Approaches

The perception of environmental and intracellular stimuli or signals is an essential feature of living systems. When such signals are received, they need to be processed and transmitted in a temporally and spatially controlled manner so that the cells can respond appropriately in the short term, for instance with change of ion channel activities, and/or adapt with an altered transcriptional and translational program. An ever-increasing body of research centers around the question of signal processing and transduction in biological systems. The methods used in these studies have greatly advanced in the last few decades both at the level of the study of structural and functional features of single components or interactions between individual molecules as well as at the systems level. NMR and X-ray crystallography studies as well as molecular imaging have and will yield increasingly detailed information about the molecular mechanism of signaling processes while bioinformatic approaches and system-level studies, particularly phosphoproteomic, will continue to enable the study of signaling processes at

Chris Gehring (ed.), *Cyclic Nucleotide Signaling in Plants: Methods and Protocols*, Methods in Molecular Biology, vol. 1016,
DOI 10.1007/978-1-62703-441-8_1,

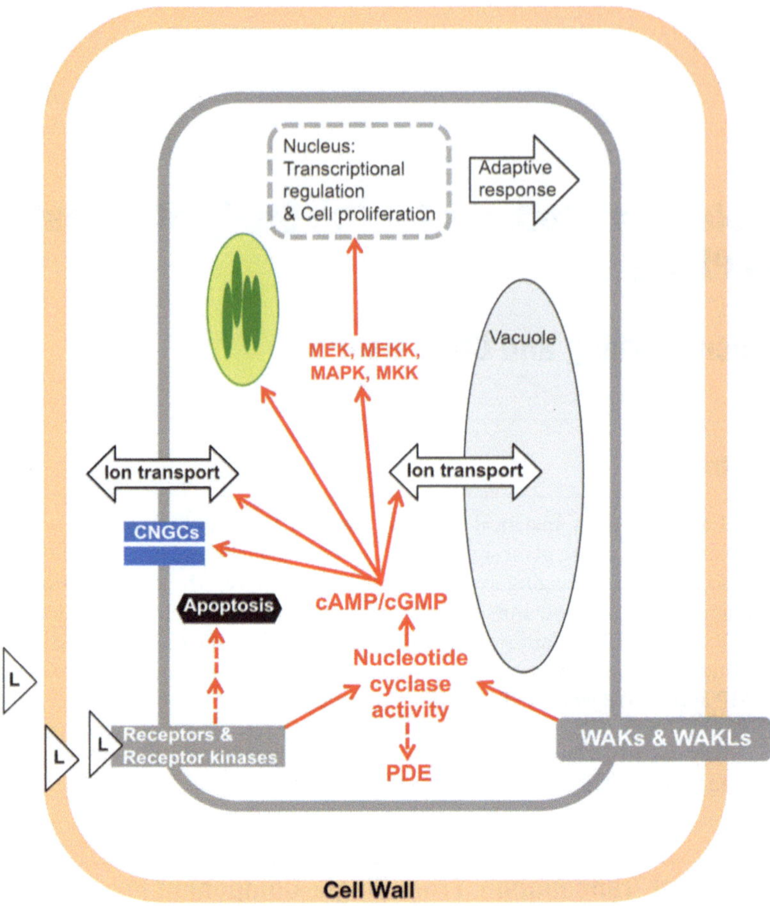

Fig. 1 Overview of nucleotide cyclase (NC)- and cyclic nucleotide (CN)-dependent pathways

the whole cell, tissue, or organ level. Given that signaling is a dynamic process, in vivo imaging has an important role in signal transduction research, and real-time resolution of second messengers such as calcium (Ca^{2+}) and cyclic nucleotides continues to provide new insights into complex signaling events.

Here we give a brief overview of two key second messenger systems, cytosolic Ca^{2+} and the cyclic nucleotides (CNs) cGMP and cAMP since they are critical to and cooperative in many plant responses including reactions to hormones and biotic and abiotic stresses (Fig. 1). We also briefly review adenylate cyclases (ACs) and guanylate cyclases (GCs), the enzymes that catalyze the reaction from ATP to cAMP and GTP to cGMP, respectively, and argue that these nucleotide cyclases (NCs), many of which are part of multi-domain receptor molecules, are increasingly recognized as having a key role in plant signal transduction. The emphasis of this overview is on technical advances that have and will continue to enable the molecular characterization of these messenger systems.

2 Ca^{2+}, the Classic Messenger in Plant Signal Transduction: Methods and Challenges

The last three decades have seen a revolution in the understanding of Ca^{2+} signaling in plants mainly because of the establishment of improved imaging systems and biotechnological reporter systems. Calcium is distributed throughout the plant and is an essential nutrient for plant growth with a major role in stabilizing cell walls where it binds pectin and calcium levels vary greatly in the different cellular compartments [1]. The vacuole forms the main storage center of cellular calcium and levels of total calcium can reach 80 mM where most is stored as bound calcium (e.g., calcium malate, citrate, oxalate, etc.) while vacuolar free calcium is in the range of 0.2–5 mM. Calcium is also relatively high in cell walls (~1 mM total and 0.33 mM free Ca^{2+}) [1]. A large amount of total Ca^{2+} (~15 mM) can also be stored in chloroplasts where it is mainly bound to thylakoid membranes and stromal proteins, but free Ca^{2+} is ~100,000-fold lower at ~150 nM to enable photosynthesis [1]. Free Ca^{2+} levels in the cytoplasm, nucleus, and mitochondria are also in this order of magnitude at ~100, 100, and 200 nM, respectively [1]. Based on functional homology with other eukaryotes it is likely that the protein calreticulin in endoplasmic reticulum binds Ca^{2+} bringing total Ca^{2+} to mM levels [1]. Obviously, the maintenance of these diverse and often largely different levels of Ca^{2+} across cellular compartments requires coordinated regulation of calcium-binding proteins and, more to the point, intact membranes containing Ca^{2+}-permeable channels and transporters. Hirschi [2] elegantly illustrated the dichotomy between Ca^{2+} levels required as a nutrient for plant growth and development and those required for its role in signal transduction. Changes in Ca^{2+} are transient and typically raise cytoplasmic Ca^{2+} two- to fivefold before returning to basal levels as prolonged exposure to high cytoplasmic Ca^{2+} triggers cell death [2]. Cytoplasmic Ca^{2+} increases are stimulated by a wide range of biotic and abiotic environmental stresses and in part mediated by increases in hormonal levels such as abscisic acid, auxin and gibberellic acid, etc. Due to the range of responses stimulating increases in cytoplasmic Ca^{2+} it is now recognized that various signature profiles of Ca^{2+} increases are occurring from single to repeated spikes (oscillations) to plateaus, in addition to changes restricted to specific cellular locations [3, 4]. This challenging situation has driven the development of a variety of methods to measure Ca^{2+} from monitoring ion movement across membranes to mapping the changes in the cytoplasm and visualizing changes in the whole plant which are briefly reviewed below.

Ion movement has been monitored by measuring fluxes in $^{45}Ca^{2+}$ but this is fraught with problems due to Ca^{2+} binding to cell wall components (mainly pectin) [5]. Ion-sensitive microelectrodes have been used to measure changes in intracellular calcium in epidermal cells and root hairs [6, 7]. In addition, electrophysiological approaches including patch clamping have resulted in the characterization of many Ca^{2+}-permeable channels in the plasma, vacuolar, and other membranes [3, 4]. Although these procedures result in characterization of channels in their native membranes, they are restricted to single-cell measurement without the normal constraints of the cell wall. Fluxes of several ions including Ca^{2+} can be followed simultaneously in intact tissue such as roots or leaf surfaces using the Microelectrode Ion Flux Electrode (MIFE) system and this can also be coupled to whole-cell patch clamping of isolated protoplasts [8, 9]. Intracellular changes in Ca^{2+} have been detected by fluorescent dyes such as Fluo-3 but it is challenging to get these dyes past the cell wall and plasma membrane and they also can easily enter different intracellular compartments.

These constraints have led to the development of various reporter systems designed for the visualization of changes in cytoplasmic Ca^{2+} in real time using genetically encoded calcium indicators that typically involve either aequorin or fluorescent protein (FP) derivatives and these recombinant methods have also been accompanied by improved microscopy and image acquisition systems. Aequorin was first isolated from luminescent jelly fish (e.g., *Aequorea victoria*) and is a luminescent photoprotein that emits blue light (emission peak: 470 nm) due to nM amounts of Ca^{2+} triggering oxidation of its chromophore coelenterazine. Knight and colleagues have pioneered the use of aequorin-expressing transgenic plants to report cytoplasmic changes in Ca^{2+} following exposure to various environmental triggers that can be detected in whole seedlings using luminometers [10, 11]. One of the advantages of using aequorin reporters is that it overcomes some of the problems associated with autofluorescent compounds such as lignin, chlorophyll, phenolics, and callose found in plant tissues. More recently, they have developed a compatible set of aequorin expression vectors that target the reporter to specific subcellular regions such as the cytosol, nucleus, mitochondria, plasma membrane, and various parts of chloroplasts [12]. Since these vectors code for YFP–aequorin fusions, the subcellular localization of the reporters can be verified using fluorescent microscopy [12]. YFP is usually excited at ~490–515 nm and since chlorophyll fluorescence is also stimulated at these wavelengths, it is important to apply emission filters that block chlorophyll emission (e.g., 520–550 nm will detect YFP fluorescence while chlorophyll fluorescence is seen at >650 nm). FP sensors are usually derivatives of green FP (GFP) which have a distinctive β-barrel structure that protects the chromophore. In general, the FP sensors are either single FP-based or fluorescence resonance energy transfer (FRET)-based sensors and the uses and limitations of such sensors in plants have recently been comprehensively reviewed [13, 14]. Single FP sensors contain one

FP that is sometimes circularly permuted to have a sensor protein/peptide sequence (e.g., a calcium-binding domain) inserted in the middle of the β-barrel sequence with suitable linkers so that the barrel can reassemble. Ca^{2+} binding to the calcium-binding domain causes allosteric changes in protein structure that stimulates changes in fluorescence [13, 15]. FRET-based sensors on the other hand contain a donor–acceptor pair and a ligand-binding domain such as calmodulin for Ca^{2+} sensors. Allosteric conformational changes occur upon ligand binding that result in changes in FRET efficiency that can be detected. The most commonly used FRET-based Ca^{2+} biosensor in plants are members of the yellow cameleon (YC) group which contain cyan FP (CFP) linked via calmodulin and mammalian myosin light-chain kinase that binds calmodulin to YFP to form a single protein with a Ca^{2+}-binding hinge in the middle [13, 14]. Such FRET-based FP sensors have been used to great effect in high-resolution imaging to reveal changes in cytoplasmic Ca^{2+} beginning with the resolution of Ca^{2+} signatures controlling guard cell closure [4, 14, 16]. Importantly, more detailed analysis of Ca^{2+} dynamics will be achievable using new vector constructs that target the YC reporters to specific subcellular locations [17].

3 A Brief History of Cyclic Nucleotide Research in Plants

While in lower eukaryotes and animals cAMP and ACs have long been established as key components and second messengers in many signaling pathways, in plants, both the presence and biological role of cAMP have been a matter of ongoing debate and some controversy [18] particularly since CN transients in plants appear to be about an order of magnitude smaller than in animal systems.

Arguably the most convincing data for a specific signaling role for cAMP came from whole-cell patch-clamp recordings from *Vicia faba* mesophyll protoplasts that revealed that outward K^+ current increased in a dose-dependent fashion following intracellular application of cAMP but not AMP, cGMP, or GMP [19] and cAMP-dependent up-regulation of a calcium-permeable conductance activated by hyperpolarization [20]. In addition, plants have also been shown to have cyclic nucleotide-gated channels (CNGCs) [21, 22], one of which (AtCNGC2) may play a key role in innate immunity by facilitating Ca^{2+} currents linking them to nitric oxide (NO) production which in turn is critical for the hypersensitive response (HR) [23]. CNs also have important roles in abiotic stress responses and in particular responses to NaCl stress [10, 24] and it has been demonstrated that voltage-independent channels (VICs) in *Arabidopsis thaliana* roots have open probabilities sensitive to μmolar concentrations of cAMP or cGMP at the cytoplasmic side [24]. We are also beginning to see some progress at the systems level where the first cGMP-induced Arabidopsis root transcriptome has become available [25]. In this first study monovalent cation transporter-encoding genes encoding nonselective ion channels and cation:proton antiporters were reported to be transcriptionally regulated by cGMP and thereby linking Na^+ and K^+ homeostasis to increased channel densities. More recently we have seen proteomic and phosphoproteomic analyses of rice shoot and root tonoplast-enriched and plasma membrane-enriched membrane fractions [26] which are beginning to yield further insight into tissue-specific roles of phosphorylation in the activation of transporters, ion channels, and aquaporins. Phosphoproteomics has also been employed to discover new cGMP-dependent phosphorylation targets with a view to link hormone-induced cGMP transients to hormone signal transduction pathways at the systems level [27].

4 Nucleotide Cyclases: A Complex and Growing Family of Enzymes

While the CNs are increasingly accepted as key components in many plant responses, it is somewhat surprising that the search for NCs and the functional characterization of these candidate molecules in higher plants are only just beginning, particularly so since in *Chlamydomonas reinhardtii* there are >90 annotated NCs that come in >20 different domain combinations with 13 different partners suggesting that these NCs could generate CNs and thereby transmit or modulate diverse and complex signals. However, none of the Chlamydomonas NCs has been functionally characterized, neither in vivo nor in vitro. The structural diversity and complexity of molecules with NC activity are a likely reason why BLAST searches with known NCs from lower and higher eukaryotes have not yielded candidate molecules in higher plants [28]. A search strategy based on conserved amino acid residues in the catalytic center of known NCs [28] has now opened the way to a systematic search of NCs in higher plants and has led to the discovery of a number of *A. thaliana* candidate molecules with catalytic activity in vitro and/or in vivo. These molecules include a wall-associated kinase-like protein (WAKL10) with a role in defense [29], the brassinosteroid receptor (AtBRI1) [30], the Pep1 receptor (AtPepR1) [31], and the phytosulfokine receptor (PSKR) [32] and an NO sensing molecule (AtNOGC1) [33]. PSKR belongs to a family of NCs that contains the GC catalytic center embedded within the intracellular kinase domain of leucine-rich repeat receptor-like molecules and in vitro experiments have shown that both the kinase and GC domain have catalytic activity. It was therefore proposed that kinase-GCs are examples of moonlighting proteins with dual-catalytic function [34]. The natural ligands for both the PSKR and BRI1 receptors increase intracellular cGMP levels in isolated mesophyll protoplast assays suggesting that the GC activity is functionally relevant [32, 34].

While the role of cGMP and GCs in higher plants is now firmly established, it is considerably less so for cAMP and particularly ACs [18]. To date the only experimentally confirmed AC in plants is a *Zea mays* pollen protein [35] with a role in polarized pollen tube growth. The Arabidopsis orthologue (At3g14460) is a disease resistance protein that belongs to the family of nucleotide-binding site-leucine-rich repeat (NBS-LRR) proteins used for pathogen sensing [36]. This molecule does indeed contain an AC catalytic center in the cytosolic domain and may be considered a candidate AC that could conceivably link pathogen sensing to cAMP transients.

5 Current Challenges in Cyclic Nucleotide Research

While changes in cellular CN concentrations have been associated with, e.g., responses to hormones [27], signaling peptides [37], virulent and avirulent pathogens [38], and salinity and drought stress [10], the cellular and molecular mechanisms that underlie these changes have remained a matter of speculation. The main reason for this is that the specific NCs that generate these changes in CN concentrations in response to the above triggers are not yet known or are insufficiently characterized. There is however one exception, the response to phytosulfokines (PSKs). In this system we know both the ligand, the phytosulfokines that are sulfated pentapeptides, and their receptor (PSKR) (Fig. 2). The latter is a leucine-rich repeat receptor-like kinase with a functional GC catalytic center embedded within the functional kinase domain [32]. Importantly, it was demonstrated that the sulfated PSK causes intracellular cGMP increase in isolated mesophyll protoplasts expressing native levels of PSKR suggesting that ligand-specific activation of the receptor GCs may have a direct role in cGMP-dependent signal transduction [32]. More recently, specific hormone-dependent GC activation has also been observed in the brassinosteroid receptor AtBRI1 [34]. The BRI1 receptor is relatively well studied and known to form homodimers and also to interact with other receptor-like molecules such as BRI1-associated kinase (BAK1) [39]. The model proposed in Fig. 2 has been informed by the documented dimerization events observed with BRI1 and also the inhibitory effect of cGMP on PSKR kinase activity. Given the complex structure of this class of leucine-rich repeat receptor-like kinases with GC activity, a whole host of questions have arisen. Firstly, do these two domains interact with each other and is cGMP production (auto-) regulating kinase activity and thus contributing to downstream responses? Secondly, is cGMP directly responsible for ligand-dependent transient changes in Ca^{2+}? Thirdly, are ligand-induced Ca^{2+}

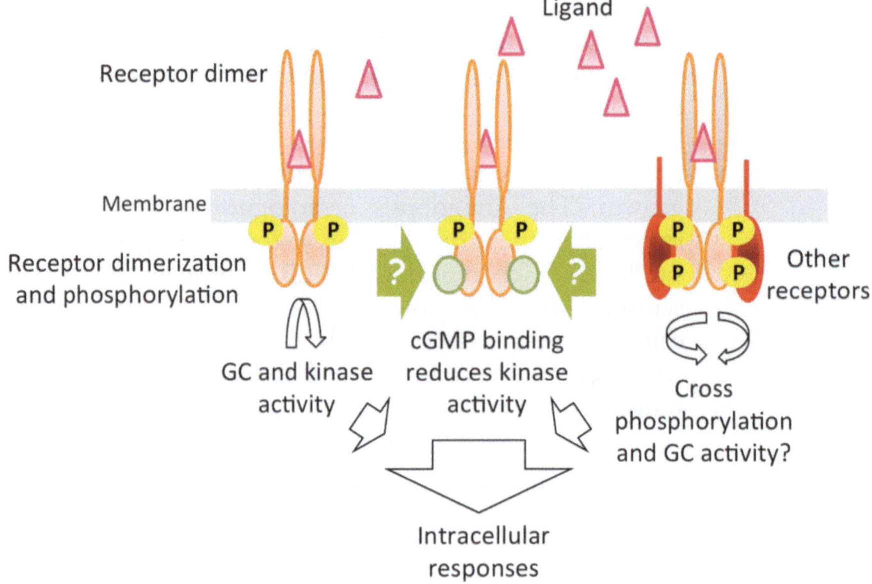

Fig. 2 Proposed model of ligand-induced cGMP-dependent signaling based on phytosulfokine and brassinosteroid receptors

transients dependent on cGMP and/or is Ca^{2+} itself affecting kinase and/or GC activity?

In addition, and much like in animals, NO has also been established as a key signaling molecule in plants [40], particularly in biotic stress responses where it has been shown to signal via cGMP [41, 42]. It has also been demonstrated recently that some NO-dependent transcriptionally activated genes show critical dependence on cGMP [43]. However, contrary to animal systems, where the molecular mechanisms of NO-dependent activation of soluble NC are well understood [44], the situation in plants is much less clear and the single identified plant GC that binds NO selectively with some NO-dependent GC activity has only been tested in vitro [33].

6 Outlook

It is likely that an increasing number of candidate NCs in higher plants will be discovered and characterized both in vitro and in vivo. In vitro studies will afford new insights into the role of Ca^{2+} on NC activity and will also allow the elucidation of the role of cAMP and/or cGMP on the auto-phosphorylation of the receptor kinases that harbor NC domains. These experiments will be complemented with NMR and X-ray crystallography studies that will yield novel insights into the structural features that underlie the function of NC-receptor kinases and other NCs. In turn, phosphoproteomics is expected to reveal new downstream targets of kinases and give new insights into the phospho-relay that links environmental stimuli to both a short-term response and a transcriptional response. Recent results have confirmed that cGMP is involved in plant hormone signaling and that it does so by altering protein phosphorylation in *A. thaliana* root [27]. It is noteworthy that rapidly phosphorylated proteins include a member of the plasma membrane intrinsic protein family (PIP) that functions as an aquaporin in salt-responses as well as signaling molecules themselves such as a leucine-rich repeat protein kinase and a calcium-dependent protein kinase (CDPK19) [27]; the former presumably enables a rapid physiological response by changing membrane H_2O permeability while the latter suggests that an intermediary signaling step may have been activated.

In vivo signal transduction studies will benefit from new and/or improved indicator dyes that can visualize and quantitate second messenger transients in response to, e.g., hormones and cellular perturbations in a nondestructive way. Recently a delta-FlincG reporter system has been developed to visualize NO- and gibberellic acid-dependent cGMP responses in living plant cells [45] and it is likely that a similar system will be developed for the analysis of cAMP-generating responses in plant cells.

Finally, the Arabidopsis model system is an extensive resource for plant signal transduction research for a number of reasons, not least the availability of a large amount of well-characterized mutants that allow for the dissection of many components of many signal transduction pathways including those linked to the classical hormones and the increasing number of peptidic signaling molecules.

References

1. Stael S, Wursinger B, Mair A, Mehlmer N, Vothknecht UC, Teige M (2012) Plant organellar calcium signalling: an emerging field. J Exp Bot 63:1525–1542
2. Hirschi KD (2004) The calcium conundrum. Both versatile nutrient and specific signal. Plant Physiol 136:2438–2442
3. Sanders D, Pelloux J, Brownlee C, Harper JF (2002) Calcium at the crossroads of signaling. Plant Cell 14(Suppl):S401–S417
4. Kudla J, Batistic O, Hashimoto K (2010) Calcium signals: the lead currency of plant information processing. Plant Cell 22:541–563
5. Reid RJ, Smith FA (1992) Measurement of calcium fluxes in plants using ^{45}Ca. Planta 186:558–566
6. Felle HH, Kondorosi E, Kondorosi A, Schultze M (1999) Elevation of the cytosolic free [Ca^{2+}] is indispensible for the transduction of the nod factor signal in alfalfa. Plant Physiol 121:273–279
7. Felle H (1988) Auxin causes oscillations of cytosolic free calcium and pH in *Zea mays* coleoptiles. Planta 174:495–499
8. Newman IA (2001) Ion transport in roots: measurement of fluxes using ion-selective microelectrodes to characterize transporter function. Plant Cell Environ 24:1–14
9. Gilliham M, Sullivan W, Tester M, Tyerman SD (2006) Simultaneous flux and current measurement from single plant protoplasts reveals a strong link between K^+ fluxes and current, but no link between Ca^{2+} fluxes and current. Plant J 46:134–144
10. Donaldson L, Ludidi N, Knight MR, Gehring C, Denby K (2004) Salt and osmotic stress cause rapid increases in *Arabidopsis thaliana* cGMP levels. FEBS Lett 569:317–320

11. Knight MR, Campbell AK, Smith SM, Trewavas AJ (1991) Transgenic plant aequorin reports the effects of touch and cold-shock and elicitors on cytoplasmic calcium. Nature 352:524–526
12. Mehlmer N, Parvin N, Hurst CH, Knight MR, Teige M, Vothknecht UC (2012) A toolset of aequorin expression vectors for *in planta* studies of subcellular calcium concnetration in *Arabidopsis thaliana*. J Exp Bot 63: 1751–1761
13. Okumoto S (2012) Quantitative imaging using genetically encoded sensors for small molecules in plants. Plant J 70:108–117
14. Choi WG, Swanson SJ, Gilroy S (2012) High-resolution imaging of Ca^{2+}, redox status, ROS and pH using GFP biosensors. Plant J 70:118–128
15. Souslova EA, Chudakov DM (2007) Genetically encoded intracellular sensors based on fluorescent proteins. Biochemstry (Mosc) 72:683–697
16. Allen GJ, Kwak JM, Chu SP, Llopas J, Tsien RY, Harper JF et al (1999) Cameleon calcium indicator reports cytoplasmic calcium dynamics in Arabidopsis guard cells. Plant J 19:735–747
17. Krebs M, Held K, Binder A, Hashimoto K, Den Herder G, Parniske M et al (2012) FRET-based genetically encoded sensors allow high-resolution live cell imaging of Ca^{2+} dynamics. Plant J 69:181–192
18. Gehring C (2010) Adenyl cyclases and cAMP in plant signaling—past and present. Cell Commun Signal 8:15
19. Li WW, Luan S, Schreiber SL, Assmann SM (1994) Cyclic-AMP stimulates K^{+} channel activity in mesophyll-cells of *Vicia faba*. Plant Physiol 106:957–961
20. Lemtiri-Chlieh F, Berkowitz GA (2004) Cyclic adenosine monophosphate regulates calcium channels in the plasma membrane of Arabidopsis leaf guard and mesophyll cells. J Biol Chem 279:35306–35312
21. Leng Q, Mercier RW, Yao W, Berkowitz GA (1999) Cloning and first functional characterization of a plant cyclic nucleotide-gated cation channel. Plant Physiol 121:753–761
22. Zelman AK, Dawe A, Gehring C, Berkowitz GA (2012) Evolutionary and structural perspectives of plant cyclic nucleotide-gated cation channels. Front Plant Sci 3:95
23. Ali R, Ma W, Lemtiri-Chlieh F, Tsaltas D, Leng Q, von Bodman S et al (2007) Death don't have no mercy and neither does calcium: Arabidopsis CYCLIC NUCLEOTIDE GATED CHANNEL2 and innate immunity. Plant Cell 19:1081–1095
24. Maathuis FJ, Sanders D (2001) Sodium uptake in Arabidopsis roots is regulated by cyclic nucleotides. Plant Physiol 127:1617–1625
25. Maathuis FM (2006) cGMP modulates gene transcription and cation transport in Arabidopsis roots. Plant J 45:700–711
26. Whiteman SA, Nuhse TS, Ashford DA, Sanders D, Maathuis FJ (2008) A proteomic and phosphoproteomic analysis of *Oryza sativa* plasma membrane and vacuolar membrane. Plant J 56:146–156
27. Isner JC, Nuhse T, Maathuis FJ (2012) The cyclic nucleotide cGMP is involved in plant hormone signalling and alters phosphorylation of *Arabidopsis thaliana* root proteins. J Exp Bot 63:3199–3205
28. Ludidi N, Gehring C (2003) Identification of a novel protein with guanylyl cyclase activity in *Arabidopsis thaliana*. J Biol Chem 278: 6490–6494
29. Meier S, Ruzvidzo O, Morse M, Donaldson L, Kwezi L, Gehring C (2010) The Arabidopsis wall associated kinase-like 10 gene encodes a functional guanylyl cyclase and is co-expressed with pathogen defense related genes. PLoS One 5:e8904
30. Kwezi L, Meier S, Mungur L, Ruzvidzo O, Irving H, Gehring C (2007) The *Arabidopsis thaliana* brassinosteroid receptor (AtBRI1) contains a domain that functions as a guanylyl cyclase in vitro. PLoS One 2:e449
31. Qi Z, Verma R, Gehring C, Yamaguchi Y, Zhao Y, Ryan CA et al (2010) Ca^{2+} signaling by plant *Arabidopsis thaliana* Pep peptides depends on AtPepR1, a receptor with guanylyl cyclase activity, and cGMP-activated Ca^{2+} channels. Proc Natl Acad Sci U S A 107: 21193–21198
32. Kwezi L, Ruzvidzo O, Wheeler JI, Govender K, Iacuone S, Thompson PE et al (2011) The phytosulfokine (PSK) receptor is capable of guanylate cyclase activity and enabling cyclic GMP-dependent signaling in plants. J Biol Chem 286:22580–22588
33. Mulaudzi T, Ludidi N, Ruzvidzo O, Morse M, Hendricks N, Iwuoha E et al (2011) Identification of a novel *Arabidopsis thaliana* nitric oxide-binding molecule with guanylate cyclase activity in vitro. FEBS Lett 585: 2693–2697
34. Irving HR, Kwezi L, Wheeler JI, Gehring C (2012) Moonlighting kinases with guanylate cyclase activity can tune regulatory signal networks. Plant Signal Behav 7:201–204

35. Moutinho A, Hussey PJ, Trewavas AJ, Malho R (2001) cAMP acts as a second messenger in pollen tube growth and reorientation. Proc Natl Acad Sci U S A 98:10481–10486
36. DeYoung BJ, Innes RW (2006) Plant NBS-LRR proteins in pathogen sensing and host defense. Nat Immunol 7:1243–1249
37. Gehring CA, Irving HR (2003) Natriuretic peptides—a class of heterologous molecules in plants. Int J Biochem Cell Biol 35:1318–1322
38. Meier S, Madeo L, Ederli L, Donaldson L, Pasqualini S, Gehring C (2009) Deciphering cGMP signatures and cGMP-dependent pathways in plant defence. Plant Signal Behav 4:307–309
39. Clouse SD (2011) Brassinosteroid signal transduction: from receptor kinase activation to transcriptional networks regulating plant development. Plant Cell 23:1219–1230
40. Durner J, Klessig DF (1999) Nitric oxide as a signal in plants. Curr Opin Plant Biol 2:369–374
41. Durner J, Wendehenne D, Klessig DF (1998) Defense gene induction in tobacco by nitric oxide, cyclic GMP, and cyclic ADP-ribose. Proc Natl Acad Sci U S A 95: 10328–10333
42. Clarke A, Desikan R, Hurst RD, Hancock JT, Neill SJ (2000) NO way back: nitric oxide and programmed cell death in *Arabidopsis thaliana* suspension cultures. Plant J 24:667–677
43. Pasqualini S, Meier S, Gehring C, Madeo L, Fornaciari M, Romano B et al (2009) Ozone and nitric oxide induce cGMP-dependent and -independent transcription of defence genes in tobacco. New Phytol 181:860–870
44. Boon EM, Huang SH, Marletta MA (2005) A molecular basis for NO selectivity in soluble guanylate cyclase. Nat Chem Biol 1:53–59
45. Isner JC, Maathuis FJ (2011) Measurement of cellular cGMP in plant cells and tissues using the endogenous fluorescent reporter FlincG. Plant J 65:329–334

Chapter 2

Recombinant Expression and Functional Testing of Candidate Adenylate Cyclase Domains

Oziniel Ruzvidzo, Bridget T. Dikobe, David T. Kawadza, Grace H. Mabadahanye, Patience Chatukuta, and Lusisizwe Kwezi

Abstract

Adenylate cyclases (ACs) are enzymes capable of converting adenosine-5′-triphosphate to cyclic 3′, 5′-adenosine monophosphate (cAMP). In animals and lower eukaryotes, ACs and their product cAMP have firmly been established as important signalling molecules with important roles in several cellular signal transduction pathways. However, in higher plants, the only annotated and experimentally confirmed AC is a *Zea mays* pollen protein capable of generating cAMP. Recently a number of candidate AC-encoding genes in *Arabidopsis thaliana* have been proposed based on functionally assigned amino acids in the catalytic center of annotated and/or experimentally tested nucleotide cyclases in lower and higher eukaryotes. Here we detail the cloning and recombinant expression of functional candidate AC domains using, as an example, the *A. thaliana* pentatricopeptide repeat-containing protein (AtPPR-AC; At1g62590). Through a complementation test, in vivo adenylate cyclase activity of candidate recombinant molecules can be prescreened and promising candidates can subsequently be further evaluated in an in vitro AC immunoassay.

Key words *Arabidopsis thaliana*, Pentatricopeptide (PPR), Adenylate cyclase (AC), Adenosine-5′-triphosphate (ATP), Cyclic 3′,5′-adenosine monophosphate (cAMP), Lactose fermenters, Enzyme immunoassay

1 Introduction

By the mid 1970s, the molecule 3′,5′-cyclic adenosine monophosphate (cAMP) had been firmly established as an important signalling chemical and a second messenger in both animals and lower eukaryotes [1–3]. It was also understood that adenylate cylases (ACs) are the enzymes responsible for the generation of this cAMP from adenosine-5′-triphosphate (ATP) hydrolysis, and that the cAMP can affect many different physiological and biochemical processes including the activity of kinases [1]. Given the growing realization of the importance of ACs and cAMP, it is not surprising that plant scientists were keen to learn if this signalling system was universal

Chris Gehring (ed.), *Cyclic Nucleotide Signaling in Plants: Methods and Protocols*, Methods in Molecular Biology, vol. 1016, DOI 10.1007/978-1-62703-441-8_2,

and therefore operating in plants too. The major reasons why cAMP functions were more difficult to establish in plants were firstly that the levels of cAMP detected in plants appeared to be very low (<20 pmol/g fresh weight) [4] compared to those found in animals (>250 pmol/g wet weight) [5] and secondly that the vagaries of the assay systems used in plants were not conducive to reach firm conclusions [6]. Nonetheless, specific pathogen-induced signalling at lower cyclic nucleotide concentrations has been reported in plants [7].

Furthermore, cell-permeant 8-Br-cAMP and the stimulation of *albeit* unknown ACs with forskolin were shown to elicit concentration-dependent and time-dependent plant biological responses such as increases in Ca^{2+} influx across the plasma membrane [8] and biochemical evidence also suggests that crude alfalfa (*Medicago sativa* L.) root extracts show calmodulin-dependent AC activity [9]. Arguably, the most convincing data for a specific signalling role for cAMP came from whole-cell patch-clamp recordings from *Vicia faba* mesophyll protoplasts, which revealed that outward K^+ current could increase in a dose-dependent fashion as a result of an intracellular application of cAMP but not AMP, cGMP, or GMP [10].

To date, the only annotated and experimentally confirmed AC in plants is a *Zea mays* pollen protein [11] and this molecule has a role in polarized pollen tube growth. An Arabidopsis orthologue of this protein (At3g14460) is annotated as disease resistance protein and belongs to the nucleotide-binding site-leucine-rich repeat (NBS-LRR) family used in pathogen sensing and with a role in defense responses and apoptosis [12]. Considering that cyclic nucleotides have important and diverse roles in plant signalling via cyclic nucleotide-responsive protein kinases, nucleotide-binding proteins, and nucleotide-gated ion channels [13], it is unlikely that a single AC or GC can account for all cAMP- and cGMP-dependent processes in higher plants. In line with this hypothesis is the fact that a number of Arabidopsis molecules with different domain organizations and experimentally confirmed GC activity in vitro and/or in vivo have recently been reported [14, 15] and it is likely that more and structurally diverse candidate ACs will in time be identified. One such candidate identified with an AC catalytic center motif search [16] is a pentatricopeptide repeat-containing protein (PPR; At1g62590) responsible for interacting with RNA and facilitating its processing [17–19]. Here we detail the cloning and expression of the AC catalytic center containing fragment of the PPR gene and demonstrate the use of a cAMP-deficient *Escherichia coli* strain (*cya*A mutant) to screen for AC function in vivo as well as the use of an assay kit to verify AC activity in vitro.

2 Materials

Prepare all solutions using sterile distilled water (sterilized by autoclaving at 121 °C for 15 min) and analytical grade chemicals. Prepare all solutions and reagents in sterile containers under laminar flow conditions and store them at room temperature unless otherwise stated. All buffers and/or solutions requiring sterilization must be filtered with a 0.22 or a 0.45 μm membrane system. Follow all waste disposal regulations when disposing of waste materials.

1. Luria Bertani (LB) agar plates: Add 20 g of LB powder and 15 g of bacteriological agar powder into a 1 l universal bottle. Add water to the 1 l mark and mix well. Sterilize by autoclaving. Allow to cool to 45–50 °C, add 10 ml of 1 M filter-sterilized $MgSO_4$, and mix well. Pour 15–20 ml of the medium into individual plates in a sterile hood and allow them to dry. Store at 4 °C (*see* **Note 1**).
2. 1 M $MgSO_4$: Weigh 24.08 g of $MgSO_4$ salt into a 250 ml universal bottle. Add water to the 200 ml mark and dissolve completely. Filter-sterilize and store at 4 °C.
3. 34 mg/ml Chloramphenicol: Weigh 0.034 g of chloramphenicol powder into an Eppendorf tube. Add 1 ml absolute ethanol and dissolve completely. Store at −20 °C.
4. 100 mg/ml Ampicillin: Weigh 0.1 g of ampicillin powder into an Eppendorf tube. Add 1 ml of water and dissolve completely. Filter-sterilize and store at −20 °C.
5. Double-strength yeast-tryptone (2YT) medium: Add 16 g of tryptone powder (1.6 % m/v), 10 g of yeast extract (1 % m/v), 5 g of NaCl (0.5 % m/v), and 4 g of glucose (0.4 %) into a 1 l universal bottle. Add water to the 1 l mark and mix well. Sterilize by autoclaving. Store at 4 °C (*see* **Note 1**).
6. 1 M Isopropyl-β-D-thiogalactopyranoside (IPTG): Weigh 0.24 g of IPTG salt into an Eppendorf tube. Add 2 ml of water and dissolve completely. Filter-sterilize and store at −20 °C.
7. Lysis buffer: 10 mM Tris-HCl (pH 8.0), 150 mM NaCl, 10 mM imidazole, 1 mM DTT, 10 μg/ml lysozyme, 0.5 mM phenylmethanesulfonylfluoride (PMSF), and 7.5 % v/v glycerol. Weigh 0.6 g of Tris–HCl, 4.39 g of NaCl, 0.34 g of imidazole salt, 0.08 g of DTT, 0.01 g of lysozyme powder, and 0.04 g of PMSF (*see* **Note 2**) into a 1 l universal bottle. Add 37.5 ml of 100 % glycerol, and top up with water to the 500 ml mark. Dissolve completely by stirring and adjust pH to 8.0 with 1 M NaOH solution. Filter-sterilize and store at 4 °C.
8. Wash buffer: 10 mM Tris-HCl (pH 8.0), 150 mM NaCl, 20 mM imidazole, 1 mM DTT, 0.5 mM PMSF, and 7.5 % v/v

glycerol. Weigh 0.6 g of Tris–HCl, 4.39 g of NaCl, 0.68 g of imidazole salt, 0.08 g of DTT, and 0.04 g of PMSF (*see* **Note 2**) into a 1 l universal bottle. Add 37.5 ml of 100 % glycerol, and top up with water to the 500 ml mark. Dissolve completely by stirring and adjust pH to 8.0 with 1 M NaOH solution. Filter-sterilize and store at 4 °C.

9. Elution buffer: 50 mM Tris-HCl (pH 8.0), 200 mM NaCl, 250 mM imidazole, 0.5 mM PMSF, and 20 % v/v glycerol. Weigh 5.84 g of NaCl, 3.0 g of Tris–HCl, 8.5 g of imidazole, and 0.04 g of PMSF (*see* **Note 2**) into a 1 l universal bottle. Add 100 ml of 100 % glycerol and top up with water to the 500 ml mark. Dissolve completely by stirring and adjust pH to 8.0 with 1 M NaOH solution. Filter-sterilize and store at 4 °C.
10. 1 M NaOH solution: Weigh 0.4 g of NaOH pellets into a 50 ml universal tube. Add water to the 10 ml mark and dissolve completely. Filter-sterilize and store at 4 °C.
11. LB broth: Add 20 g of LB powder into a 1 l universal bottle. Add water to the 1 l mark and mix well. Sterilize by autoclaving. Allow to cool to 45–50 °C, add 10 ml of 1 M filter-sterilized $MgSO_4$, and mix well. Store at 4 °C (*see* **Note 3**).
12. 15 mg/ml Kanamycin: Weigh 0.015 g of kanamycin powder into an Eppendorf tube. Add 1 ml of water and dissolve completely. Filter-sterilize and store at −20 °C.
13. Transformation Buffer 1 (TFB1): 30 mM KAc, 100 mM RbCl, 10 mM $CaCl_2$, 50 mM $MnCl_2$, 15 % v/v glycerol. Add 0.294 g KAc, 0.990 g $MnCl_2$, 1.209 g RbCl, 0.147 g $CaCl_2$, 15 ml of 100 % glycerol, and 85 ml of water into a 250 ml universal bottle. Mix well to dissolve contents and adjust pH to 5.8 with 1 M KOH solution. Filter-sterilize and store at 4 °C.
14. Transformation Buffer 2 (TFB2): 10 mM MOPS, 75 mM $CaCl_2$, 10 mM RbCl, and 15 % v/v glycerol. Add 0.209 g MOPS, 0.121 g RbCl, 1.103 g $CaCl_2$, 15 ml of 100 % glycerol, and 85 ml of water into a 250 ml universal bottle. Mix well to dissolve contents and adjust pH to 6.8 with 1 M KOH solution. Filter-sterilize and store at 4 °C.
15. 1 M KOH solution: Weigh 0.56 g of KOH pellets into a 50 ml universal tube. Add water to the 10 ml mark and dissolve completely. Filter-sterilize and store at 4 °C.
16. SOC broth: 2 % (m/v) Tryptone, weigh 20 g of powder, 0.5 % (m/v) yeast extract, weigh 5 g, 8.56 mM NaCl, weigh 0.5 g and 2.5 mM KCl, weigh 0.186 g of salt into a 1 l universal bottle. Add water to the 1 l mark and mix well to dissolve contents. Adjust pH to 7.0 with 1 M NaOH solution and sterilize by autoclaving. Allow to cool to 45–50 °C, add 20 ml of 1 M filter-sterilized glucose, and mix well. Store at 4 °C (*see* **Note 4**).

17. MacConkey Agar: 5 % (m/v) MacConkey agar, weigh 5 g of powder into a 250 ml universal bottle. Add water to the 100 ml mark and mix well. Boil whilst stirring until completely dissolved. Sterilize by autoclaving. Allow to cool to 45–50 °C, add 10 μl of 1 M filter-sterilized IPTG, and mix well. Pour 15–20 ml of the medium into individual plates in a sterile hood and allow them to dry. Store at 4 °C (*see* **Note 3**).
18. 50 mM Isobutyl methylxanthine (IBMX): Weigh 0.004 g of IBMX salt into an Eppendorf tube. Fill to the 1 ml mark with absolute ethanol and dissolve completely. Store at -20 °C.
19. 100 mMMgCl_2 (Hexahydrate): Weigh 0.02 g of MgCl_2 salt into a 50 ml universal tube. Add water to the 10 ml mark and dissolve completely. Filter-sterilize and store at 4 °C.
20. 10 mM ATP: Weigh 0.006 g of ATP salt into an Eppendorf tube. Add water to the 1 ml mark and dissolve completely. Filter-sterilize and store at –20 °C.
21. 100 mM EDTA: Weigh 0.7 g of EDTA salt into a 50 ml universal tube. Add water to the 20 ml mark and dissolve completely. Filter-sterilize and store at 4 °C.
22. *E. coli* cells, SP850 *cya*A mutant were obtained from Coli Genetic Stock Center, Yale, USA.

3 Methods

3.1 Preparation of the Recombinant Protein

1. Extract total RNA from 3-week-old *Arabidopsis thaliana* ecotype Columbia-0 (Col-0) seedlings using any plant RNA extraction kit, in combination with DNase 1 treatment, and in accordance with the manufacturer's instructions (*see* **Note 5**).
2. Synthesize the AtPPR-AC fragment from the total RNA using any one-step RT-PCR kit (hot- or cold-start) and in accordance with the manufacturer's instructions (*see* **Note 6**).
3. Clone the amplified PCR product into a pCRT7/NT-TOPO expression vector (*see* **Note 7**) using any cloning kit and according to the manufacturer's instructions to make a pCRT7/NT-TOPO:AtPPR-AC fusion expression construct with an N-terminal His purification tag (*see* **Note 8**).
4. Transform competent BL21 (DE3) Star pLysS *E. coli* cells with the pCRT7/NT-TOPO:AtPPR-AC fusion expression construct using any cloning kit and in accordance with the manufacturer's instructions.
5. Streak transformed cells onto LB agar plates supplemented with 34 μg/ml chloramphenicol and 100 μg/ml ampicillin (*see* **Note 1**). Incubate overnight at 37 °C.

6. Inoculate 5 ml of fresh 2YT broth containing 34 μg/ml chloramphenicol and 100 μg/ml ampicillin with a single colony from the LB plates (*see* **Note 1**). Incubate overnight at 37 °C in a shaking incubator at 200 rpm.
7. In a 50 ml universal tube containing 15 ml fresh and pre-warmed 2YT broth and supplemented with 34 μg/ml chloramphenicol and 100 μg/ml ampicillin, add 300 μl of the overnight culture. Incubate at 37 °C in a shaking incubator at 200 rpm until the OD_{600} has reached 0.5.
8. Split the culture into two equal volumes of 5 ml. Add 5 μl of sterile water to one tube and 5 μl of 1 M IPTG to the other tube. Incubate the cultures at 37 °C in a shaking incubator at 200 rpm for a further 3 h.
9. Harvest the cell by centrifugation for 5 min and analyze part of the pellet by sodium dodecyl sulfate-polyacrylamide gel electrophoresis (SDS-PAGE) (Fig. 1a, [20]).
10. Resuspend the pellet into a lysis buffer at a ratio of 1 g pellet weight per 5 ml buffer. Incubate the contents on ice for 30 min. Using a benchtop centrifuge, spin at 10,000 × *g* for 10 min and collect the supernatant.
11. Transfer the supernatant into an empty column (of a minimum internal diameter of 1.0 cm and a minimum length of 5 cm) pre-loaded with 2 ml of 50 % nickel-nitrilotriacetic acid (Ni-NTA) slurry that has been pre-equilibrated with 10 ml of lysis buffer. Allow the mixture to settle and discard the flow-through.
12. Add 10 ml wash buffer into the column and allow the flow-through to go out. Repeat this step five more times.
13. Elute the protein in 2 ml of elution buffer. Desalt and concentrate the protein fraction by spinning on a benchtop centrifuge at 5,000 × *g* for 4 h at 4 °C in filtration devices with a molecular weight cutoff point of 3,000 Da.
14. Determine the protein concentration with a Bradford assay (*see* ref. 21) or using a nano-drop and evaluate the protein quality by SDS-PAGE (Fig. 1c). Store the recombinant protein at −20 °C.

3.2 Complementation Testing

1. Streak SP850 *cya*A mutant *E. coli* cells onto LB agar plates supplemented with 15 μg/ml kanamycin (*see* **Note 3**) and incubate at 37 °C overnight.
2. Inoculate 10 ml of fresh LB broth containing 15 μg/ml kanamycin (*see* **Note 3**) with a single colony of the SP850 *cya*A mutant cells. Incubate at 37 °C overnight in a shaking incubator at 200 rpm.

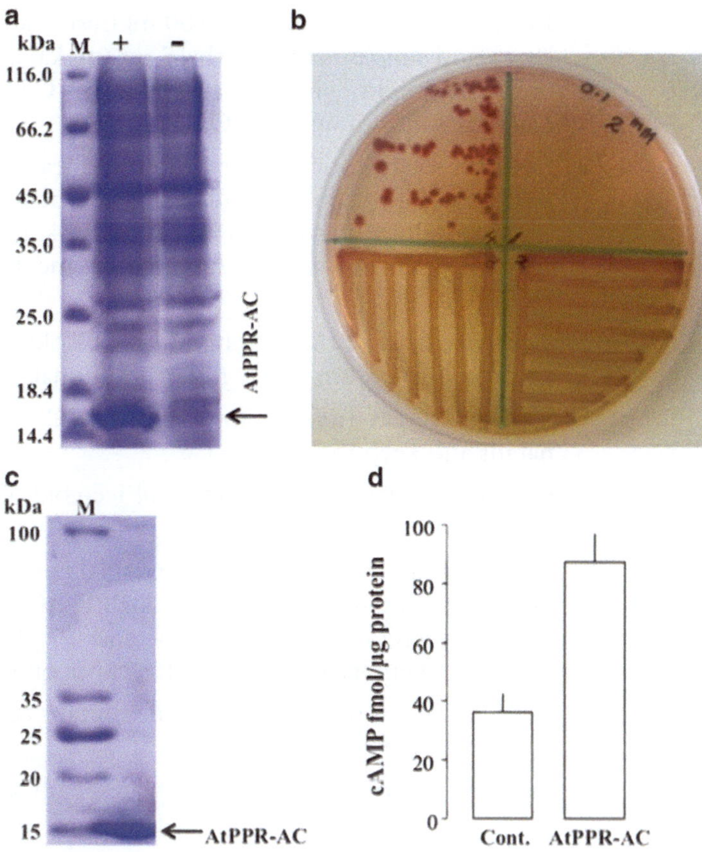

Fig. 1 Recombinant expression and functional characterization of the AtPPR-AC protein. (**a**) SDS-PAGE of protein fractions expressed in BL21 (DE3) Star pLysS cells transformed with the pCRT7/NT-TOPO:AtPPR-AC fusion construct, where *M* represents a low-molecular-weight marker while (*plus*) and (*minus*) represent protein fractions from the induced and un-induced cultures, respectively. The *arrow* marks the expressed recombinant AtPPR-AC fusion product in the induced culture. (**b**) Functional complementation of the SP850 *cya*A mutant *Escherichia coli* cells by the AtPPR-AC gene fragment whereby three different types of mutant SP850 *cya*A cells were plated onto MacConkey agar supplemented with 0.1 mM IPTG for 40 h at 37 °C followed by their analysis for possible lactose-fermenting characteristic. Section 1 of the plate contains no cells, section 2 contains non-transformed mutant cells, section 3 contains the mutant cells transformed with the pCRT7/NT-TOPO empty vector while section 4 contains the mutant cells transformed with the pCRT7/NT-TOPO:AtPPR-AC fusion construct. Cells in sections 2 and 3 are both non-lactose fermenters and produce white or yellowish colonies. Cells in section 4 have picked a deep-purple phenotype signifying a lactose-fermenting phenotype (*see* **Note 8**). (**c**) SDS-PAGE of the purified AtPPR-AC where *M* represents a low-molecular-weight marker and the *arrow* marks the purified recombinant fusion product. (**d**) In vitro adenylate cyclase activity of the recombinant AtPPR-AC, whereby 1 mM ATP, 5 mM $MgCl^{2+}$, and 2 mM IBMX were incubated in 50 mM Tris-HCl (pH 8.0) for 20 min at room temperature in the presence (AtPPR-AC) or the absence (Cont) of 5 µg of the recombinant protein followed by measurement of the generated cAMP levels by enzyme immunoassay. Experiments were carried out in triplicates and the error bars represent standard errors of the means ($n=3$)

3. In a 250-ml flask containing 100 ml fresh and pre-warmed LB broth and supplemented with 15 μg/ml kanamycin, add 1 ml of the overnight culture. Incubate at 37 °C in a shaking incubator at 200 rpm until the OD_{600} is 0.5.
4. Cool the culture on ice for 5 min and transfer it to a sterile round-bottom centrifuge tube. Using a benchtop centrifuge, spin the culture at 4,000 × *g* for 5 min at 4 °C in order to harvest the cells. Discard the supernatant and keep the cells on ice.
5. Gently resuspend the cells in 30 ml ice-cold TFB1 buffer and keep the suspension on ice for an additional 90 min.
6. Spin the cells at 4,000 × *g* for 5 min at 4 °C. Discard the supernatant and keep the cells on ice.
7. Gently resuspend the cells in 4 ml ice-cold TFB2 buffer and keep the suspension on ice.
8. Cool two Eppendorf tubes on ice for 10 min. In one tube, add 10 μl of the pCRT7/NT-TOPO:AtPPR-AC fusion expression construct while in the other tube add 10 μl of the pCRT7/NT-TOPO empty vector. Keep the tubes on ice.
9. In each tube, transfer 100 μl of the mutant cells and mix carefully. Incubate on ice for 20 min.
10. Transfer the tubes to a 42 °C water bath and incubate for 90 s.
11. To each tube add 500 μl of SOC broth and incubate at 37 °C for 60–90 min in a shaking incubator at 200 rpm (*see* **Note 4**).
12. Prepare a single plate of MacConkey agar supplemented with 15 μg/ml kanamycin (*see* **Note 3**) and 0.1 mM IPTG.
13. Mark four equal sectors on the agar plate with a permanent marker.
14. Streak one sector of the plate with the *cyaA* mutant cells transformed with the pCRT7/NT-TOPO:AtPPR-AC fusion expression construct, another sector with the *cyaA* mutant cells transformed with the pCRT7/NT-TOPO empty vector, and one other sector with the non-transformed *cyaA* mutant cells. Leave one section of the plate un-streaked (*see* **Note 9**).
15. Invert the plate and incubate at 37 °C for 18–40 h.
16. Analyze the phenotypes of all cells growing in the different sectors of the agar plate and check if only the color of those cells harboring the AtPPR-AC gene fragment turns deep purple while the colors of the rest of the other cells remain yellowish or colorless (Fig. 1b) (*see* **Note 9**).

3.3 Measurement of AC Activity In Vitro

1. Collect 11 Eppendorf tubes, label them 0–10, and place on ice.
2. To tube 0, add 100 μl of 50 mM Tris–HCl (pH 8.0), 10 μl of 20 mM IBMX, 10 μl of 50 mM $MgCl_2$ (*see* **Note 10**), and

10 μl of 10 mM ATP. To the other ten tubes, add all reaction components except the recombinant protein.

3. To tube 1, add 1 μg of the purified recombinant AtPPR-AC; to tube 2, add 2 μg of the recombinant protein and continue in that order until tube 10 has 10 μg of the protein. Incubate all tubes at room temperature for 20 min.
4. At time intervals of 0, 5, 10, 15, and 20 min, collect 20 μl from each reaction tube into a fresh Eppendorf tube and terminate the reaction by adding 10 μl of 10 mM EDTA followed by boiling for 5 min.
5. Spin all the tubes at full speed on a benchtop centrifuge for 10 min and collect the supernatants into fresh Eppendorf tubes.
6. Measure the cAMP content in the supernatants of all the tubes using a cAMP enzyme immunoassay kit following the acetylation protocol and according to instructions by the supplier's manual (*see* **Note 11**).
7. Compare the cAMP content in the control tube (tube 0) to the cAMP contents in all the other ten experimental tubes and take note of the optimum protein quantity and incubation time necessary for the generation of the highest levels of cAMP in vitro (Fig. 1d).
8. Using the optimized protein content and incubation time, further determine the other necessary and important parameters for the recombinant AtPPR-AC (*see* **Notes 12–17**).

4 Notes

1. Each time the BL21 (DE3) Star pLysS *E. coli* cells transformed with the pCRT7/NT-TOPO:AtPPR-AC expression construct are cultured, 1 ml of the 34 mg/ml chloramphenicol stock and 1 ml of the 100 mg/ml ampicillin stock need to be added to every 1 l of the medium.
2. PMSF is very difficult to dissolve in nonorganic solvents and always tends to precipitate in most buffers when added. In order to overcome this problem, weigh 0.05 g PMSF into a 50 ml universal tube and dissolve it in 25 ml of absolute ethanol to make a 10 mM stock for further use. Furthermore, PMSF, DTT, and lysozyme solutions should not be kept for any further uses but must always be prepared and used fresh.
3. For selection purposes of the SP850 *cya*A mutant *E. coli* cells, each time the cells are grown, there should always be an addition of 1 ml of the 15 mg/ml kanamycin stock to every 1 l of culture medium.

4. Since SP850 *cya*A mutant *E. coli* cells need to recover after transformation with the pCRT7/NT-TOPO:AtPPR-AC expression construct, they should not be immediately exposed to antibiotics, hence the use of SOC medium free of antibiotics during the first 90 min of culture growth.
5. We have noted that elution of the spin column with water pre-warmed to 65 °C yields more than twice the amount of RNA as compared to elution with water at room temperature. The yield is even further improved if the spin column membrane is first incubated for at least 1 min with this pre-warmed water.
6. The AtPPR-AC gene (At1g62590) has intronic sequences (*see* www.arabidopsis.org) and therefore, its cDNA sequence was used as a template to derive its gene-specific primers. Primers were designed to amplify between positions 3132 and 3638 of the gene. This translates to a protein product that is inclusive of the catalytic center, position 100–115, that is flanked by 48 and 50 amino acids on the amino-termini and carboxyl-termini, respectively. The following forward and reverse primers were, respectively, designed to specifically prime and amplify this region: (Fwd.) 5′-CG**GGATCC**GATGGGTGGCAGTG GTG-3′ and (Rev.) 5′-TCCA**GAATTC**TCAAGCAACTTTT AAATGT-3′. The bold and underlined regions are Bam H1 (forward) and EcoR1 (reverse). Two to four bases on the 5′ side of each primer were added to provide a scaffold for restriction enzymes and also to protect the restriction site from mutation by *Taq* polymerase during the insertion of a poly-A tail.
7. For the cloning purpose of the amplified AtPPR-AC, we preferred to use the pCRT7/NT-TOPO vector over the other expression vectors because it adds only a very small and short fragment (6× His tag) to the recombinant fusion protein, thereby having very minimal interferences to both the soluble expression and native purification of the recombinant product. However, other expression vectors may be used as long as they do not interfere with the soluble expression and native purification of the recombinant protein product.
8. For the ligation of the AtPPR-AC insert into the pCRT7/NT-TOPO expression vector, the optimal quantity of vector DNA should be in the range of 20–100 ng and the most effective ligation occurs when the molar concentration of the insert DNA to the vector DNA is in the range of 1:1 to 5:1. Further, a preheating of these two samples at 55 °C for 5 min before mixing them together helps melting any sticky ends that could have improperly self-annealed during the low-temperature storages of −20 °C. We also found that an additional longer incubation at low temperatures of 4–16 °C during the ligation process significantly improves efficiency.

9. In order to demonstrate the ability of the AtPPR-AC gene product to produce cAMP from ATP, a *cya*A mutant host strain (*E. coli* SP850) was complemented with the expression construct and screened for the selectable marker. The *E. coli* SP850 is a *cya*A deletion mutant that lacks the only AC in this organism and, because of this, cannot use lactose (Lac$^-$). Consequently, it produces colorless or yellowish colonies on MacConkey agar instead of the deep-purple colonies [11, 22]. Here, the AC-encoding region of *AtPPR-AC* cloned in the pCRT7/NT-TOPO expression vector was transformed into the SP850 *cya*A *E. coli* cells and the recombinant colonies were then screened on MacConkey agar containing 0.1 mM IPTG. When compared with the colorless or yellowish colonies produced by either the non-transformed bacteria or bacteria transformed with the empty pCRT7/NT-TOPO expression vector, colonies expressing *AtPPR-AC* stained deep-purple, signifying a rescued cAMP-dependent lactose-fermentation.
10. The truncated recombinant AtPPR-AC construct contains the cytoplasmic domain of the protein and should contain the N-terminal part with an aspartic acid residue [D] at the -33 from the catalytic center, a key residue in metal binding. As reported previously [23, 24], some adenylate cyclases may exhibit no inherent preference of either Mn^{2+} or Mg^{2+} metal ion as a cofactor for enzymatic activity. In the presence of either of these metal ions, the adenylate cyclase activity increases in a dose-dependent manner [23, 24].
11. The cAMP enzyme immunoassay Biotrak (EIA) kit was used because of its high sensitivity and specificity for the signalling molecule, cAMP. The anti-cAMP antibody in this kit is highly specific for cAMP at fento-molar levels and has approximately 10^6 times lower affinity for cGMP. However, it is also very important and advisable to verify and validate the results obtained by enzyme immunoassay through mass spectrometry as an additional method. In this method, the system is first calibrated by extracting mass chromatograms at *m/z* 328 [M-1] of the cAMP molecule before actual levels of cAMP in the experimental samples are measured. Generally, samples are introduced into the machine with a UPLC at a flow rate of 180 μl/min and separation is then achieved by a Phenomenex Synergi 4 μm Fusion-RP (250 × 2.0 mm) column. A gradient of solvent "A" (0.1 % v/v formic acid) and solvent "B" (100 % acetonitrile) over 18 min is normally applied. During the first 7 min, the solvent composition is kept at 100 % "A" followed by a linear gradient of over 3 min to 80 % "B" and then a re-equilibration to the initial conditions. Electrospray ionization in the negative mode is always used at a cone voltage of 35 V.

12. *Standard curve*: Set the reaction with increasing concentrations of ATP from 0 to 2 mM at 0.2 mM intervals and plot the standard curve of ATP concentration (*x*-axis) and generated cAMP (*y*-axis).
13. *Cofactor preference*: Set two reactions with equimolar concentrations (5 mM) of $MgCl_2$ and $MnCl_2$ and compare their levels of generated cAMP. The cofactor with higher levels of cAMP would be the preferred one.
14. *Substrate preference*: Set two reactions with equimolar concentrations (1 mM) of ATP and GTP and compare the levels of the generated cAMP and cGMP, respectively. A substrate with the higher levels of cAMP would be the preferred one.
15. *Co-substrate effect*: Set one reaction with equimolar concentrations (1 mM) of ATP and GTP and assess if the generation of cAMP is enhanced, reduced, or unaltered.
16. *Product feedback*: Set the reaction in the presence of 100 μM cAMP or its analogue 8-Br-cAMP and check if its presence results in an enhanced or a diminished protein activity.
17. *Modulator effect*: Set the reaction in the presence of either 100 μM $CaCl_2$ [25] or 50 mM $NaHCO_3$ [26] and determine if any of these molecules can enhance the activity of the recombinant protein.

Acknowledgement

This material is based upon work supported financially by the National Research Foundation but any opinion, findings and conclusions or recommendations expressed in this material are those of the author(s) and therefore the NRF does not accept any liability in regard thereto.

References

1. Robison GA, Butcher RW, Sutherland EW (1968) Cyclic AMP. Annu Rev Biochem 37:149–174
2. Goodman DB, Rasmussen H, DiBella F et al (1970) Cyclic adenosine 3′:5′-monophosphate-stimulated phosphorylation of isolated neurotubule subunits. Proc Natl Acad Sci USA 67:652–659
3. Gerisch G, Hülser D, Malchow D et al (1975) Cell communication by periodic cyclic-AMP pulses. Philos Trans R Soc Lond B Biol Sci 272:181–192
4. Ashton AR, Polya GM (1978) Cyclic adenosine 3′:5′-monophosphate in axenic rye grass endosperm cell cultures. Plant Physiol 61: 718–722
5. Butcher RW, Baird CE, Sutherland EW (1968) Effects of lipolytic and antilipolytic substances on adenosine 3′,5′-monophosphate levels in isolated fat cells. J Biol Chem 243:1705–1712
6. Amrhein N (1977) The current status of cyclic AMP in higher plants. Annu Rev Plant Physiol 28:123–132
7. Meier S, Ruzvidzo O, Morse M et al (2010) The Arabidopsis wall-associated kinase-like 10 gene encodes a functional guanylyl cyclase and is co-expressed with pathogen defense related genes. PLoS One 5:e8904
8. Kurosaki F, Nishi A (1993) Stimulation of calcium influx and calcium cascade by cyclic AMP in cultured carrot cells. Arch Biochem Biophys 302:144–151

9. Carricarte VC, Bianchini GM, Muschietti JP et al (1988) Adenylate cyclase activity in a higher plant, alfalfa (*Medicago sativa*). Biochem J 249:807–811
10. Li W, Luan S, Schreiber SL et al (1994) Cyclic AMP stimulates K^+ channel activity in mesophyll cells of *Vicia faba L.* Plant Physiol 106: 957–961
11. Moutinho A, Hussey PJ, Trewavas AJ et al (2001) cAMP acts as a second messenger in pollen tube growth and reorientation. Proc Natl Acad Sci USA 98:10481–10486
12. DeYoung BJ, Innes RW (2006) Plant NBS-LRR proteins in pathogen sensing and host defense. Nat Immun 7:1243–1249
13. Newton RP, Smith CJ (2004) Cyclic nucleotides. Phytochemistry 65:2423–2437
14. Kwezi L, Meier S, Mungur L et al (2007) The *Arabidopsis thaliana* brassinosteroid receptor (AtBRI1) contains a domain that functions as a guanylyl cyclase *in vitro*. PLoS One 2:7
15. Mulaudzi T, Ludidi N, Ruzvidzo O et al (2011) Identification of a novel *Arabidopsis thaliana* nitric oxide-binding molecule with guanylate cyclase activity *in vitro*. FEBS lett 585:2693–2697
16. Gehring C (2010) Adenyl cyclases and cAMP in plant signaling—past and present. Cell Commun Signal 8:15
17. Nakamura T, Schuster G, Sugiura G et al (2004) Chloroplast RNA-binding and pentatricopeptide repeat proteins. Biochem Soc Trans 32:571–574
18. Kotera E, Tasaka M, Shikanai T (2005) A pentatricopeptide repeat protein is essential for RNA editing in chloroplasts. Nature 433: 326–330
19. Schmitz-Linneweber C, Williams-Carrier R, Barkan A (2005) RNA immunoprecipitation and microarray analysis show a chloroplast pentatricopeptide repeat protein to be associated with the 5′ region of mRNAs whose translation it activates. Plant Cell 17:2791–2804
20. Kurien BT, Scofield RH (2009) Nonelectrophoretic bidirectional transfer of a single SDS-PAGE gel with multiple antigens to obtain 12 immunoblots. Methods Mol Biol 536:55–65
21. Bradford MM (1976) A rapid and sensitive method for the quantitation of microgram quantities of protein utilizing the principle of protein-dye binding. Anal Biochem 72:248–254
22. Tang WJ, Stanzel M, Gilman AG (1995) Truncation and alanine-scanning mutants of type I adenylyl cyclase. Biochemistry 34: 14563–14572
23. Tesmer JJ, Dessauer CW, Sunahara RK et al (2000) Molecular basis for P-site inhibition of adenylyl cyclase. Biochemistry 39: 14464–14471
24. Geng W, Wang Z, Zhang J et al (2005) Cloning and characterization of the human soluble adenylyl cyclase. Am J Physiol 288:C1305–C1316
25. Oh M-H, Kim HS, Wu X et al (2012) Calcium/calmodulin inhibition of the Arabidopsis BRASSINOSTEROID-INSENSITIVE 1 receptor kinase provides a possible link between calcium and brassinosteroid signalling. Biochem J 443:515–523
26. Chen Y, Cann MJ, Litvin TN et al (2000) Soluble adenylyl cyclase as an evolutionarily conserved bicarbonate sensor. Science 289: 625–628

Chapter 3

Quantification of Cyclic Dinucleotides by Reversed-Phase LC-MS/MS

Heike Burhenne and Volkhard Kaever

Abstract

Cyclic dinucleotides such as bis-(3′,5′)-cyclic dimeric adenosine monophosphate (c-di-AMP) and bis-(3′,5′)-cyclic dimeric guanosine monophosphate (c-di-GMP) represent important *second messengers* in bacteria. Although their synthesis has not been described in plants so far, they may be involved in the regulation of bacterial phytopathogen–plant interactions as well as rhizobium plant symbiosis. Here, we describe a sensitive and specific quantification method for c-di-AMP and c-di-GMP by HPLC-coupled tandem mass spectrometry. Additional linear dinucleotide metabolites and mononucleotides, as well as cyclic mononucleotides, can be simultaneously determined by this method.

Key words Cyclic dinucleotides, HPLC, Tandem mass spectrometry

1 Introduction

Cyclic mononucleotides such as 3′,5′-cyclic adenosine monophosphate (cAMP) and 3′,5′-cyclic guanosine monophosphate (cGMP) are well-known *second messengers* in prokaryotes and eukaryotes. In various bacterial species additional nucleotides likewise acting as signalling molecules have been described. Besides the linear nucleotides guanosine-bis-diphosphate (ppGpp) and diadenosine tetraphosphate (Ap4A), the cyclic dinucleotides bis-(3′,5′)-cyclic dimeric adenosine monophosphate (c-di-AMP) and bis-(3′,5′)-cyclic dimeric guanosine monophosphate (c-di-GMP) have gained increased significance within the last years. Very recently we could suggest a function of c-di-AMP in controlling cell size and cell wall stress in *Staphylococcus aureus* [1]. On the other hand, a clear role for c-di-GMP as bacterial lifestyle switch regulator, i.e., in controlling the transition from planktonic to biofilm-forming species, and as virulence factor has been demonstrated [2]. Extracellular cyclic dinucleotides may influence eukaryotic cell functions via specific membrane receptors. For example, in a mouse model exogenously administered c-di-GMP acts as an immunomodulator and protects against a bacterial infection [3].

Chris Gehring (ed.), *Cyclic Nucleotide Signaling in Plants: Methods and Protocols*, Methods in Molecular Biology, vol. 1016,
DOI 10.1007/978-1-62703-441-8_3,

To our knowledge, nothing is known as yet about the existence and function of c-di-AMP or c-di-GMP in plants. However, c-di-GMP signalling may be involved in bacterial pathogenesis of plants [4, 5] or in the regulation of bacteria and plant symbiosis [6].

In 2009 a method for quantitative determination of c-di-GMP concentrations in bacterial extracts was described [7]. However, the applied mass spectrometric method (MALDI-TOF) suffers from marginal linearity of the standard calibrator curve. In addition, this method requires a tedious separate chromatographic workup for the isolation of c-di-GMP. We have, therefore, established a versatile HPLC-coupled tandem mass spectrometry (LC-MS/MS) method for sensitive and reliable quantification of c-di-GMP [8]. In this protocol we describe an upgrade of this method, in which the simultaneous determination of the cyclic dinucleotides c-di-AMP and c-di-GMP, the dinucleotide metabolites 5′-phosphoadenylyl-3′,5′-adenosine (pApA) and 5′-phosphoguanylyl-3′-5′-guanosine (pGpG), and the mononucleotides AMP and GMP, as well as cyclic mononucleotides cAMP and cGMP, is included.

2 Materials

The HPLC (Series 200, Perkin-Elmer) and MS/MS (API 3000, ABSciex) instrumentation is listed in Table 1.

All solutions for sample preparation and LC-MS/MS analysis are prepared using at least HPLC or, even better, HPLC-MS-grade solvents (water, methanol, acetonitrile) (*see* **Note 1**).

2.1 Sample Preparation

1. Vials: 1.5 mL Safe-Lock (Eppendorf), 2.0 mL Safe-Seal (Sarstedt), 15 mL Polypropylene tubes (Falcon) (*see* **Note 2**).
2. Extraction solution: Acetonitrile/methanol/water, 2/2/1, v/v/v. We use acetonitrile ultra gradient HPLC grade and methanol and water HPLC grade specification.

2.2 Liquid Chromatography

1. The HPLC instrumentation is specified in Table 1. The configuration used is consisting of a binary pump system (micro pumps) and a 150 μL binary TEE high-pressure mixer (e.g., PE N2911206) (*see* **Note 3**). An auto-sampler fitted with a six-port valve (Rheodyne series 7125) with a 100 μL sample loop and an HPLC-column oven is used.
2. LC solvents A and B are specified in Table 1. For solvent A (10 mM ammonium acetate/0.1 % acetic acid) dissolve 1.54 g of ammonium acetate (NH_4OAc) in 2 L HPLC-grade water and mix it with 2 mL acetic acid (*see* **Note 4**). Store at 4 °C. Solvent B consists of pure HPLC-grade methanol.

Table 1
HPLC and MS/MS parameters of standard method for quantification of cyclic dinucleotides and additional nucleotides

Instrumentation	
HPLC	Series 200 (Perkin Elmer)
LC Column	EC 50/3 Nucleodur C18 Pyramid 3 µ, 50 x 3 mm (#760263.30, Macherey-Nagel)
Security Guard	C18, 4 x 2 mm (#AJO-4286, Phenomenex)
Column Saver	0.5 µm (#55214-U, Supelco)
Mass Spectrometer	API 3000 (ABSciex)

<table>
<tr><th colspan="2">HPLC parameters</th></tr>
<tr><td>Sample Solvent</td><td>H_2O (HPLC-grade)</td></tr>
<tr><td>Injection Volume</td><td>50 µL</td></tr>
<tr><td>Flow Rate</td><td>0.4 mL/min</td></tr>
<tr><td>Eluent A</td><td>10 mM NH_4OAc / 0,1 % HAc</td></tr>
<tr><td>Eluent B</td><td>MeOH</td></tr>
<tr><td>Injection Needle Flushing</td><td>AcN/H2O, 80/20, v/v</td></tr>
<tr><td>Temperature (Column Oven)</td><td>30°C</td></tr>
<tr><td>Temperature (Autosampler)</td><td>Ambient Temperature</td></tr>
<tr><td>Maximal Column Backpressure</td><td>1160 PSI</td></tr>
<tr><td></td><td></td></tr>
<tr><td>Analytes</td><td>GMP, cGMP, c-di-GMP, pGpG, AMP , cAMP, c-di-AMP, pApA</td></tr>
<tr><td>Calibrators</td><td>0.08/0.21/0.52/1.3/3.3/8.2/20.5/51.2/128/320/800/2000 / 5000 nM</td></tr>
<tr><td>Internal Standards</td><td>$^{13}C^{15}N$ c-di-GMP, $^{13}C^{15}N$ c-di-AMP</td></tr>
<tr><td>Analysis Time / Sample</td><td>16 min (Injection to Injection)</td></tr>
<tr><td>Volume Eluent A / Sample</td><td>6,4 mL</td></tr>
<tr><td>Volume Eluent B / Sample</td><td>0,75 mL</td></tr>
<tr><td>LC Gradient Method</td><td>
<table>
<tr><th>Total Time [min]</th><th>Flow Rate [µL/min]</th><th>A [%]</th><th>B [%]</th></tr>
<tr><td>0,0</td><td>400</td><td>100</td><td>0</td></tr>
<tr><td>5,0</td><td>400</td><td>100</td><td>0</td></tr>
<tr><td>6,3</td><td>400</td><td>90</td><td>10</td></tr>
<tr><td>8,3</td><td>400</td><td>90</td><td>10</td></tr>
<tr><td>11,0</td><td>400</td><td>70</td><td>30</td></tr>
<tr><td>11,1</td><td>400</td><td>100</td><td>0</td></tr>
<tr><td>15,0</td><td>400</td><td>100</td><td>0</td></tr>
</table>
</td></tr>
<tr><td>Average Retention Times</td><td>
<table>
<tr><td>AMP</td><td>3.0 min</td><td>GMP</td><td>1.4 min</td></tr>
<tr><td>cAMP</td><td>11.0 min</td><td>cGMP</td><td>7.8 min</td></tr>
<tr><td>c-di-AMP</td><td>9.9 min</td><td>c-di-GMP</td><td>8.2 min</td></tr>
<tr><td>pApA</td><td>10.5 min</td><td>pGpG</td><td>7.8 min</td></tr>
<tr><td>$^{13}C^{15}N$ c-di-AMP</td><td>9.9 min</td><td>$^{13}C^{15}N$ c-di-GMP</td><td>8.2 min</td></tr>
</table>
</td></tr>
</table>

Table 1
(continued)

MS/MS-Method	
Ionisation Mode	Positive
Ion Source Parameters	Nebulizer Gas: 6; Curtain Gas: 15; Collision Gas: 10; Ion Spray Voltage: 5500; Temperature: 350 °C; Ion Source Gas 1: -; Ion Source Gas 2: 7000 mL/min; Interface Heater: -
Analyte Parameters	(see below)

Analyte	m/z Precursor	m/z Fragments	Dwell Time [ms]	DP [V]	FP [V]	EP [V]	CE [V]	CXP [V]
AMP	348,048	136,100	40	71	280	10	29	8
	348,048	331,200	40	71	280	10	11	22
GMP	364,009	152,200	40	66	270	10	21	8
	364,009	135,100	40	66	270	10	65	8
cAMP	330,132	136,202	40	41	140	10	37	8
	330,132	119,134	40	41	140	10	77	8
cGMP	346,066	152,115	40	46	160	10	33	8
	346,066	135,224	40	46	160	10	63	8
c-di-AMP	659,231	136,215	40	101	340	10	63	8
	659,231	330,165	40	101	340	10	29	22
	659,231	524,044	40	101	340	10	33	16
c-di-GMP	691,135	135,207	40	101	370	10	123	8
	691,135	152,118	40	101	370	10	61	8
	691,135	248,168	40	101	370	10	39	16
pGpG	709,240	152,200	40	76	350	10	47	10
	709,240	135,200	40	76	350	10	125	8
pApA	677,135	136,100	40	66	240	10	61	8
	677,135	410,200	40	66	240	10	29	26
^{13}C ^{15}N-c-di-AMP	689,033	146,100	40	96	370	10	61	8
^{13}C ^{15}N-c-di-GMP	721,198	162,056	40	101	370	10	61	8

2.3 Mass Spectrometry

1. The MS/MS instrumentation is specified in Table 1. The API 3000 triple quadrupole mass spectrometer is equipped with a TurboIonspray source (ESI) (*see* **Note 5**).
2. Vials: 2 mL injection vials with 200 μL micro glass inserts and screw caps N9 (Macherey-Nagel).
3. Nitrogen gas 5.0 (supplied from liquid nitrogen) (*see* **Note 6**).
4. Calibrators: All nucleotides were obtained from BIOLOG Life Science Institute.
5. $^{13}C_{20}{}^{15}N_{10}$ c-di-AMP and $^{13}C_{20}{}^{15}N_{10}$ c-di-GMP are used as isotope-labelled internal standards.

2.4 Software

Control of the HPLC and MS/MS systems as well as data generation was done by the Analyst software (version 1.4, ABSciex).

3 Methods

3.1 Extraction of Nucleotides from Bacterial Liquid Culture

The described protocol works well with Gram-negative bacteria. In case of Gram-positive species a modified extraction method has already been published [1]. Tissue species can be homogenized and lysed by a special instrumentation, e.g., FastPrep® system (MP Biomedicals).

1. Incubate the liquid bacterial culture until the desired optical density is reached (*see* **Note 7**).

   All procedures described in **steps 2–7** should be performed on ice or at 4 °C.
2. Take 1–5 mL of bacterial suspension and put it into the 15 mL tube (*see* **Note 8**).
3. Centrifuge for 20 min at 4 °C at 2,500 ×*g*. Discard supernatant (*see* **Note 9**).
4. Resuspend bacterial pellet with 2 × 500 μL culture medium and transfer suspension into 1.5 mL vials.
5. Centrifuge for 20 min at 4 °C at 2,500 ×*g*. Discard supernatant (*see* **Note 9**).
6. Resuspend bacterial pellet with 300 μL extraction solution (*see* **Note 10**).
7. Incubate suspension on ice for 15 min (*see* **Note 11**).
8. Heat extracted suspension for 10 min at 95 °C and then cool again on ice (*see* **Note 12**).
9. Centrifuge for 10 min at 4 °C at 20,800 ×*g* (*see* **Note 13**) and then transfer supernatant into 2.0 mL vial.
10. Repeat extraction (**step 6**) twice with 200 μL extraction solution but omit heating at 95 °C (**step 8**).
11. Combine supernatant fluids of the three extraction steps (about 700 μL) and store at −20 °C overnight (*see* **Note 14**).
12. Centrifuge for 10 min at 4 °C at 20,800 ×*g* and then transfer supernatant into new 2.0 mL vial. The final extracts can be stored at −20 °C or directly be evaporated to dryness at 40 °C by a gentle nitrogen stream or by using an evaporation system (*see* **Note 15**).
13. For exact quantification of cyclic dinucleotides the protein content of the respective bacterial culture should be determined (BCA protein assay) (*see* **Note 16**).

3.2 Chromatographic Separation of Nucleotides

1. Calibrators: Stock solutions of all compounds were prepared in HPLC-grade water and stored in polypropylene tubes at −20 °C in 10 μL aliquots. Working solutions for calibration

were prepared by dilution of the stock solutions with HPLC-grade water. Final concentrations of the calibrators are listed in Table 1.

2. Internal Standards: $^{13}C_{20}{}^{15}N_{10}$ c-di-AMP and $^{13}C_{20}{}^{15}N_{10}$ c-di-GMP were prepared from $^{13}C_{10}{}^{15}N_{5}$ ATP (Sigma) or $^{13}C_{10}{}^{15}N_{5}$ GTP (Sigma), respectively, using recombinant diadenylate cyclase DisA from *Bacillus subtilis* [1] or recombinant diguanylate cyclase PleD*, which is a constitutively active mutant of PleD from *Caulobacter crescentus* [8] (*see* **Note 17**). The internal standards were stored in polypropylene tubes at −20 °C at a concentration of 400 ng/mL.
3. Preparation of calibrators for LC-MS/MS analysis: Mix 10 μL of each calibrator working solution with 40 μL of HPLC-grade water and centrifuge for 10 min at 4 °C at 20,800×*g*. Mix 40 μL of each calibrator sample with an equal volume of the internal standard directly in the micro insert of the injection vial. Avoid air bubbles (*see* **Note 18**).
4. Preparation of extracted samples for LC-MS/MS analysis: Reconstitute dried sample extracts with 200 μL of HPLC-grade water by intensive vortexing for at least 10 s. Centrifuge for 10 min at 4 °C at 20,800×*g*. Mix 40 μL of each sample with an equal volume of the internal standard directly in the micro insert of the injection vial. Avoid air bubbles (*see* **Note 18**).
5. 50 μL of each calibrator or biological sample are automatically injected into the HPLC system and separated on the HPLC column according to the applied LC gradient method (*see* Table 1) (*see* **Notes 19–21**).

3.3 Analysis of Nucleotides by Tandem Mass Spectrometry

1. The applied tandem mass spectrometer is operated in the positive ionization mode. The ion source parameters (electrospray ionization), mass-to-charge ratios (*m/z*) for precursor ions and specific fragments (selected reaction monitoring), and mass spectrometer-specific settings are listed in Table 1 (*see* **Note 22**). The most intensive mass transitions are used as quantifiers, whereas additional fragments serve as qualifiers (*see* Fig. 2).
2. Calibration curves are constructed for each analyte. Data interpretation of the MS/MS signals is carried out by calculating the ratios of the peak areas of the calibrators and samples in relation to the respective peak areas of the internal standard (*see* **Notes 23** and **24**).
3. Beware of misinterpretation of false-positive c-di-AMP or c-di-GMP peaks by recording specific quantifier and qualifier mass transitions (*see* **Note 25**).

4 Notes

1. All laboratory glassware should be intensely rinsed with deionized water after cleaning with dish-washing liquid. Plastic labware should not be reused.
2. All vials should be checked for extractable residues or prewashed with the extraction solution.
3. A versatile HPLC system that comprises at least two high-pressure solvent pumps and a binary mixer for exact gradient mixture is needed to separate cyclic dinucleotides from interfering metabolites. A further improvement of analyte separation may also be achieved by UPLC methods.
4. Preparation of solvent A should be performed under a fume hood. No pH adjustment is necessary.
5. A sensitive tandem mass spectrometer (triple quadrupole) system is recommended.
6. As alternative to liquid nitrogen a nitrogen generator can be used.
7. Biological triplicates are recommended.
8. Increased culture volumes will lead to declined chromatographic separation and unwanted matrix effects in mass spectrometry.
9. Take care that the supernatant fluid is completely removed.
10. If culture volumes >5 mL are chosen as starting sample the amount of extraction solution should be adapted accordingly.
11. Following resuspension the prolonged incubation ensures proper extraction of the nucleotides.
12. Put an additional heavy cover on the lids in order to avoid spilling. At this step phosphodiesterases are inactivated.
13. The high centrifugation force leads to sedimentation of precipitated proteins.
14. Overnight storage of the extraction solution at −20 °C leads to enhanced protein precipitation.
15. Dried extracts are stable and can be stored at ambient temperatures or at 4 °C.
16. For determination of the protein content resuspend the bacterial pellet in 800 μL of 0.1 N NaOH and heat at 95 °C for 15 min. If the bacterial pellet is not completely solved after 15 min the incubation should be prolonged. In some cases, a centrifugation step will be necessary. Perform a BCA protein assay.

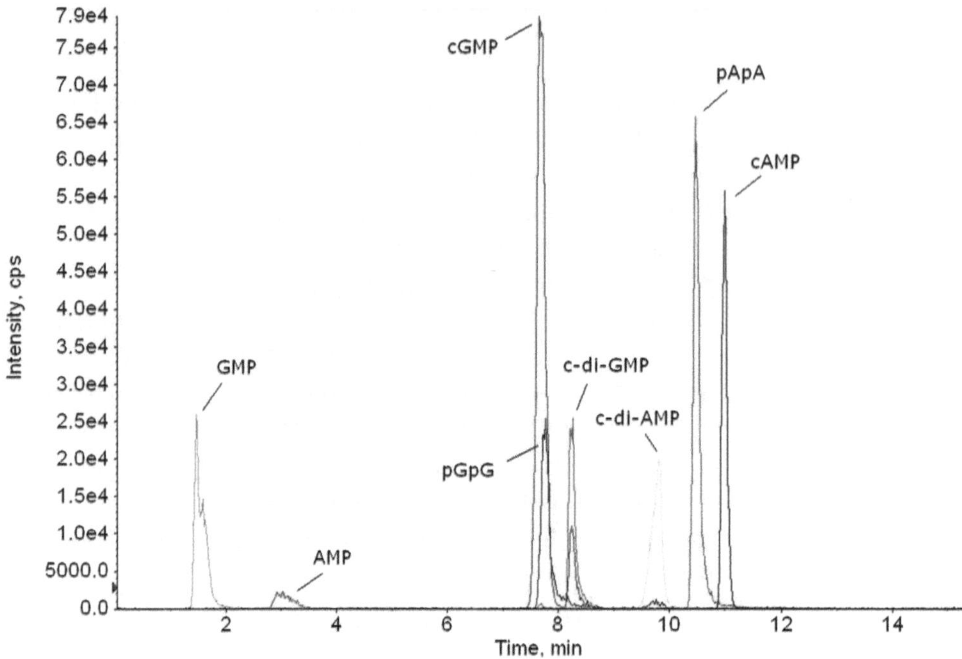

Fig. 1 Representative chromatogram of nucleotide standards (0.5 μM each) of GMP (1.4 min), AMP (3.0 min), cGMP (7.8 min), pGpG (7.8 min), c-di-GMP (8.2 min), c-di-AMP (9.9 min), pApA (10.5 min), and cAMP (11.0 min) with the respective retention times specified in parentheses

17. Methods for large-scale enzymatic production of c-di-GMP have already been described [9, 10]. Usage of recombinant dinucleotide cyclases lacking product inhibition is recommended.
18. In case of air bubbles in the micro insert of the injection vial correct injection of the sample could be hampered. Therefore, the vials should be vortexed or slightly flipped with a finger.
19. MS-"friendly" solvents and additives have to be applied. Prior degassing is only indicated if no degassing unit is integrated in the HPLC system. Several blank samples and an appropriate test mix should be analyzed to ensure trouble-free working of all instruments before the biological samples are measured.
20. A representative chromatogram of authentic standard nucleotides is shown in Fig. 1. Note that by MS/MS analysis a second peak for cGMP is found at the retention time of c-di-GMP and a second peak for cAMP is seen at the retention time of c-di-AMP (*see* **Note 23**).
21. This HPLC method could also be applied for purification of c-di-AMP or c-di-GMP after chemical or enzymatic synthesis. In this case a simple UV detector (254 nm) would be sufficient [11].

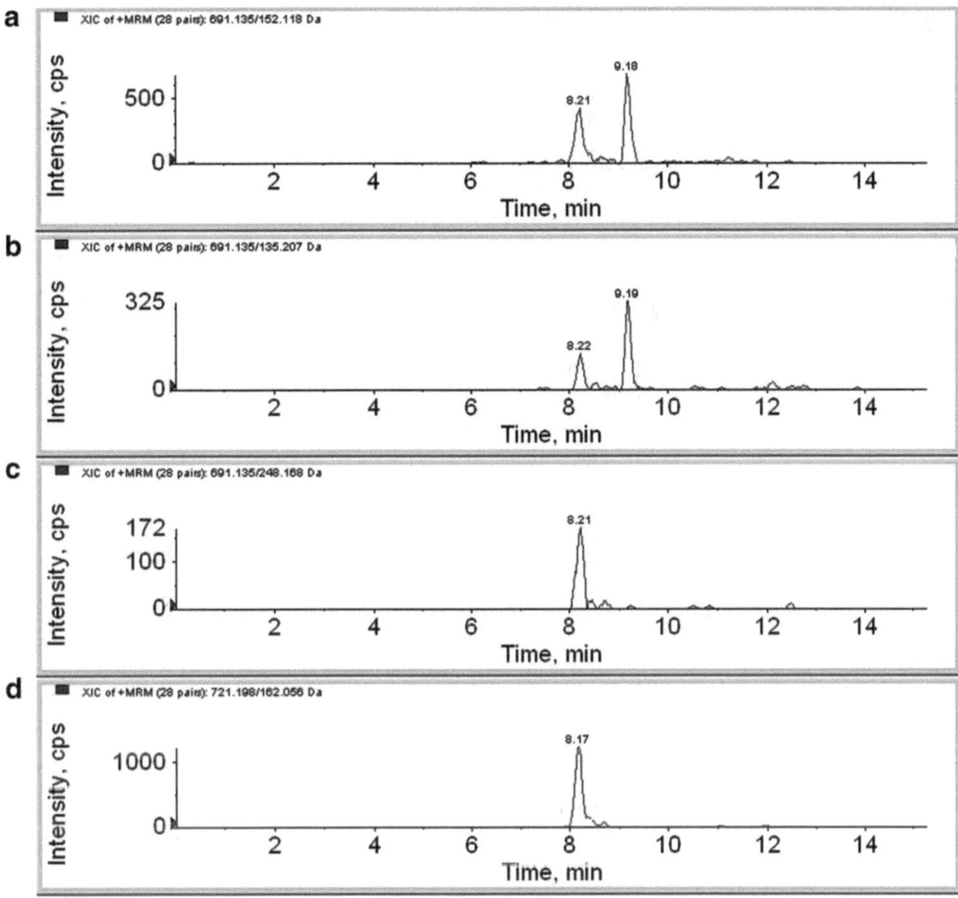

Fig. 2 Determination of c-di-GMP extracted from a bacterial culture (*E. coli*). (**a**) Quantifier SRM (*m/z* 691/152) of c-di-GMP. (**b**) First qualifier SRM (*m/z* 691/135) of c-di-GMP. (**c**) Second qualifier SRM (*m/z* 691/248) of c-di-GMP. (**d**) Quantifier SRM (*m/z* 721/162) of the internal standard $^{13}C^{15}N$-c-di-GMP

22. The instrument-specific settings will vary between mass spectrometry systems of different vendors and have to be adapted by the respective operator.
23. The additional peaks for cAMP and cGMP at the retention times of c-di-AMP and c-di-GMP, respectively, seen in Fig. 1 are due to partial in-source transformation of the cyclic dinucleotides in the ion source. For correct data interpretation a summation of c-di-NMP and cNMP peak areas (at identical retention times) has to be implemented.
24. The established method should be validated in terms of precision and accuracy. The lower limit of detection (LOD) is defined at a signal-to-noise (S/N) ratio of >3. The lower limit of quantification (LLOQ) is specified by an S/N ratio of >10.
25. In Fig. 2a–c the chosen mass transition signals for c-di-GMP from an extracted *E. coli* sample are presented. A first peak is

obvious at the retention time of authentic c-di-GMP (8.2 min); however, an additional second peak appears at a retention time of about 9.2 min at the quantifier mass transition (*m/z* 691/152, Fig. 2a) and the first qualifier mass transition (*m/z* 691/135, Fig. 2b). Checking of the exact precursor and fragment masses by an LC-quadrupole/time-of-flight MS hybrid system (TripleTOF 5600, ABSciex) with extremely higher mass accuracy compared to triple quadrupole systems revealed that peak 1 (8.2 min) indeed represents c-di-GMP whereas the second peak (9.2 min) represents a yet unknown bacterial metabolite. This is also confirmed through the recording of a second qualifier mass transition (*m/z* 691/248, Fig. 2c), which is specific for c-di-GMP. The internal standard $^{13}C_{20}{}^{15}N_{10}$ c-di-GMP also elutes at a retention time of 8.2 min (Fig. 2d). These findings emphasize the importance of the application of appropriate internal standards in LC-MS/MS analyses and the establishment of efficient HPLC separation methods.

Acknowledgements

We gratefully acknowledge the skillful technical assistance of Annette Garbe.

References

1. Corrigan RM, Abbott JC, Burhenne H, Kaever V, Gründling A (2011) c-di-AMP is a new second messenger in *Staphylococcus aureus* with a role in controlling cell size and envelope stress. PLoS Pathogens 7:e1002217
2. Hengge R (2009) Principles of c-di-GMP signalling in bacteria. Nat Rev Microbiol 7: 263–273
3. Zhao L, KuoLee R, Harris G, Tram K, Yan H, Chen W (2011) c-di-GMP protects against intranasal *Acinetobacter baumannii* infection in mice by chemokine induction and enhanced neutrophil recruitment. Int Immunopharmacol 11:1378–1383
4. Dow JM, Fouhy Y, Lucey JF, Ryan RP (2006) The HD-GYP domain, cyclic di-GMP signalling, and bacterial virulence to plants. Mol Plant Microbe Interact 19:1378–1384
5. Perez-Mendoza D, Coulthurst SJ, Humphris S, Campbell E, Welch M, Toth IK, Salmond GP (2011) A multi-repeat adhesin of the phytopathogen, *Pectobacterium atrosepticum*, is secreted by a type I pathway and is subject to complex regulation involving a non-canonical diguanylate cyclase. Mol Microbiol 82: 719–733
6. Wang Y, Xu J, Chen A, Wang Y, Zhu J, Yu G, Xu L, Luo L (2010) GGDEF and EAL proteins play different roles in the control of *Sinorhizobium melitoli* growth, motility, exopolysaccharide production, and competitive nodulation on host alfalfa. Acta Biochim Biophys Sin 42:410–417
7. Simm R, Morr M, Remminghorst U, Andersson M, Römling U (2009) Quantitative determination of cyclic diguanosine monophosphate concentrations in nucleotide extracts of bacteria by matrix-assisted laser desorption/ionization-time-of-flight mass spectrometry. Anal Biochem 386:53–58
8. Spangler C, Böhm A, Jenal U, Seifert R, Kaever V (2010) A liquid chromatography-coupled tandem mass spectrometry method for quantitation of cyclic di-guanosine mono-

phosphate. J Microbiol Methods 81: 226–231

9. Rao F, Pasunooti S, Ng Y, Zhuo W, Lim L, Liu AW, Liang Z-X (2009) Enzymatic synthesis of c-di-GMP using a thermophilic diguanylate cyclase. Anal Biochem 389:138–142

10. Spehr V, Warrass R, Höcherl K, Ilg T (2011) Large-scale production of the immunomodulator c-di-GMP from GMP and ATP by an enzymatic cascade. Appl Biochem Biotechnol 165:761–775

11. Schmidt AJ, Ryjenkov DA, Gomelsky M (2005) The ubiquitous protein domain EAL is a cyclic diguanylate-specific phosphodiesterase: enzymatically actice and inactive EAL domains. J Bacteriol 187:4774–4781

Chapter 4

Determination of ADP-Ribosyl Cyclase Activity, Cyclic ADP-Ribose, and Nicotinic Acid Adenine Dinucleotide Phosphate in Tissue Extracts

Richard M. Graeff and Hon Cheung Lee

Abstract

Cyclic ADP-ribose (cADPR) is a novel second messenger that releases calcium from intracellular stores. Although first shown to release calcium in the sea urchin egg, cADPR has been shown since to be active in a variety of cells and tissues, from plant to human. cADPR stimulates calcium release via ryanodine receptors although the mechanism is still not completely understood. cADPR is produced enzymatically from NAD by ADP-ribosyl cyclases; several of these proteins have been identified including one isolated from *Aplysia californica*, two types found in mammals (CD38 and CD157), and three forms in sea urchin. A cyclase activity has been measured in extracts from *Arabidopsis thaliana* although the protein is still unidentified. Nicotinic acid adenine dinucleotide phosphate (NAADP) is another novel messenger that releases calcium from internal stores and is produced by these same enzymes by an exchange reaction. NAADP targets lysosomal stores whereas cADPR releases calcium from the endoplasmic reticulum. Due to their importance in cell signaling, cADPR and NAADP have been the focus of numerous investigations over the last 25 years. This chapter describes several assay methods for the measurements of cADPR and NAADP concentration and cyclase activity in extracts from cells.

Key words Cyclic ADP ribose, NAADP, CD38, ADP-ribosyl cyclase

Abbreviations

ADP-ribose	Adenine diphosphate-ribose
cADPR	Cyclic adenine diphosphate-ribose
NAD	Nicotinamide adenine dinucleotide
NADP	Nicotinamide adenine dinucleotide phosphate
NAADP	Nicotinic acid adenine dinucleotide phosphate
NGD	Nicotinamide guanine dinucleotide
NHD	Nicotinamide hypoxanthine dinucleotide
PBS	Phosphate-buffered saline
PCA	Perchloric acid
TFA	Trifluoroacetic acid
wt	Wild type

Chris Gehring (ed.), *Cyclic Nucleotide Signaling in Plants: Methods and Protocols*, Methods in Molecular Biology, vol. 1016, DOI 10.1007/978-1-62703-441-8_4, © Springer Science+Business Media New York 2013

1 Introduction

Cyclic ADP-ribose (cADPR) was first identified as the product of nicotinamide adenine dinucleotide (NAD) that had calcium release activity in sea urchin egg homogenate [1]. At that time, the sea urchin egg was shown to be sensitive to both NAD and NADP although the details of how these pyridine nucleotides produced their effects were unknown [2]. It was subsequently shown that NAD was converted enzymatically to cADPR by the sea urchin homogenate and that the commercial NADP was contaminated with NAADP, a derivative of NADP which has a nicotinic acid group instead of nicotinamide [3]. Although it became apparent that a variety of tissues could produce cADPR from NAD, the identity of the protein(s) responsible for the enzyme activity was unknown [4]. The first protein shown to produce cADPR was named "ADP-ribosyl cyclase" and was purified from extracts of *Aplysia californica* [5]. Subsequently, CD38, a human lymphocyte antigen, was shown to have cyclase activity as well [6]. These two proteins have been expressed in yeast and an abundance of details about the crystal structures and enzyme mechanisms have been reported [7–9]. The information about NAADP has lagged somewhat behind that of cADPR. However, besides their effects on calcium release in sea urchin, another interesting phenomenon that links these two compounds is that both are produced enzymatically by ADP-ribosyl cyclases, including CD38. The reaction to produce cADPR from NAD involves release of the nicotinamide group and attachment of the terminal ribose to adenine through a linkage at N1 of the adenine ring [10]. The reaction to produce NAADP is through the exchange of nicotinic acid for the nicotinamide group of NADP [11]. Given that plants have also been shown to be responsive to both cADPR and NAADP [12, 13] and that these molecules play an important role in plant signaling, it is necessary to be able to accurately resolve temporal, spatial, and stimulus specific changes in their concentrations.

2 Materials

All buffers and reagents can be purchased from chemical supply companies with the exception of the ADP-ribosyl cyclase and NADase. Diaphorase (from *Clostridium kluyveri*), alcohol dehydrogenase, and glucose-6-phosphate dehydrogenase should be of high quality that is suitable for cycling assays. NADase can be prepared as described below. The ADP-ribosyl cyclase is the recombinant protein prepared as described previously [14]. Other sources of ADP-ribosyl cyclase that have been used for the cycling assay described below include the protein purified from *A. californica* and one prepared from the sponge *Axinella polypoides* [15, 16].

Table 1
Composition of Vogel's salt solution

Minimal media (2 % Glucose) (1 l):	
Vogel's 50×	20 ml
Glucose	20 g
dH_2O	980 ml
Autoclave or filter sterilize	
Vogel's 50× salt solution (1 l):	
$Na_3citrate \cdot 2H_2O$	125 g
KH_2PO_4	250 g
NH_4NO_3	100 g
$MgSO_4 \cdot 7H_2O$	10 g
$CaCl_2 \cdot 2H_2O$	5 g
Trace element solution	5 ml
Biotin solution (0.1 mg/ml)	2.5 ml
H_2O to 1 l and 2 ml chloroform	
Trace element stock solution: Prepared by adding the following ingredients successively, with stirring, to 95 ml of distilled water	
Citric acid$\cdot H_2O$	5.00 g
$ZnSO_4 \cdot 7H_2O$	5.00 g
$Fe(NH_4)_2(SO_4)_2 \cdot 6H_2O$	1.00 g
$CuSO_4 \cdot 5H_2O$	0.25 g
$MnSO_4 \cdot H_2O$	0.05 g
H_3BO_3	0.05 g
$Na_2MoO_4 \cdot 2H_2O$	0.05 g
The final volume is adjusted to 100 ml, 1 ml of chloroform is added as a preservative, and the solution is stored at room temperature	

2.1 Buffers, Solutions, and Reagents

1. Conditions for HPLC analysis: 10×220 mm column packed with AG MP-1 resin (Bio-Rad), flow rate 3 ml/min, gradient elution with buffer A: H_2O, buffer B: 150 mM TFA.
2. Perchloric acid (PCA): 11.6 M stock.
3. Trifluoroacetic acid (TFA): 13.1 M stock.
4. C/T extraction mixture: Chloroform/tri-*n*-octylamine (3:1).
5. Vogel's solution, for culture of *Neurospora crassa* (*see* Table 1).

6. NAD, NADP, NAADP, NGD, cADPR, cGDPR: 10 mM of each compound dissolved in H_2O and stored frozen in aliquots.
7. Nicotinamide: 1 M solution dissolved in H_2O and stored at 4 °C.
8. Resazurin: 5 mM stock solution dissolved in H_2O and stored frozen in aliquots.
9. Flavin mononucleotide (FMN): 5 mM stock solution dissolved in H_2O and stored frozen in aliquots.
10. Glucose-6-phosphate: 100 mM stock solution dissolved in H_2O and stored frozen in aliquots.
11. cADPR reagent, continuous assay: 100 mM disodium phosphate, 1 % ethanol, 5 μM resazurin, 5 μM FMN, 50 mM nicotinamide, 150 μg/ml alcohol dehydrogenase, and 7.5 μg/ml diaphorase (*see* **Note 1**).
12. cADPR reagent, discontinuous assay: Step 1: 5 mM disodium phosphate, 2 μg/ml ADP-ribosyl cyclase, and 100 mM nicotinamide. Step 2: 100 mM disodium phosphate, 2 % ethanol, 10 μM resazurin, 10 μM FMN, 200 μg/ml alcohol dehydrogenase, and 10 μg/ml diaphorase.
13. Etheno-NAD reagent, used to measure the activity of NADase: 100 μM etheno-NAD, 25 mM Tris–HCl, pH 8, and 2 mM $MgCl_2$.
14. NAD reagent, for reverse cycling assay or sample extract: 100 mM disodium phosphate, 1 % ethanol, 5 μM resazurin, 5 μM FMN, 100 μg/ml alcohol dehydrogenase, and 5 μg/ml diaphorase.
15. NADP reagent: 50 mM disodium phosphate, 1 mM glucose-6-phosphate, 5 μM resazurin, 5 μM FMN, 1 U/ml of glucose-6-phosphate dehydrogenase, and 5 μg/ml diaphorase.
16. NAADP reagent: **Step1**: 5 mM disodium phosphate, 10 μg/ml of E98G, and 100 mM nicotinamide. **Step 2**: 50 mM disodium phosphate, 2 mM glucose-6-phosphate, 10 μM resazurin, 10 μM FMN, 2 U/ml glucose-6-phosphate dehydrogenase, and 10 μg/ml diaphorase.
17. NGD reagent: 20 mM Tris–HCl, pH 8, 100 μM NGD.

2.2 Extraction Media for ADP-Ribosyl Cyclase Activity Assay

1. Homogenization medium: 10 mM Tris–HCl, pH 7.4, and 5 mM $MgCl_2$.
2. Sonication medium: 40 mM HEPES, pH 7, 100 mM NaCl, 5 mM $MgCl_2$, 4 % glycerol, 0.05 % NP-40, 10 mM NaF, 20 mM B-glycerophosphate, 2 mM sodium vanadate, 5 mM DTT, 1 mM PMSF, 30 nM okadaic acid, and complete EDTA-free protease inhibitors.

3 Methods

3.1 Measurement of ADP-Ribosyl Cyclase Activity

1. Preparation of extracts: Cells or tissues are washed with cold PBS, suspended in 10 volumes of a buffer containing homogenization medium, and disrupted by homogenization with 50 strokes in a Dounce glass homogenizer. The homogenate is centrifuged for 5 min at 1,000 × *g* to remove debris (*see* **Note 2**).
2. Membranes are prepared by centrifugation at 105,000 × *g* for 15 min and used in assays for ADP-ribosyl cyclase activity [17]. Alternatively, the tissue is flash frozen in liquid nitrogen, pulverized to a fine powder, and extracted in a buffer containing detergent. A complete buffer utilizing this technique is detailed in Sanchez et al. [18]. The powder is re-suspended in a cold sonication medium, sonicated for 20 s, and centrifuged at 12,000 × *g* for 10 min. The supernatant is centrifuged at 100,000 × *g* for 1 h. The resulting pellet is used in assays of cyclase activity.
3. HPLC analysis: Plant extracts have been shown to have ADP-ribosyl cyclase activity [12, 18]. Originally, the assays were conducted by incubating the extracts with NAD and following the production of cADPR by using a bioassay of calcium release in sea urchin homogenate [12] (*see* **Note 3**). Although this method is sensitive and reliable, it is limited by the need to have readily available sea urchin homogenate and low background calcium in samples and buffers.
4. Alternative HPLC-based method: The sample is incubated with 1 mM NAD and the cADPR product is identified as a peak on a chromatographic separation. Typically the peaks that can be identified include the NAD substrate and nicotinamide, cADPR, and ADPR products. An HPLC tracing of reaction products for an incubation of CD38 with NAD is shown in Fig. 1a. Details of the analysis are given in the figure legend. Since cADPR is often the least abundant peak, its identification and quantification can be difficult. However, low amounts of the cADPR product that cannot be identified by HPLC can be quantified by the cycling assay for cADPR described below. As shown in Fig. 1b, ADP-ribosyl cyclase from *Aplysia* produces a sufficient amount of cADPR that can be identified and quantified by HPLC. If needed, a method employing ^{32}P-NAD can also be used (*see* **Note 4**).
5. Assay with NGD as substrate: These problems of separating substrate from products can be avoided by using an alternative substrate, NGD, which has a guanine ring instead of the adenine ring of NAD. NGD is also a substrate for enzymes that have ADP-ribosyl cyclase activity. However, whereas the

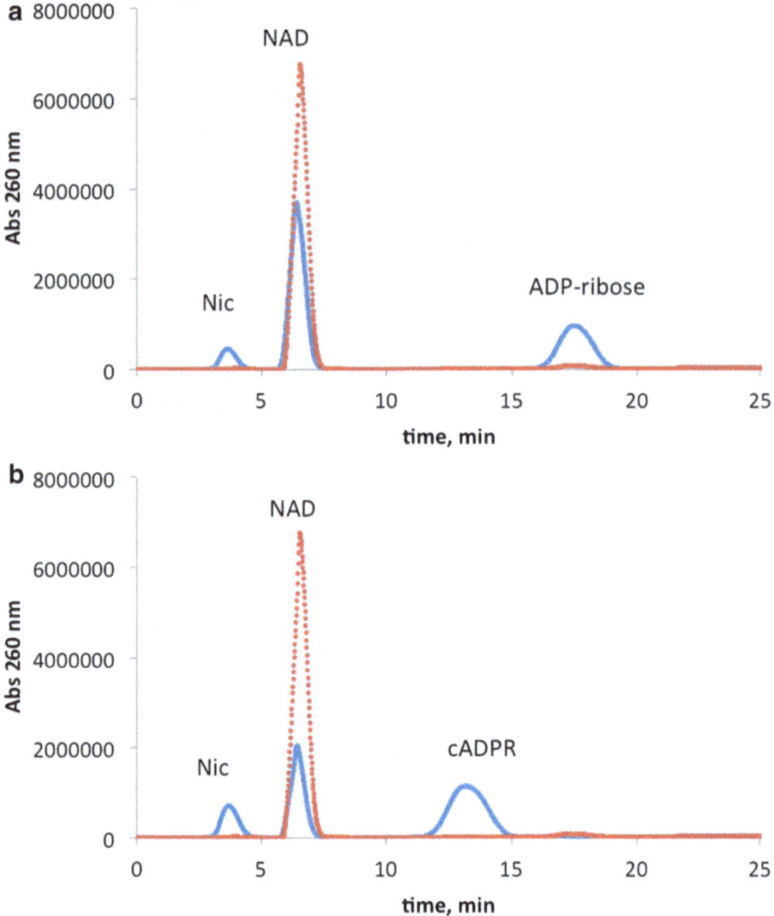

Fig. 1 HPLC separation of reaction products after incubation with NAD. (**a**) Incubation with CD38. CD38, 1 μg/ml, is incubated with 1 mM NAD in 20 mM Tris–HCl, pH 8, in a volume of 100 μl for 5 min and the reaction stopped by adding 0.1 % SDS. The products are analyzed by HPLC on a 10 × 220 mm column packed with AG MP-1 (Bio-Rad) and eluted with a gradient of trifluoroacetic acid from 0 to 100 % over 30 min at a flow rate of 3 ml/min [9]. Solvent A is water and solvent B is 150 mM trifluoroacetic acid and the gradient is increased linearly to 1 % from 0 to 1 min, increased linearly to 2 % from 1 to 2 min, increased linearly to 4 % from 2 to 5 min, increased linearly to 8 % from 5 to 9 min, increased linearly to 16 % from 9 to 13 min, increased linearly to 32 % from 13 to 17 min, increased linearly to 100 % from 17 to 18 min, held at 100 % from 18 to 22 min, decreased to 0 % from 22 to 22.1 min, and held at 0 % from 22.1 to 30 min. The products of the reaction include nicotinamide (Nic), and ADP-ribose. The cADPR produced by CD38 cannot be seen in this trace. The time zero trace for NAD is shown as a *dashed line*. (**b**) ADP-ribosyl cyclase, 1 μg/ml, is incubated with 1 mM NAD in 20 mM Tris–HCl for 2 min and the reaction stopped by adding 0.1 % SDS. Conditions for HPLC elution are the same as described above. The products include nicotinamide (Nic) and cADPR. The time zero trace for NAD is shown as a *dashed line*. The retention times are 3.6 min for Nic, 6.6 min for NAD, 13.0 min for cADPR, and 17.4 min for ADPR

terminal ribose of cADPR is linked to the N1 position of the adenine ring, the linkage in cGDPR is to the N7 position of the guanine ring [19]. The difference in these linkages is reflected in the fact that cGDPR is fluorescent whereas cADPR is not (*see* **Note 5**). The cyclase reaction with NGD can be monitored in a plate reader by the increase in fluorescence using excitation wavelength of 300 nm and emission wavelength of 410 nm. A typical reaction contains 5–50 μg of protein extract and NGD reagent in a 200 μl reaction volume. The reactions can be initiated by adding the solution containing NGD to the protein extract and measuring the increase in fluorescence with time. A reaction with NGD and recombinant CD38 protein is shown in Fig. 2a. If the background absorbance of the extract is too high, the fluorescence signal of cGDPR might be quenched and no increase in fluorescence seen. In this case, a discontinuous assay can be used. The sample is incubated with 100 μM NGD for various times and the reaction is stopped by adding 0.6 M PCA. The precipitated protein is removed by centrifugation for 10 min at 10,000 × *g* and the supernatant is recovered. The PCA is removed by extraction with the C/T mixture [20]. Although the original method used 1,1,2-trichlorotrifluoroethane, we have found that chloroform is a suitable substitute. Four parts of C/T organic extraction solution are added to one part of sample containing PCA. The sample is mixed well by vortexing for 1 min and the aqueous and organic phases are allowed to separate from one another. The neutralized extract is recovered from the upper aqueous phase, and the fluorescence of the sample is measured (excitation 300 nm, emission 410 nm). A time course should show that the fluorescence increases with time of incubation. This method has been employed by some investigators [21], and an example is shown in Fig. 2b. The fluorescence signal can be calibrated by using authentic cGDPR or by adding ADP-ribosyl cyclase to convert any remaining NGD to cGDPR. By using a time zero blank and cGDPR standard a rate of cGDPR production can be calculated.

6. Reverse cycling assay: All enzymes with cyclase activity can produce cADPR from NAD, but the percentage of product that is cADPR varies, depending on the source of enzyme. For example, CD38 produces less than 1 % of its product as cADPR from NAD whereas the ADP-ribosyl cyclase from *Aplysia* produces essentially 100 % cADPR [22]. The other product of NAD utilization by CD38 is ADP-ribose. Despite these differences, however, it is possible to use the reverse reaction to detect activity. The tissue extract (10–100 μg/ml) is incubated with 10 μM cADPR in the presence of 10 mM nicotinamide in 20 mM Tris–HCl, pH 8, in a volume of 100 μl. This reaction produces NAD which can be detected by a cycling assay, as

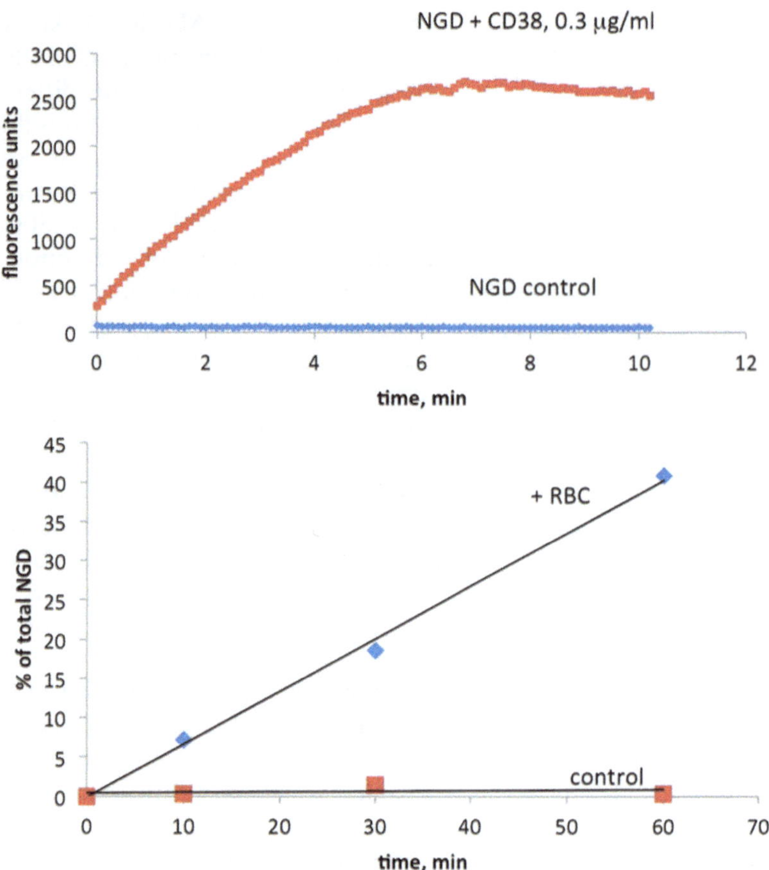

Fig. 2 Assay of CD38 activity with NGD as substrate. *Upper panel*: Continuous assay. 100 μM NGD is incubated with CD38, 1 μg/ml, in 20 mM Tris–HCl, pH 8, and the fluorescence measured at excitation of 300 nm and emission of 410 nm. *Lower panel*: Discontinuous assay. A sample, such as a suspension of red blood cells (40 %) is incubated with 100 μM NGD in HBSS. At the times indicated, the reaction is stopped by adding 0.6 M perchloric acid. The protein is removed by centrifugation at 10,000 × *g* for 10 min. The supernatant is recovered and the acid removed by extraction with 4 volumes of chloroform/tri-*n*-octylamine as described in the text. The neutral aqueous phase is recovered and fluorescence measured at 300 nm excitation and 410 nm emission. The fluorescence signal is quantified by comparing to cGDPR standards

described below. Reactions can be incubated for various times (hours if necessary) and stopped by adding 100 mM excess HCl. The samples are filtered on Immobilon-P plates (Millipore) and recovered with vacuum. The pH is adjusted with 1 M Tris–HCl, pH 8. In this example, the tissue extract serves the role of ADP-ribosyl cyclase, i.e., the reaction is conducted with and without extract. The presence of activity is indicated by accumulation of NAD with time. Prior to the cycling reaction, the sample is diluted 10- to 100-fold in 5 mM disodium phosphate. The assay for NAD includes 10 μl of diluted sample and 200 μl of NAD reagent. Fluorescence is detected in a plate reader at 544 nm excitation and 590 nm emission.

7. NAADP production: Assuming that it is possible to detect ADP-ribosyl cyclase activity by one of the above methods, it should also be possible to detect NAADP production. Reactions include 10–100 μg/ml of extract protein, 1 mM NADP, and 50 mM nicotinic acid, pH 5 [11]. Reaction volumes are usually 0.1–1 ml. The isolation and quantification of NAADP can be accomplished by HPLC as described [11]. The products are analyzed on a 10 × 220 mm column packed with AG MP-1 (Bio-Rad) and eluted with a gradient of trifluoroacetic acid from 0 to 100 % over 30 min at a flow rate of 3 ml/min [9]. Solvent A is water and solvent B is 150 mM trifluoroacetic acid and the gradient is increased linearly to 1 % from 0 to 1 min, increased linearly to 2 % from 1 to 2 min, increased linearly to 4 % from 2 to 5 min, increased linearly to 8 % from 5 to 9 min, increased linearly to 16 % from 9 to 13 min, increased linearly to 32 % from 13 to 17 min, increased linearly to 100 % from 17 to 18 min, held at 100 % from 18 to 22 min, decreased to 0 % from 22 to 22.1 min, and held at 0 % from 22.1 to 30 min. Under these conditions NADP elutes at 14.9 min and NAADP at 19.8 min. The NAADP product should accumulate with time and the NADP substrate decrease with time.

3.2 Measurement of Cyclic ADP Ribose

1. Preparing acid extracts: cADPR can be measured in acid extracts prepared from cells. At the appropriate time of treatment, the samples are either quickly frozen in liquid nitrogen or immersed in cold 0.6 M PCA and homogenized. The frozen sample is ground to a powder in liquid nitrogen and the powder is extracted with 1–5 volumes of cold 0.6 M PCA and incubated on ice for 20 min.
2. The insoluble protein is removed by centrifugation for 10 min at 10,000 × *g* at 4 °C (*see* **Note 6**). The acid is subsequently removed from the supernatant by extraction with C/T. Four parts of C/T (3:1) are added to one part extract and vortexed for 1 min. The phases will separate spontaneously or the samples can be centrifuged for 5 min at 1,000 × *g*. Typically, 75 % of the upper aqueous phase can be recovered. If needed, the sample is centrifuged at 10,000 × *g* for 10 min to remove any remaining debris. Generally, samples are clear after centrifugation.
3. cADPR can also be measured in samples from reactions of cyclase activity. Prior to the cycling assay for cADPR, the NAD is removed by treatment with NADase; otherwise the cycling assay itself does not distinguish between cADPR and NAD.

Following treatment with NADase, cADPR is measured by the cycling assay [20].

4. NADase preparation: NADase has been available from commercial sources but may be difficult to obtain.

5. If it is not available, an extract of *N. crassa* can be prepared. The NADase from *N. crassa* has been shown to have no ADP-ribosyl cyclase activity and converts NAD to ADPR [22]. In addition, the NADase does not have to be a pure protein for it to effectively remove NAD or NADP. The *N. crassa* spores can be obtained from commercial sources and the mold is grown in liquid culture in minimal media: 50× Vogel's solution is diluted to 1× and supplemented with 2 % glucose. *See* Table 1 for the composition of 50× Vogel's solution.

6. The 1× medium can be autoclaved or filtered to prepare a sterile stock. The mold can be grown in 150 mm culture plates in 50 ml of medium. After several days, the mold reaches confluency and spreads out over the surface of the liquid. The remaining liquid is removed, which causes the culture to produce conidia over the next few days and develop a characteristic orange color. After several days, the conidia are carefully scraped from the dish and transferred to a 50 ml polypropylene tube. Any liquid that is transferred along with the mold is removed. The NADase, an ectoenzyme, is extracted by adding 10 volumes of 0.17 M KCl and gently shaking the mixture for several minutes [23] and any particulate material is removed by centrifugation at 10,000 × *g* for 10 min. The final solution containing NADase should be clear. This extract of NADase can be stored frozen in aliquots and retains activity. However, activity is lost with several freeze/thaw cycles.

7. The activity of the crude NADase can be tested by using etheno-NAD as substrate. The reaction is started by adding 10 or 20 μl of NADase to 200 μl of etheno-NAD reagent and fluorescence is monitored in a plate reader at room temperature, excitation 300 nm and emission 410 nm. As shown in Fig. 3, the reaction is completed by 10 min. The specific activity of a typical NADase preparation using these conditions is 32 μmol/mg/min. This preparation of NADase can be used at a dilution of 1/20 (NADase volume/sample volume) to treat samples to remove NAD and NADP. Overnight treatment at 37 °C is sufficient to remove as much as 20 μM NAD from an extract. After incubation with NADase, the sample is filtered on 96-well PEI plates (Millipore) and collected in 8-well pcr strips (*see* **Note 7**).

8. cADPR assay: The assay for cADPR is typically performed in 96-well plates. The scheme for the assay is shown in Fig. 4. The assay can be conducted in a continuous or a discontinuous

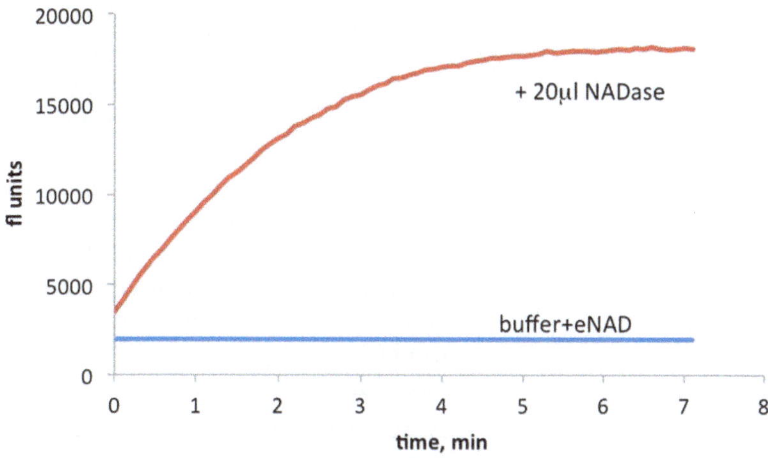

Fig. 3 Measurement of NADase activity by fluorescence assay with etheno-NAD. The NADase is prepared from *N. crassa* extracts as described in the text. 100 μM etheno-NAD is added to 20 μl of extract in 200 μl of 25 mM Tris–HCl, pH 8, and 2 mM $MgCl_2$ and fluorescence measured at 300 nm excitation and 410 nm emission

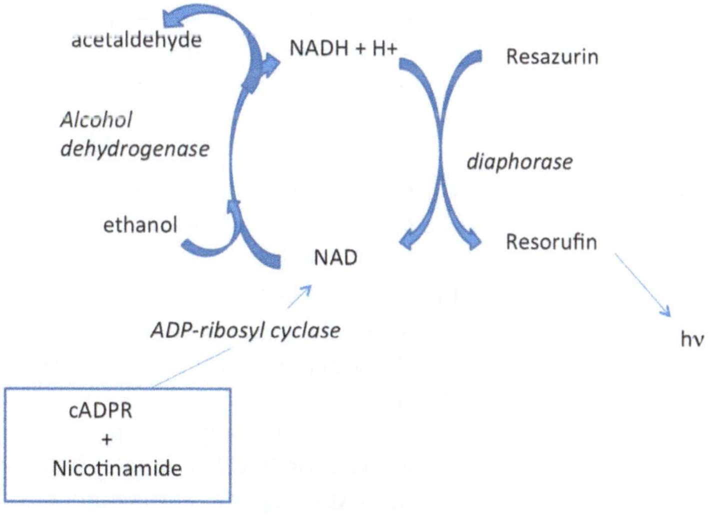

Fig. 4 Components of the cycling assay for cADPR. The assay for cADPR depends on the conversion of cADPR to NAD in the presence of nicotinamide and ADP-ribosyl cyclase. Since extracts from tissue samples contain NAD, the NAD is first removed by treatment with NADase from *N. crassa*. Assays for cADPR are conducted in the absence and presence of ADP-ribosyl cyclase. For comparison, cADPR standards are also extracted with perchloric acid and treated with NADase prior to the cycling assay

manner. In the continuous assay, 35 μl of sample is added per well. Each sample is assayed in a paired manner by adding 65 μl of cADPR reagent for continuous assay in the absence or the presence of ADP-ribosyl cyclase, 2 μg/ml. After addition of reagent, the fluorescence is monitored for several hours in a plate reader (excitation 544 nm, emission 590 nm). The difference in the rate of fluorescence increase represents the change due to cADPR. Ideally, the background fluorescence in the absence of cyclase is minimal and the rate of increase for cADPR is significantly higher in the presence of cyclase. A typical example of an assay of cADPR standards is shown in Fig. 5a (*see* **Note 8**).

9. A standard curve can be constructed from the slopes of fluorescence/min versus cADPR concentration, as shown in Fig. 5b. For the continuous assay, the lag in fluorescence increase is between 10 and 30 min (Fig. 5c). After this lag time, cADPR standards and NAD standards have virtually identical slopes. The assay is linear in the range of 1–20 nM cADPR. For the discontinuous assay, the sample is first treated with nicotinamide and cyclase before adding reagent. Typically, 40 μl of sample are added to 40 μl of step 1 reagent. After 30 min, 80 μl of **step 2** reagent is added. Fresh reagent is prepared just prior to the assay and diaphorase is added last. Some commercial preparations of diaphorase contain NAD, which may contribute to a gradual increase in fluorescence with time. This problem can be minimized by charcoal treatment [20] or by reducing the amount of diaphorase (*see* **Note 9**).
10. Confirmation that a signal is specific for cADPR should be done by pretreatment with nucleotide pyrophosphatase (*see* **Note 10**).
11. NAD assay: The cycling assay can also be used to measure NAD. Typically, NAD is about 1,000× more abundant than cADPR. Recovery of NAD from the acid extraction and neutralization steps is essentially 100 %. A fraction of the neutralized extract can be diluted appropriately (typically 1,000–10,000-fold) and assayed for NAD. The assay includes 10 μl of sample and 200 μl of NAD reagent. Fluorescence is detected in a plate reader at 544 nm excitation and 590 nm emission.
12. NADP assay: A cycling assay for NADP can be done on the extracts as well; the alcohol dehydrogenase is replaced with glucose-6-phosphate dehydrogenase (1 U/ml) and 1 % ethanol is replaced with 1 mM glucose-6-phosphate. Following acid extraction and neutralization, the sample is diluted (10- to 100-fold) in 5 mM disodium phosphate. The assay includes 10 μl of sample and 200 μl of NADP reagent. Fluorescence is detected in a plate reader at 544 nm excitation and 590 nm emission.

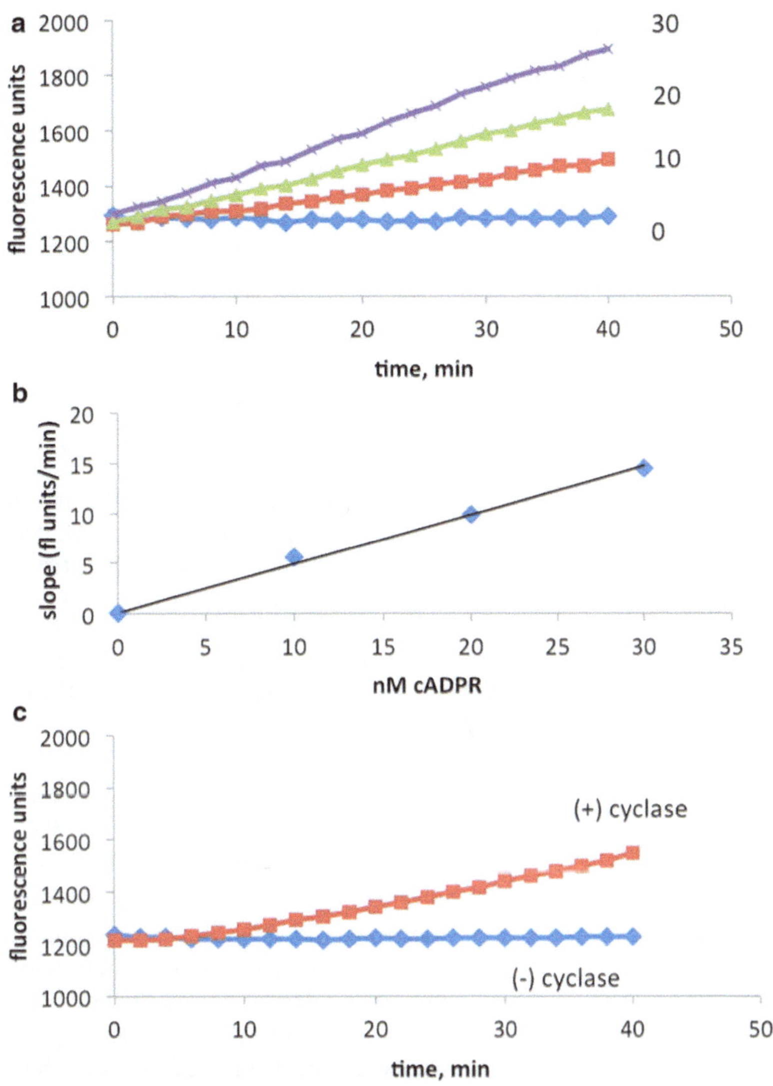

Fig. 5 Cycling assay for cADPR. (**a**) Time-dependent increases in fluorescence for cADPR standards at 544 nm excitation and 590 nm emission. The numbers to the right refer to nM concentration of cADPR. (**b**) Standard curve constructed from the slopes of increase in fluorescence versus cADPR concentration. (**c**) Time-dependent increase in fluorescence for a sample prepared from red blood cells

3.3 Measurement of NAADP

1. NAADP assay: The measurement of NAADP in extracts presents a greater challenge than that for cADPR. NAADP concentrations in tissues are generally lower than those of cADPR, and therefore it is more likely that some components can interfere with the assay. To date, the most sensitive assay utilizes ^{32}P-NAADP binding to sea urchin homogenate [24–26]. This method requires preparation of ^{32}P-NAADP, which has a

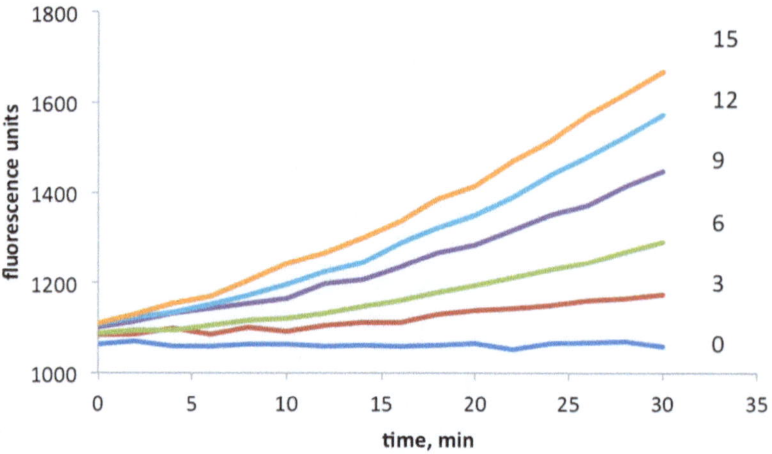

Fig. 6 Cycling assay for NAADP. NAADP standards are prepared in buffer at four times the final concentration, extracted with PCA, neutralized, and treated overnight at 37 °C with NADase. The NADase is removed by filtration on Immobilon-P filters. The assay for NAADP is conducted in the absence and presence of E98G, a mutant form of ADP-ribosyl cyclase. Wt cyclase does not catalyze the exchange reaction under these conditions. The *numbers* on the *right side* of the figure refer to the final concentration of NAADP in nM. Fluorescence is measured at 544 nm excitation and 590 nm emission

limited lifetime, and access to sea urchin homogenate with good NAADP-binding properties.

2. Alternative assays for NAADP have utilized various cycling methods [27–29]. A new cycling assay for NAADP has been developed which may eliminate some of the difficulties of the previously described assays. The basis of the assay is the finding that some mutants of ADP-ribosyl cyclase catalyze the base-exchange reaction at pH greater than 7 [9] (*see* **Note 11**). The assay is set up in a similar manner to that for cADPR except that E98G replaces wt cyclase. Samples are extracted with PCA, neutralized with the C/T mixture, and treated with NADase as described above. Due to the characteristics of the various components, the discontinuous assay works best for NAADP. Similar to the above condition, 40 μl of sample are added to 40 μl of step 1 reagent. After 1 h at room temperature, 80 μl of step 2 reagent are added and the fluorescence is monitored in a plate reader. The assay is conducted in the absence and presence of E98G and the difference in the rate of fluorescence represents the change due to NAADP. A typical standard curve is shown in Fig. 6. As described above for the assay for cADPR, there are circumstances where a false positive for the NAADP assay has been detected (*see* **Note 12**).

4 Notes

1. It is convenient to prepare 100× stock solutions of alcohol dehydrogenase and glucose-6-phosphate; 1,000× stock solutions of resazurin, FMN, and glucose-6-phosphate dehydrogenase; and a 5,000× stock solution of diaphorase. They can all be stored at −20 °C in aliquots. Resazurin is readily soluble in H_2O when it is prepared in polypropylene tubes. Nicotinamide can be prepared as a 1 M stock in H_2O and is stable for a few months at 4 °C.
2. Most of the methods used to measure ADP-ribosyl cyclase activity and cADPR concentration have been developed for cells or animal tissues. However, these methods have been used for plant extracts as well and can be adapted as needed. Homogenization might require additional reagents and equipment.
3. For a complete description of the preparation and use of sea urchin homogenate *see* ref. 30. Briefly, eggs of female *Strongylocentrotrus purpuratus* are obtained by stimulating ovulation with a 1 ml injection of 0.5 M KCl, washed with artificial seawater once, and twice in Ca^{2+}-free seawater. The eggs are homogenized in homogenization buffer at 25 % (v/v). The composition of homogenization buffer is 250 mM *N*-methylglucamine, 250 mM potassium gluconate, 20 mM HEPES, pH 7.2, 1 mM $MgCl_2$,10 μg/ml leupeptin, 10 μg/ml aprotinin, 50 μg/ml soybean trypsin inhibitor, 2 U/ml creatine kinase, 4 mM phosphocreatine, and 0.5 mM ATP. The eggs are gently homogenized with five strokes in a Dounce homogenizer. The homogenate can be stored frozen at −80 °C before use. After thawing at 17 °C for 30 min, the homogenate is diluted to 5 %, then 2.5 %, and finally 1.25 % at 30-min intervals. The dilution buffer is composed of 250 mM *N*-methylglucamine, 250 mM potassium gluconate, 20 mM HEPES, pH 7.2, 1 mM $MgCl_2$, 2 U/ml creatine kinase, 4 mM phosphocreatine, 0.5 mM ATP, and 2 μM fluo3, potassium salt. Calcium release is measured in a plate reader at excitation of 490 nm and emission of 535 nm. Sample volumes of 2–10 μl are tested with 200 μl of diluted homogenate.
4. HPLC analysis of reaction products: If needed, a method employing ^{32}P-NAD can also be used. In such a case, 100 μM NAD containing 100,000 dpm ^{32}P-NAD is incubated with the extract for various times. The reaction is stopped and the products are separated and identified by HPLC or thin-layer chromatography [31]. Analysis by HPLC can be very useful and quantitative. However, controls that should be used include incubating for different times to show that the product accumulates with time and that the NAD substrate correspondingly decreases.

5. cGDPR has unique UV and fluorescence spectral properties that distinguish it from NGD and GDPR. The fluorescence of cGDPR is linear up to 200 μM at neutral pH. If NGD is not available, NHD can also be used to measure cyclase activity [19].
6. After extraction with PCA and before incubation with NADase, the samples are kept on ice. Samples can be frozen at any step after extraction. However, samples in PCA should be stored at −80 °C to avoid breakdown of cADPR and NAADP. The NADase treatment is most effective when the samples are incubated overnight at 37 °C. The fluorescence assays are conducted at room temperature.
7. In order for the cycling assay for cADPR to work effectively, the NADase treatment must remove all of the original NAD in the sample. NADase can be removed by filtration on Immobilon plates or by ultrafiltration with 10,000 MWCO filters. When the NAD is removed, the background fluorescence of the sample should be similar to that of the control. Nucleotide pyrophosphatase can be used instead of NADase to treat samples prior to the assay of cADPR but it cannot be used before the NAADP assay because NAADP is a substrate for nucleotide pyrophosphatase [20].
8. The cycling assays for cADPR and NAD (or NADP and NAADP) are sensitive in the low nM range and can also be used to measure very low activities of cyclase that might be present in extracts. It is critical that each sample be measured as a pair of data points, i.e., in the presence and absence of cyclase or nicotinamide. Standards should be treated like samples and taken through the steps of acid extraction and NADase treatment before the cycling assay.
9. Charcoal treatment is performed on stock solutions of diaphorase that range in protein concentration from 1 to 10 mg/ml. A 2 % charcoal suspension is prepared in 20 mM sodium phosphate, pH 7, and 20 μl of enzyme is added to 60 μl of charcoal suspension and incubated for 30 min at 37 °C. The charcoal is removed by centrifugation at 10,000 × *g* for 5 min and the diaphorase is recovered in the supernatant.
10. An important feature of the cycling assay for cADPR is that the cyclase has specific recognition of cADPR and converts it to NAD in the presence of nicotinamide. However, there are some rare circumstances where analogs of NAD are resistant to NADase treatment but undergo exchange with nicotinamide in the presence of ADP-ribosyl cyclase. The measurement of such molecules results in a false positive in the cycling assay for cADPR. In order to avoid this possibility, an important control is treatment with nucleotide pyrophosphatase. Since cADPR is

the only nucleotide that resists both NADase and nucleotide pyrophosphatase, treatment with both enzymes is an important control that eliminates this false positive in the cycling assay.

11. By changing the charge on the Glu at position 98 of the cyclase, the nicotinic acid group of NAADP gains access to the active site. One of the mutants has a Gly residue at position 98 and it catalyzes the exchange reaction at pH 8 at a rate 71 times that of the wt cyclase.
12. Although the identity of the molecule is unknown at this time, a characteristic feature is that it does not co-chromatograph with NAADP. Therefore, one way of assuring that the cycling assay detects NAADP is to run the sample on an HPLC chromatographic system using the conditions described above for the exchange reaction. The cycling assay can be used to assay fractions collected from the column. Prior to the assay, the pH of the fractions is adjusted with 1 M Tris–HCl, pH 8. The positive fractions must co-chromatograph with a known standard of NAADP for correct identification of NAADP in a sample.

Acknowledgments

This work was supported in part by an RCGAS seed grant for Basic Research #201105159001 to R.M.G. This work was also supported by grants from the Council of Hong Kong (Nos. 769107, 768408, 769309, and 770610) and the National Natural Science Foundation of China/the Research Grants Council of Hong Kong (No. N_HKU 722/08).

References

1. Lee HC, Walseth TF, Bratt GT, Hayes RN, Clapper DL (1989) Structural determination of a cyclic metabolite of NAD^+ with intracellular Ca^{2+}-mobilizing activity. J Biol Chem 264:1608–1615
2. Clapper DL, Walseth TF, Dargie PJ, Lee HC (1987) Pyridine nucleotide metabolites stimulate calcium release from sea urchin egg microsomes desensitized to inositol trisphosphate. J Biol Chem 262:9561–9568
3. Lee HC, Aarhus R (1995) A derivative of NADP mobilizes calcium stores insensitive to inositol trisphosphate and cyclic ADP-ribose. J Biol Chem 270:2152–2157
4. Rusinko N, Lee HC (1989) Widespread occurrence in animal tissues of an enzyme catalyzing the conversion of NAD^+ into a cyclic metabolite with intracellular Ca^{2+}-mobilizing activity. J Biol Chem 264:11725–11731
5. Lee HC, Aarhus R (1991) ADP-ribosyl cyclase: an enzyme that cyclizes NAD^+ into a calcium-mobilizing metabolite. Cell Regul 2:203–209
6. Howard M, Grimaldi JC, Bazan JF, Lund FE, Santos-Argumedo L, Parkhouse RM, Walseth TF, Lee HC (1993) Formation and hydrolysis of cyclic ADP-ribose catalyzed by lymphocyte antigen CD38. Science 262:1056–1059
7. Love ML, Szebenyi DM, Kriksunov IA, Thiel DJ, Munshi C, Graeff R, Lee HC, Hao Q (2004) ADP-ribosyl cyclase; crystal structures reveal a covalent intermediate. Structure 12:477–486

8. Liu Q, Kriksunov IA, Graeff R, Munshi C, Lee HC, Hao Q (2005) Crystal structure of human CD38 extracellular domain. Structure 13:1331–1339
9. Graeff R, Liu Q, Kriksunov IA, Hao Q, Lee HC (2006) Acidic residues at the active sites of CD38 and ADP-ribosyl cyclase determine nicotinic acid adenine dinucleotide phosphate (NAADP) synthesis and hydrolysis activities. J Biol Chem 281:28951–28957
10. Lee HC, Aarhus R, Levitt D (1994) The crystal structure of cyclic ADP-ribose. Nat Struct Biol 1:143–144
11. Aarhus R, Graeff RM, Dickey DM, Walseth TF, Lee HC (1995) ADP-ribosyl cyclase and CD38 catalyze the synthesis of a calcium-mobilizing metabolite from NADP. J Biol Chem 270:30327–30333
12. Wu Y, Kuzma J, Marechal E, Graeff R, Lee HC, Foster R, Chua NH (1997) Abscisic acid signaling through cyclic ADP-ribose in plants. Science 278:2126–2130
13. Navazio L, Bewell MA, Siddiqua A, Dickinson GD, Galione A, Sanders D (2000) Calcium release from the endoplasmic reticulum of higher plants elicited by the NADP metabolite nicotinic acid adenine dinucleotide phosphate. Proc Natl Acad Sci U S A 97:8693–8698
14. Munshi CB, Fryxell KB, Lee HC, Branton WD (1997) Large-scale production of human CD38 in yeast by fermentation. Methods Enzymol 280:318–330
15. Lee HC, Graeff RM, Munshi CB, Walseth TF, Aarhus R (1997) Large-scale purification of Aplysia ADP-ribosylcyclase and measurement of its activity by fluorimetric assay. Methods Enzymol 280:331–340
16. Zocchi E, Carpaneto A, Cerrano C, Bavestrello G, Giovine M, Bruzzone S, Guida L, Franco L, Usai C (2001) The temperature-signaling cascade in sponges involves a heat-gated cation channel, abscisic acid, and cyclic ADP-ribose. Proc Natl Acad Sci USA 98: 14859–14864
17. Higashida H, Yokoyama S, Hashii M, Taketo M, Higashida M, Takayasu T, Ohshima T, Takasawa S, Okamoto H, Noda M (1997) Muscarinic receptor-mediated dual regulation of ADP-ribosyl cyclase in NG108-15 neuronal cell membranes. J Biol Chem 272:31272–31277
18. Sanchez JP, Duque P, Chua NH (2004) ABA activates ADPR cyclase and cADPR induces a subset of ABA-responsive genes in Arabidopsis. Plant J 38:381–395
19. Graeff RM, Walseth TF, Hill HK, Lee HC (1996) Fluorescent analogs of cyclic ADP-ribose: synthesis, spectral characterization, and use. Biochemistry 35:379–386
20. Graeff R, Lee HC (2002) A novel cycling assay for cellular cADP-ribose with nanomolar sensitivity. Biochem J 361:379–384
21. Sun L, Adebanjo OA, Koval A, Anandatheerthavarada HK, Iqbal J, Wu XY, Moonga BS, Wu XB, Biswas G, Bevis PJ, Kumegawa M, Epstein S, Huang CL, Avadhani NG, Abe E, Zaidi M (2002) A novel mechanism for coupling cellular intermediary metabolism to cytosolic Ca^{2+} signaling via CD38/ADP-ribosyl cyclase, a putative intracellular NAD^{+} sensor. FASEB J 16:302–314
22. Graeff RM, Walseth TF, Fryxell K, Branton WD, Lee HC (1994) Enzymatic synthesis and characterizations of cyclic GDP-ribose. A procedure for distinguishing enzymes with ADP-ribosyl cyclase activity. J Biol Chem 269:30260–30267
23. Menegus F, Pace M (1981) Purification and some properties of NAD-glycohydrolase from conidia of *Neurospora crassa*. Eur J Biochem 113:485–490
24. Aarhus R, Dickey DM, Graeff RM, Gee KR, Walseth TF, Lee HC (1996) Activation and inactivation of Ca^{2+} release by $NAADP^{+}$. J Biol Chem 271:8513–8516
25. Churchill GC, O'Neill JS, Masgrau R, Patel S, Thomas JM, Genazzani AA, Galione A (2003) Sperm deliver a new second messenger: NAADP. Curr Biol 13:125–128
26. Churamani D, Carrey EA, Dickinson GD, Patel S (2004) Determination of cellular nicotinic acid-adenine dinucleotide phosphate (NAADP) levels. Biochem J 380:449–454
27. Graeff R, Lee HC (2002) A novel cycling assay for nicotinic acid-adenine dinucleotide phosphate with nanomolar sensitivity. Biochem J 367:163–168
28. Gasser A, Bruhn S, Guse AH (2006) Second messenger function of nicotinic acid adenine dinucleotide phosphate revealed by an improved enzymatic cycling assay. J Biol Chem 281:16906–16913
29. Kim BJ, Park KH, Yim CY, Takasawa S, Okamoto H, Im MJ, Kim UH (2008) Generation of nicotinic acid adenine dinucleotide phosphate and cyclic ADP-ribose by glucagon-like peptide-1 evokes Ca^{2+} signal that is essential for insulin secretion in mouse pancreatic islets. Diabetes 57:868–878
30. Walseth TF, Wong L, Graeff RM, Lee HC (1997) Bioassay for determining endogenous levels of cyclic ADP-ribose. Methods Enzymol 280:287–294
31. Graeff RM, Franco L, De Flora A, Lee HC (1998) Cyclic GMP-dependent and -independent effects on the synthesis of the calcium messengers cyclic ADP-ribose and nicotinic acid adenine dinucleotide phosphate. J Biol Chem 273:118–125

Chapter 5

In Vivo Imaging of cGMP in Plants

Jean-Charles Isner and Frans J.M. Maathuis

Abstract

The cyclic nucleotide 3′,5′-cyclic guanyl monophosphate (cGMP) has been implicated in the regulation of important plant processes. To unravel its physiological role further, accurate recording of dynamic changes in cGMP concentration is necessary. Fluorescent sensors based on biological molecules for "live imaging" are ideal for this since they have high specificity, a sensitivity that is in the range of biologically relevant concentrations, high spatial and dynamic resolution, and measurements with such sensors are nondestructive. In this chapter we describe the use of the cGMP FlincG sensor in plant materials that either transiently or stably express this sensor.

Key words cGMP, 3′,5′-Cyclic guanyl monophosphate, Imaging, FlincG, Microscopy, Sensor

1 Introduction

The cyclic nucleotide 3′,5′-cyclic guanyl monophosphate (cGMP) has been implicated in the regulation of important plant processes such as stomatal functioning [1], monovalent and divalent cation fluxes [2–5], chloroplast development [6], gibberellic acid signalling [7], pathogen response [8, 9], and gene transcription [3].

Using a range of different techniques, cGMP has been detected in various tissues of a number of plant species, including barley, tobacco, and *Arabidopsis* [7, 8, 10]. To unravel its physiological role, accurate recording of cGMP concentration and fluctuations therein is necessary. Several factors hamper this endeavor; firstly, plants have many secondary metabolites which requires more sophisticated analytical approaches to detect the signalling active 3′,5′ isomer of cGMP [1] and secondly the overall tissue concentrations in plants are lower compared to animal cells. The estimates for cytoplasmic cGMP range from ~10 nM to 1 μM though local concentrations may be much higher. Thus, whereas in animal preparations cAMP and cGMP are routinely determined using ELISA-based kits, these often give inconsistent results in plant tissues.

Chris Gehring (ed.), *Cyclic Nucleotide Signaling in Plants: Methods and Protocols*, Methods in Molecular Biology, vol. 1016,
DOI 10.1007/978-1-62703-441-8_5,

An analytically more robust alternative would be the use of HPLC-MS-based techniques. However, these are labor intensive, only work with large amounts of tissue, require high levels of expertise, and are very costly in terms of infrastructure. Furthermore, both ELISA and MS methods are highly invasive and require tissues to be destroyed.

To circumvent such disadvantages, several groups developed fluorescence-based sensors for both cAMP and cGMP [3–5]. These are typically based on mammalian cyclic nucleotide-dependent kinases (e.g., PKA and PKG) or on EPACs, proteins that undergo a conformational change when cAMP or cGMP is bound. The uses of such sensors for "live imaging" are manifold and include: (1) the biological basis of the reporters endows a very high specificity for and selectivity between cAMP and cGMP; (2) the sensitivity (Kd) is in the range of biologically relevant concentrations; (3) cyclic nucleotide levels can be recorded live with high spatial and dynamic resolution; (4) measurements are nondestructive. Recently we adapted one of these sensors (δ-FlincG) for use in plants by cloning it into the binary vector pART7 for transient expression and subsequently into pGREEN for stable expression in plants [11]. In plants, the sensor has a Kd for cGMP of approximately 200 nM giving it a dynamic range of around 20–2,000 nM. This chapter gives a detailed, step-by-step protocol for usage of this cGMP sensor either via transient expression or by using tissues or cells from plants that stably express the sensor.

2 Materials

2.1 Plant Material

1. *Arabidopsis thaliana*, of any developmental stage, or rice seedlings (*see* **Note 1**) were routinely used in our lab to isolate protoplasts for transient expression but protoplasts from other plants can also be used as long as they are easily transformed.
2. Instead of transient transformation, plants that stably express FlincG can be used (*see* **Notes 2** and **3**).

2.2 Protoplast Isolation

1. Enzymes for cell wall digestion: Macerozyme R10, Cellulase RS (Scientific Laboratory Supplies LTD.), Hemicellulase (Sigma) (*see* **Note 4**).
2. Stock solutions (filter sterilized and kept frozen):
   (a) 1 M 2-(*N*-morpholine)-ethanesulfonic acid (MES)-KOH pH 5.6 (50 mL).
   (b) 1 M Mannitol (50 mL).
   (c) 1 M $CaCl_2$ (50 mL).
   (d) 1 M KCl (50 mL).
   (e) 5 M NaCl (50 mL).

3. Working solutions:
   (a) Plasmolysis buffer (50 mL): 750 mM mannitol (6.825 g in 50 mL), 1 mM $CaCl_2$, 15 mM MES-KOH pH 5.6.
   (b) Enzyme solution (~10 mL/g tissue): 1.5 % cellulase RS, 0.75 % macerozyme, 0.2 % Hemicellulase, 0.6 M mannitol, 10 mM MES-KOH pH 5.6, 1 mM $CaCl_2$, 0.1 % BSA. Incubate for 10 min at 55 °C to inactivate proteases.
   (c) Protoplast incubation medium (PIM) (50 mL): 0.6 M Mannitol, 4 mM MES (pH 5.6), 4 mM KCl, 3 mM $CaCl_2$.
4. 40 mm diameter petri dishes.
5. Desiccator.
6. Double sided razor blades.
7. Vacuum or aspirator pump.
8. Plastic 50 mL tubes.
9. Centrifuge with swing-out rotor suitable for 50 mL plastic tubes.
10. Water bath set to 55 °C.
11. 40 μm mesh sieves (e.g., Fisherbrand Cell Strainers 40 μm).
12. 0.20 μm syringe filters.

2.3 Protoplast Transformation

1. Stock solutions (filter sterilized and kept frozen):
   (a) 1 M $Ca(NO_3)_2$ (50 mL).
   (b) BSA 10 mg/mL (5 mL).
   (c) MaMg solution (50 mL): 0.6 M mannitol, 15 mM $MgCl_2$, 4 mM MES-KOH pH 5.6.
   (d) W5 solution (50 mL): 125 mM $CaCl_2$, 154 mM NaCl, 2 mM MES-KOH pH 5.6, 5 mM KCl.
   (e) PEG-CMS solution (prepare fresh each time): 1 g PEG 4000, 1 mL mannitol (1 M), 0.375 mL water, 0.250 mL $Ca(NO_3)_2$ (1 M).
2. Plasmid midi prep of FlincG-pART7(Amp^R) construct (Fig. 1) at a concentration of at least 1 μg/μL.
3. 10–15 mL plastic tubes with round bottom.
4. 25–28 °C Dark Incubators for protoplasts incubation.

2.4 Imaging and Analysis

1. Microscope. We use an inverted epifluorescence microscope (Diaphot-TMD; Nikon, http://www.nikon.com) with a 40× air objective and a Hamamatsu Orca ER CCD camera (Hamamatsu City, Japan). Emission and excitation filters were 480/20 and 520/40 nm. Clearly, any good quality epifluorescence microscope with a cooled camera having a high signal/noise ratio is suitable (*see* **Note 5**).

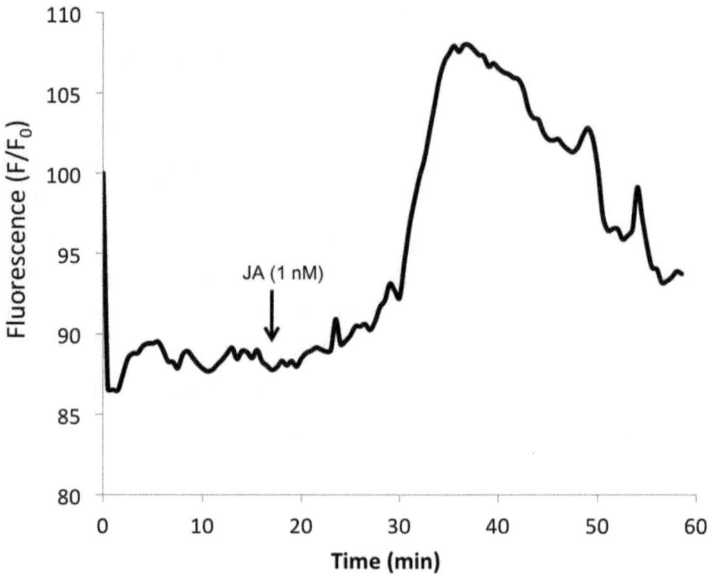

Fig. 1 δ-FlincG plant expression vector. δ-FlincG was subcloned in the plant expression vector pART7 under the control of the CaMV 35S promoter. The vector can easily be subcloned into any binary vector by using NotI restriction sites for stable expression

2. Software to acquire images and analyze fluorescence intensities. We use Simple PCI 6.1.2 imaging software (Compix Imaging Systems, http://www.compix.com).
3. 1.5–2 cm dish with a coverslip glued at the bottom.
4. Poly-L-lysine (MW >300 kDa) at a concentration of around 1 mg/mL.

3 Methods

3.1 Plant Growth

There are many suitable growth protocols for *Arabidopsis*, rice, and other plants. Any protocol that produces tissue suitable for protoplast extraction is fine.

3.2 Protoplast Isolation from Arabidopsis Leaves (See Note 6)

1. Cut 2–5 rosettes leaves and place them in a small dish (40 mm diameter) in the presence of 3 mL plasmolysis buffer. A fresh razor blade is used to slice the leaves perpendicularly in 1–2 mm strips (*see* **Note** 7).
2. Remove plasmolysis buffer and wash leaves with another 3 mL plasmolysis buffer.
3. Cover leaf slices with enzyme solution and place in desiccator under vacuum for 2 min.

4. Release vacuum slowly and incubate protoplasts in enzyme solution for 2 h in the dark at 25–28 °C.
5. Swirl dish gently to release the remaining protoplast into the buffer and filter through a small sieve (40–60 μM mesh) placed on top of a 50 mL tube.
6. Wash tissue with 2 mL plasmolysis buffer.
7. Centrifuge collected flow-through for 3 min at around 200×*g*.
8. Resuspend the protoplast pellet in PIM (if protoplasts are extracted from stably transformed lines) or W5 medium if they are to be used for a transient transformation.

3.3 Protoplast Transient Transformation (See Note 8)

1. Place protoplasts (**step 8** of Subheading 3.2) in the dark at 4 °C for 30 min.
2. Centrifuge protoplasts in W5 medium at around 200×*g* for 3 min.
3. Remove W5 medium carefully by pipetting and redissolve protoplasts in 1 mL MaMg medium.
4. Place 100 μL protoplast suspension in a round bottom tube and add 10 μL of FlincG-pArt7 plasmid (1 μg/μL), homogenize by gently tapping the tube.
5. Add 110 μL of PEG-CMS solution and homogenize by gently tapping the tube.
6. Place protoplasts in the dark for 15 min in a 25 °C incubator.
7. Add PIM buffer stepwise: Add 0.1 mL of PIM to the protoplasts and gently rotate the tube to mix. Subsequently, add 0.2, 0.4, 0.8, 1, and 2 mL of PIM, making sure the solution is mixed each time.
8. Place tube in the dark for a minimum of 6 h without exceeding 12–18 h (*see* **Note 9**).

3.4 Cell and Tissue Preparation for Imaging

1. To contain the tissues and/or cells to be imaged, we use teflon dishes with microscope coverslips glued to the bottom (*see* **Notes 10** and **11**).
2. To immobilize protoplasts, coat a coverslip with a solution of poly-l-lysine (1 mg/mL prepared in distilled water) for 10 min. Then immerse coverslip in distilled water, rinse for 5 min, and dry at room temperature or in a suitable incubator (<50 °C).
3. Swirl the tube containing protoplasts expressing FlincG to homogenize the solution.
4. Pipette 0.1–0.3 mL of protoplasts into the imaging dish/chamber and leave for 5–10 min for protoplasts to adhere to the coverslip.

5. For imaging of intact roots (*see* **Note 12**) or leaves, tissue can be taped to the microscope chamber (e.g., using a bit of tape with hole in the middle for access), or alternatively, plasticine or medical glue can be used.

3.5 Image Acquisition

We use a Hamamatsu Orca ER CCD camera, Simple PCI 6.1.2 imaging software, and a 40× air objective. Protocols may be different for other types of equipment and software.

1. If necessary install correct filters which should be suitable to detect GFP such as 480/20-nm excitation and 520/40-nm emission filter (*see* **Notes 13** and **14**).
2. Define a region of interest (ROI) on a protoplast or a region of a tissue.
3. Set the exposure time to about 1 s to test and subsequently adjust it in order to obtain around 80 % of signal saturation from the most fluorescent pixels of the ROI (*see* **Notes 15** and **16**).
4. Monitor signal until it stabilizes, typically by acquiring an image every 30 s (*see* **Note 17**).
5. When using SimplePCI software, tick on "return to idle position" and "exposure protection" in the "Filter Menu" in order to minimize photobleaching.
6. After signal stabilization, treatment can be applied, for example, by addition to the bath or via perfusion.

3.6 Signal Analysis

The description below is based on off-line signal analysis with SimplePCI, Compix (*see* **Note 18**), but generally valid for all types of software.

1. Define a ROI on a protoplast or a region of a tissue expressing FlincG (FlincG-ROI).
2. For background subtraction, define a "Background-ROI" of a nontransformed protoplast or on an area in the tissue without signal (*see* **Note 19**).
3. Analyze the average (noted Grey Average in SimplePCI) of the pixel intensity of FlincG-ROI and Background-ROI at each time point (*see* **Note 20**).
4. Export values to a spreadsheet type package.
5. Subtract the Background-ROI intensity from FlincG-ROI intensity at each time point (*see* **Note 20**).
6. Calculate the relative fluorescence F/F_0 (with F being the varying fluorescence signal and F_0 being the initial fluorescence signal) as a function of time (*see* **Note 21**).

7. A procedure for semiquantitative calibration of F/F_0 versus [cGMP] concentration is described in ref. 11 and its outcome can be used as an indication of fold changes in cellular cGMP levels. If quantitative values are required, ratiometric imaging methodology can be applied (*see* **Note 22**).

4 Notes

1. To improve rice protoplast yield we grow plants for less than 7–9 days in the dark at 28 °C.
2. Seed of *Arabidopsis* that expresses FlincG is available from the corresponding author and from the NASC seed stock center (http://arabidopsis.info/).
3. A stable transformed line of any *Arabidopsis* genotype can be made using *Agrobacterium*-mediated transformation [12] using the binary vector FlincG-pGreen0229.
4. For protoplast extraction of *Arabidopsis*, cellulysin (or other types of cellulase) can be used instead of Cellulase RS, and Hemicellulase can be omitted.
5. Confocal microscopy may also be suitable. In our experience laser scanning systems led to rapid FlincG bleaching but spinning-disk systems may be appropriate using a solid state excitation laser at 488 nm with detection at 510 nm.
6. The same protocol can be used for extracting protoplasts from other *Arabidopsis* tissues or from rice etiolated seedlings.
7. Chopping the leaves too finely is not recommended because it releases small tissue pieces with autofluorescence and it decreases the efficiency of transformation.
8. This protocol is based on ref. 13 with slight modifications.
9. Incubation may vary depending on cell type and plant species. If incubation time is too long, degradation of protoplasts might be observed.
10. For imaging the fluorescence of protoplasts, and especially the dynamics, the protoplasts and the stage need to be as stable as possible. We typically use a teflon chamber with a glued coverslip which is coated with poly-l-lysine. The microscope stage itself is mounted on a vibration free table.
11. Another possibility is to use a sealed perfusion recording chamber (e.g., RC-21BDW from Warner Instruments, Hamden, CT, USA). Although more demanding technically, it allows greater control over the incubation buffer for wash and treatment steps.

12. Roots usually show a lot of autofluorescence but this can be mathematically corrected by subtraction of signal from non-transformed roots.
13. The excitation and emission maxima of FlincG are 491 and 511 nm, respectively.
14. Since the fluorescence of FlincG is not very high, using other filters may reduce the signal/noise ratio considerably.
15. For FlincG detection in leaves, smaller exposure times with higher gains can be used with minor loss of sensitivity and lower photobleaching.
16. In photosynthetic tissue illumination with the blue light will degrade chlorophyll to a product which fluoresces near the GFP emission wavelength. To avoid incorrect signal analysis, ROIs should be chosen that are located away from chloroplasts and excitation should be reduced either with neutral density filters or by increasing the time interval between image acquisition.
17. Initially the signal often decreases over time. This apparent bleaching is not due to photobleaching but due to the partial photoconversion of the FlincG fluorescent core to a dark state upon illumination with blue light [14]. Therefore, FlincG fluorescence decreases until it reaches an equilibrium between the fluorescent and dark form of the chromophore.
18. A free imaging software can be used for Image analysis like IMAGEJ BUNDLE software (http://rsb.info.nih.gov/ij/).
19. A nontransformed root or leaf can be used to measure autofluorescence background, ensuring that the same measurement conditions (exposure time, gain, etc.) are used.
20. Most software packages will carry out automatic background subtraction.
21. Calculation F/F_0 is important to standardize the increase of FlincG fluorescence and allows the determination of the increase of cGMP concentration [11]. A signal increase of 10–20 % is usually observed in protoplasts exposed to Br-cGMP (100 μM) or ABA (10 μM) [11] (Fig. 2).
22. Ratiometric signal detection can be made with FlincG [15] because in the presence of cGMP, FlincG emission increases when excited at 480 nm while it decreases when excited at 410 nm. However, a common problem observed in plants is the high level of autofluorescence when tissue is excited 410 nm. In our experience no or little change in FlincG fluorescence was observed when excited at 410 nm, possibly due to a large autofluorescence signal. This problem might be alleviated to some extent by using software capable of spectral unmixing.

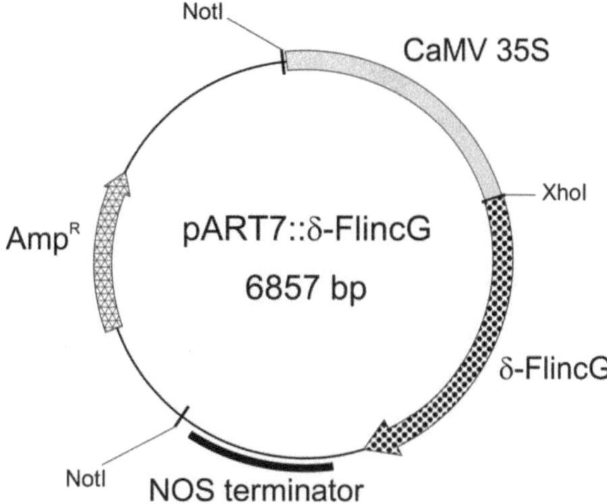

Fig. 2 Example of FlincG fluorescence detection. Fluorescence was measured in one mesophyll protoplast and standardized by dividing the FlincG signal by its initial fluorescence. After a short equilibration time, the fluorescence became stable and addition of Jasmonic acid (1 nM, *arrow*) to the protoplast buffer transiently induced a ~10 % increase of FlincG fluorescence

References

1. Newton RP, Smith CJ (2004) Cyclic nucleotides. Phytochemistry 65:2423–2437
2. Essah PA, Davenport R, Tester M (2003) Sodium influx and accumulation in *Arabidopsis*. Plant Physiol 133:307–318
3. Maathuis FJM (2006) cGMP modulates gene transcription and cation transport in *Arabidopsis* roots. Plant J 45:700–711
4. Maathuis FJM, Sanders D (2001) Sodium uptake in *Arabidopsis* roots is regulated by cyclic nucleotides. Plant Physiol 127:1617–1625
5. Rubio F, Flores P, Navarro JM, Martinez V (2003) Effects of Ca^{2+}, K^{+} and cGMP on Na^{+} uptake in pepper plants. Plant Sci 165: 1043–1049
6. Bowler C, Neuhaus G, Yamagata H, Chua NH (1994) Cyclic GMP and calcium mediate phytochrome phototransduction. Cell 77:73–81
7. Penson SP, Schuurink RC, Fath A, Gubler F, Jacobsen JV, Jones RL (1996) cGMP is required for gibberellic acid-induced gene expression in barley aleurone. Plant Cell 8: 2325–2333
8. Durner J, Wendehenne D, Klessig DF (1998) Defense gene induction in tobacco by nitric oxide, cyclic GMP, and cyclic ADP-ribose. Proc Natl Acad Sci U S A 95:10328–10333
9. Ma W, Qi Z, Smigel A, Walker RK, Verma R, Berkowitz GA (2009) Ca^{2+}, cAMP, and transduction of non-self perception during plant immune responses. Proc Natl Acad Sci USA 106:20995–21000
10. Donaldson L, Ludidi N, Knight MR, Gehring C, Denby K (2004) Salt and osmotic stress cause rapid increases in *Arabidopsis thaliana* cGMP levels. FEBS Lett 569:317–320
11. Isner JC, Maathuis FJM (2011) Measurement of cellular cGMP in plant cells and tissues using the endogenous fluorescent reporter FlincG. Plant J 65:329–334
12. Hellens RP, Edwards EA, Leyland NR, Bean S, Mullineaux PM (2000) pGreen: a versatile and flexible binary Ti vector for *Agrobacterium*-mediated plant transformation. Plant Mol Biol 42:819–832
13. Abel S, Theologis A (1994) Transient transformation of *Arabidopsis* leaf protoplasts, a versatile experimental system to study gene expression. Plant J 5:421–427
14. Sinnecker D, Voigt P, Hellwig N, Schaefer M (2005) Reversible photobleaching of enhanced green fluorescent proteins. Biochemistry 44: 7085–7094
15. Nausch LW, Ledoux J, Bonev AD, Nelson MT, Dostmann WR (2008) Differential patterning of cGMP in vascular smooth muscle cells revealed by single GFP-linked biosensors. Proc Natl Acad Sci U S A 105:365–370

Chapter 6

Characterization of Heterologously Expressed Transporter Genes by Patch- and Voltage-Clamp Methods: Application to Cyclic Nucleotide-Dependent Responses

Fouad Lemtiri-Chlieh and Rashid Ali

Abstract

The application of patch- and voltage-clamp methods to study ion transport can be limited by many hurdles: the size of the cells to be patched and/or stabbed, the subcellular localization of the molecule of interest, and its density of expression that could be too low even in their own native environment. Functional expression of genes using recombinant DNA technology not only overcomes those hurdles but also affords additional and elegant investigations such as single-point mutation studies and subunit associations/regulations. In this chapter, we give a step-by-step description of two electrophysiological methods, patch clamp and two-electrode voltage clamp (TEVC), that are routinely used in combination with heterologous gene expression to assist researchers interested in the identification and characterization of ion transporters. We describe how to (1) obtain and maintain the cells suitable for the use with each of the above-mentioned methods (i.e., HEK-293 cells and yeast spheroplasts to use with the patch-clamp methodology and *Xenopus laevis* oocytes with TEVC), (2) transfect/inject them with the gene of interest, and (3) record ion transport activities.

Key words Heterologous expression, Patch clamp, Two-electrode voltage clamp, HEK-293, Xenopus oocytes, Yeast protoplasts

1 Introduction

The combination of gene cloning followed by heterologous expression and electrophysiological assay of the translation product constitutes one of the most elegant and powerful tools to identify and characterize channel protein functions [1–4]. Having said that, every methodology, including the ones described here, has its own advantages and disadvantages and researchers are always advised to plan carefully to avoid the many caveats inherent in the use of these techniques [5–7]. To cite but a few examples, one common and potential problem inherent to the use of any kind of host system is the endogenous expression of native transporters, whether it is ion channels, pumps, or even ionotropic receptors, which could, unless

Chris Gehring (ed.), *Cyclic Nucleotide Signaling in Plants: Methods and Protocols*, Methods in Molecular Biology, vol. 1016, DOI 10.1007/978-1-62703-441-8_6, © Springer Science+Business Media New York 2013

adequate measures are taken, severely impede the detection and proper characterization of the protein of interest [8] and although it is theoretically possible to express any gene in any homo- or heterologous biological system, there may be specific drawbacks in using a particular system and/or protein, e.g., functional redundancy, presence of multiple isoforms, mislocalization, improper glycosylation, and folding or in some cases, preference of expression of certain isoforms rather than others [9]. Another major problematic factor is the relative difficulty in expressing significant amounts of nontoxic, stable, and functional proteins, which is required for their identification and characterization [10]. Researchers working on plant cyclic nucleotide-gated channels (CNGCs) are facing some of the aforementioned challenges and this presumably could explain why few significant attempts have been made to biochemically and biophysically characterize these channels [11–18]. CNGCs have been characterized as membrane proteins predominantly localized at the plasma membrane [19, 20]; however the true nature of their association with the plasma membrane is unknown due to the lack of biochemical data. Most CNGCs function as nonselective cation channels in plants [13, 21]. Based on sequence data there are 20 members of this gene family in Arabidopsis, which are phylogenetically clustered into five groups [22, 23]. These channels play an important role in various biological and physiological functions, both under normal plant growth and development as well as under biotic and abiotic stress [24, 25]. Indirect evidence suggests that CNGCs may form homo- or heteromeric proteins by interacting with identical or other isoforms [14, 24, 26]; however co-expression and functional analysis data in a heterologous system required to supporting this presumption is lacking. CNGCs have been shown to express in various heterologous systems, which can be effectively used for their functional characterization [12, 17, 27–29]. Plant CNGCs unlike their animal homologues contain both the nucleotide-binding domain (NBD) and the calmodulin-binding domain (CAM-BD) at the C-terminal as overlapping domains, a challenging characteristic which needs to be considered during the selection of a host for expression and functional characterization [14, 30]. As CNGCs are likely to be integral membrane proteins, in general it is difficult to express such proteins in heterologous system and a careful selection of a suitable host system is indispensable for two reasons: (1) some membrane proteins express well in certain host system (e.g., plant protoplasts, HEK cells, Xenopus oocytes, and yeast cells) whereas some isoforms are difficult to express specially those of membrane protein and (2) the host system should allow functional characterization of the channels with the expectation that the channel does not cause any toxicity to the host either as a result of its expression or due to its biological activity. For isoforms, which are difficult to express, codon optimization should be considered to improve heterologous expression.

Yeast is a simple and effective single-cell eukaryotic system for both the expression of heterologous genes and their functional and physiological characterization [27, 28]. Yeast is easy to grow and is genetically well characterized. The complete genome is accessible from the National Center for Biotechnology Information (NCBI) and the yeast genome database (SGD; www.yeastgenome.org). Various cloning vectors for genetic manipulations and gene expression containing both inducible and constitutive promoter are available from either commercial sources or nonprofit research institutions and laboratories [31]. In addition, approximately 6,000 deletion mutants are available as a deletion library or as independent strains from Invitrogen (tools.invitrogen.com/content/sfs/manuals/yeast_deletion_clones_man.pdf) or from American Type Culture Collection (ATCC; www.atcc.org/CulturesandProducts/Microbiology/FungiandYeast/tabid/177/Default.aspx). Although it is difficult to patch haploid yeast protoplasts due to their size (7 μm diameter), various methods and strains are available, which can yield bigger spheroplasts of reasonable size diameter (~20 μm) for patch clamping [32]. One important factor to be considered while expressing transporters in yeast for functional characterization is that the protein glycosylation (a posttranslational modification or PTM) pattern is different in yeast compared to that in plants, and this difference in glycosylation may affect the biological activity of that transporter. In cases where the long mannose chain may interfere with the biological activity, methylotrophic yeast (*Pichia pastoris*) can be used [33] (this is a patented technology by Invitrogen—http://tools.invitrogen.com/content/sfs/manuals/pich_man.pdf).

This chapter details two commonly used electrophysiological methods: (1) the patch-clamp technique that can be applied to assay from a variety of cells such as the human embryonic kidney cells (KEK-293) or yeast spheroplasts and (2) the two-electrode voltage-clamp (TEVC) technique suitable to assay for protein transport function in bigger cells such as the eggs of the African clawed frog (*Xenopus laevis*). Both techniques and maintenance protocols for three host cell types will be described in detail. In addition, we will outline the benefits of using one biological system over another, and provide some future directions.

2 Materials

2.1 HEK-293 Cells

1. Maintenance medium (MM) for HEK-293 cells based on Dulbecco's modified Eagle's medium (D-MEM): D-MEM (1×) endotoxin <0.03 %, 10 % fetal bovine serum (FBS), 1 % streptomycin solution maintained in 37 °C incubator at 5 % CO_2. HEPES-buffered saline (HBS). Freezing (or cryopreservation)

medium: 60 % FBS, 20 % MM, and 20 % DMSO. Trypsin (trypsin/EDTA).

2. Patch-clamp recording solutions: The choice of your internal and external solutions will depend on the transfected protein (channel/transporter) that you intend to study. The following example is from [12] where currents from HEK-293 cells transfected with AtCNGC2 cDNA were measured. Bath solution: 145 mM KCl, 10 mM HEPES [4-(2-hydroxyethyl)-1-piperazineethanesulfonic acid]–KOH—pH 7.4, 10 mM d-Glc, and 0.1 mM $MgCl_2$. Intracellular solution: 145 mM *N*-methyl-D-glucamine; 10 mM HEPES-KOH, pH 7.4, and 0.5 mM $MgCl_2$ (*see* **Notes 1–4**).
3. Cell culture equipment: Cell culture hood (laminar flow cabinet), CO_2 cell incubator, microscope, hemocytometer, centrifuge, water bath, freezer, osmometer (optional), and additional small items needed to work in aseptic conditions: a flame source, forceps, set of pipettes and disposable tips, serological pipettes (5, 10, and 25 ml) and pipette gun, cell culture dishes/flasks, corning tubes, cryogenic vials, waste container, and biohazard container.
4. Transfection equipment for HEK-293 cells: BioRad GenePulserII (with the Capacitance Extender Plus Module) and electroporation cuvettes.
5. Patch-clamp setup (can be used for both HEK-293 and yeast protoplasts):
   (a) Vibration isolation table and Faraday cage (*see* **Note 5**).
   (b) Inverted microscope to visualize your cells (300 times magnification; *see* **Note 6**).
   (c) Micromanipulator (*see* **Note 7**).
   (d) Gantry towers (*see* **Note 8**).
   (e) Perfusion system (*see* **Note 9**).
   (f) Amplifier (*see* **Note 10**).
   (g) Analog-to-digital (A/D) converter interface for computer data acquisition.
   (h) Software for acquisition and analysis (*see* **Note 11**).
   (i) A desktop computer running windows XP (*see* **Note 12**).
   (j) Electrical isolation transformer (*see* **Note 13**).
   (k) Pipette puller to make your patch pipettes (*see* **Note 14**).
   (l) Microforge (optional; *see* **Note 15**).
   (m) 19 in. metal electrical rack to house all electronics (also optional).

2.2 Yeast Cells

1. YEPD/YPD: An enriched medium composed of 1 % (w/v) Bacto Yeast Extract, 2 % (w/v) Bacto Peptone, and 2 % (w/v) Dextrose (glucose). The same medium is also available as premixed powder (Difco-242810; *see* **Note 16**). Synthetic Dextrose medium (SD): A synthetic medium composed of 0.17 % (w/v) Bacto Yeast Nitrogen base without amino acid and ammonium sulfate, ammonium sulfate 0.5 % (w/v), and 2 % glucose (w/v). The medium is also available as premixed powder that can be dissolved in high DI water according to the vendor's instruction (*see* **Note 17**). Synthetic complete medium (SC) prepared by autoclaving complete supplement mixture (CSM) of amino acid (as per vendor's instructions; *see* **Note 18**).
2. Restriction enzymes, T4 DNA Ligase, high-fidelity DNA polymerase, dNTP mix, gel loading buffer, DNase/RNase-free distilled water, agarose for purification of nucleic acids with agarose gel electrophoresis, 10 mg/ml ethidium bromide solution, 10× TAE buffer, and shuttle vectors for cloning and expression in yeast (*see* Subheading 3).
3. Standard laboratory incubators and incubating shakers, both set at 28–30 °C for growing yeast culture on solid or liquid medium, respectively. Microcentrifuge and PCR tubes, culture tubes and plates, cryogenic vials for culture storage, thermal cycler, agarose gel electrophoresis equipment.

2.3 *Xenopus laevis* Oocytes

1. Oocyte maintenance medium (OMM) or Barth's medium: 88 mM NaCl, 1 mM KCl, 2.4 mM $NaHCO_3$, 0.82 mM $MgSO_4$, 0.33 mM $Ca(NO_3)_2$, 0.41 mM $CaCl_2$, 5 mM HEPES, 0.1 mg/ml gentamycin, pH 7.4 (adjust pH with NaOH). The bath solution formulation for the voltage-clamp experiments will also depend on the transporter/channel that is being assayed. The following example is for KAT1 (a plant gene encoding an inward rectifier): 96 mM KCl, 1.8 mM $CaCl_2$, 1.8 mM $MgCl_2$, 10 mM HEPES, pH 7.5 (adjust pH with KOH), and a micropipette filling of 3 M KCl [12].
2. A water tank (5–10 gal.) with heavy transparent and perforated lid for housing the frogs (*see* **Note 19**).
3. Surgical tools to extract the eggs: A scalpel with new and sharp razor, a small sharp scissors, a suturing needle, surgical grade thread, and needle holder.
4. Homemade fire-polished Pasteur pipette to safely handle the eggs.
5. Incubator set at 18 °C (for oocyte storage).
6. Microinjector, glass capillaries, and the standard required when handling cRNA (RNase-free pipette tips and water) including mineral oil.

7. TEVC setup: Most patch equipment (vibration isolation table, computer, A/D interface, and software) can be used for TEVC. We will therefore only highlight some of the required items to run TEVC recording experiments (*see* **Note 20**).
   (a) Amplifier for TEVC (with two headstages).
   (b) Binocular microscope with adjustable fiber-optic lights.
   (c) Two coarse manipulators (high quality:drift-free).
   (d) Experimental chamber for oocytes.

3 Methods

3.1 HEK-293 Cell Preparation

This section details how to culture and maintain a human cell line (HEK-293) as well as how to transfect heterologously cDNAs of interest into these cells for electrophysiological recording [34]. We also describe step by step how to (1) achieve Giga seal resistances, (2) obtain whole-cell configuration of the patch-clamp technique from the initial attached cell and record whole-cell currents under voltage-clamp mode, and (3) program your Clampex software in order to run a current–voltage (*I–V*) protocol. The method described here can be easily adapted to other kinds of cells, e.g., CHO that are used for heterologously expressing proteins.

3.2 Thawing Cells for Subculturing

1. Pre-warm 10 ml of MM in one 100 mm Petri dish (do this at least 2 h in advance) (*see* **Note 21**).
2. Start the hood and spray with 70 % alcohol and wipe dry the surface.
3. Forceps and other accessories should be immersed in ethanol.
4. Have a flame ready.
5. Loosen the caps on all your vials and bottles (inside the hood).
6. Remove the cryovial of HEK-293 cells from the box stored in the nitrogen tank (should contain ~1×10^6 cells per vial and the viability should be over 80–90 %) (*see* **Note 22**).
7. Transfer the vial (still closed) to a 37 °C water bath (do not immerse in water) for just long enough time to rapidly thaw the content and check the vial to assess the thawing by observing the amount of crystals left inside.
8. When only few crystals are left, sterilize the vial by immersing it in ethanol.
9. Inside the hood, transfer the contents of the vial to a 15 ml centrifuge tube and add 5 ml MM.
10. Centrifuge in a balanced rotor for 3 min at $100 \times g$.

11. Discard as much of the supernatant as you can (which contains DMSO from the freezing medium) and be careful not to aspirate the cells (*see* **Note 23**).
12. Resuspend the pellet in 2.5 ml of MM by gently pipetting three to five times (*see* **Note 24**).
13. One could at this time perform a viable cell count. You should have at least 3×10^5 cells per ml; if this is not the case, recentrifuge and resupend in less MM medium (*see* **Note 25**).
14. Plate the cells into a 100 mm culture dish containing the pre-warmed 10 ml MM.
15. Incubate at 37 °C incubator with 5 % CO_2 and change the culture medium every 3 days or if the medium starts to go orange (indicating a change in pH).

3.3 Changing the Medium and Splitting Cell Cultures

1. Pre-warm some culture medium for 2 h in the incubator or at least 30 min in the water bath.
2. Aspirate your medium from the Petri dish or just pour the content into a waste.
3. Replace with the pre-warmed medium and incubate at 37 °C and 5 % CO_2.
4. Splitting is done when the cells are 80 % confluent and use trypsin to detach the cells from the bottom of the culture dishes (*see* **Note 26**). Have some warm culture medium ready.
5. Remove the medium and wash the cell monolayer with 5 ml of HBSS without Ca^{2+} and Mg^{2+}.
6. Discard and then wash with 5 ml of HBS with Ca^{2+} and Mg^{2+}.
7. Discard again.
8. Add ~1 ml of fresh trypsin/EDTA to just cover the cells for about 5–10 s.
9. Discard and add 3 ml of fresh trypsin/EDTA and incubate for 40–60 s (*see* **Note 27**).
10. Add 2 ml of MM to stop trypsination reaction (*see* **Note 26**).
11. Detach the cells from the bottom of the dish by squirting the medium few times.
12. Transfer the 5 ml cell suspension to a 15 ml centrifuge tube and spin at $100 \times g$ for 3 min.
13. Discard the supernatant and resuspend the cells in 3–5 ml of MM (*see* **Note 28**).
14. Plate the cells in few Petri dishes at the ratios required for your experimental load (say at 1:10 or 1:5 split ratio; *see* **Note 29**).
15. Incubate your dishes at 37 °C in 5 % CO_2.

3.4 Freezing of Cells

1. Once the cell culture in the dish reaches 80 % confluency, follow the subculturing step as described above (from **steps 5** to **13**) and resuspend the cells in a pre-thawed freezing medium (*see* Subheading 2). Perform a viable cell count and aim for cell viability of >90 % and cell density of at least 1×10^6 cells/ml. Start the gradual freeze of the cells (*see* **Note 30**).
2. Place vials in ice or at −20 °C for 1–2 h.
3. Then, transfer to an insulated container (foam ice chest), place it in −80 °C freezer overnight, and transfer the vials into liquid nitrogen.

3.5 Gene Expression in HEK-293 Cells

1. Follow the steps outlined below in the yeast section (Subheading 3.2, **step 3**) to clone the target gene in a suitable mammalian expression vector.
2. The open reading frame (ORF) can be cloned under Cytomegalovirus (CMV) immediate-early enhancer/promoter, in any commercially available mammalian expression vector such as pcDNA3.1 or pCI plasmids.
3. Once the chimeric construct is made and the cloned fragment is confirmed by sequencing, it can be introduced into HEK cells with a mammalian transfection kit (follow the protocol provided by the supplier) or by electroporation.

3.6 Electroporation Protocol

1. Split the cells 2–3 days prior to electroporation and make sure that they are 50–70 % confluent on the day of electroporation (*see* **Note 31**).
2. Place five clean sterile coverslips (*see* Subheading 3.7 below) in a plastic 35 mm dish.
3. Pipette 2 ml of MM medium into the 35 mm dish and place it in the incubator.
4. Pre-warm MM medium in two 100 mm dishes for splitting (1:5 and 1:10).
5. Remove medium from the dish containing 50–70 % confluent cells and follow the trypsination step as described above.
6. Add 2 ml of cold HBS.
7. Squirt cells a few times to detach most of the cells.
8. Collect in 15 ml conical tubes and spin down for 1 min at 1,000 rpm.
9. Resuspend cells in 1 ml of cold HBS (*see* **Note 32**).
10. Mix cells and cDNA in the electroporation cuvette: We typically use 200 ml of suspension HEK-293 cells and 1–10 mg "channel" protein encoding cDNA to which should be added 0.5 mg of CD8 plasmid (for visualization of transfected cells using anti-CD8 antibody-coated Dynabeads).

11. Homogenize the mixture gently making sure to get rid of any air bubbles on the sides of the electroporation cuvette and place on ice for 2 min.
12. Place the 4 mm gap width cuvette in its holder making sure that the electrodes face each other and are connected.
13. Set the voltage to 200 V and the capacitance to 350 μF—other cell type may require different settings—and follow instructions in the manual.
14. Switch on the GenePulser by pushing the two red buttons at the same time and release only when you hear a beep.
15. Return cuvette on ice for an additional 2 min before spreading them equally on the five round clean coverslips (see below how to clean them).
16. Before spreading, make sure to push firmly on the coverslips to make them stick to the bottom of the dish.
17. Return dish to incubator (*see* **Note 33**).

3.7 Cleaning Procedure for Microscope Cover Glasses

1. Place coverslips in a sterile beaker.
2. Fill the beaker halfway with 70 % nitric acid.
3. Cover the beaker with aluminum foil.
4. Swirl the beaker slowly so that the nitric acid flows all around the coverslips.
5. Place the beaker in a fume hood for 24–48 h.
6. Dispose of nitric acid in an appropriate container.
7. Thoroughly rinse the beaker with DI water.
8. Place coverslips in another sterile beaker and soak with 95 % ethanol.
9. Leave the coverslips for another 24 h.
10. Thoroughly rinse the beaker with DI water.
11. Place coverslips in another sterile beaker.
12. Bake the coverslips in an oven at 200 °C for 2–3 h.
13. Coverslips are ready after having cooled down to room temperature.

3.8 Patch Clamping

The patch-clamp technique is classically used for any of the cells described in this chapter (including oocytes, but removal of the vitteline membrane is required) but TEVC method is only routinely used for big cells such as oocytes and that is for technical reasons only. Both techniques require nearly the same kind of instrumentation and setup (*see* Subheading 2 above). We highly recommend that the free guide "*The Axon Guide for Electrophysiology and Biophysics Laboratory Techniques*" be downloaded and consulted.

3.9 Acquisition Protocols

Write your voltage protocols ahead of time and store them with easy-to-remember names (e.g., *I–V* curve, inactivation curve) so that you can use them on the day of the patching according to the corresponding protocol you want to use. Most recording software come with pre-written protocols as examples. One can use them as templates to start and then you can change them according to the specific kinetics of activation and inactivation characteristics of the protein or the channel studied. For instance, a neuronal outward current such as Kv2.1 (a slowly inactivating delayed rectifier K^+-channel found predominately in the CNS) will activate faster and reach steady state in much less time than an outward current originating from a plant. Instead of 200-ms long pulses required for the full activation of Kv2.1, you will need at least 600–1,000 ms and consequently the number of acquired data points or samples may need to be increased.

These are some important guidelines if you are running Clampex software:

Open Edit Protocol, in the "mode/rate" tab you can define:

1. The acquisition mode to "episodic stimulation" (one episode is made of successive voltage levels or steps A, B, C, etc. that you can choose from the "waveform" tab).
2. Set "start to start interval" to at least 5 s to leave enough time for the channel to reactivate.
3. Choose one run per trial (given that you can always rerun it).
4. Choose the sweep duration (*see* **Note 34**).
5. Number of sweeps per run will depend on the voltage span (initial and last value of testing voltages) and the interval step (or increment) between consecutive voltages. These will be set in "waveform" tab.

3.10 Making Patch Recording Pipettes

For consistent micropipettes, use a Flaming/Brown horizontal puller (e.g., from Sutter Instruments) and borosilicate capillary glass type (use of glass with filament is a personal choice). Here we follow the instructions provided by Sutter Instruments (www.sutter.com/contact/faqs/pipette_cookbook.pdf). Patch-clamp micropipette programs use four to five pulling steps with varying heat, velocity, and pull [35]. The first heat value is deduced from the built-in "heat slope program" that one can perform when receiving a new batch of glass capillaries. Decrease the heat and velocity values at the following steps. Use a "pull" step at the end if needed. Store you pipette in any homemade tight box away from the dust. To help decrease capacitance transients and reduce dielectric noise in your recordings coat your micropipettes with either Sylgard or dental wax. Here is the procedure for the latter option which requires a microforge but it is a much simpler method than Sylgarding.

1. Apply positive pressure to your micropipette.
2. Dip the tip for a brief period of time into pre-molten dental wax.
3. Place your micropipette in the microforge holder and using the microscope, position the tip about few tens of microns away from the heating element (the heating element will expand when heated).
4. Apply one or two brief heat pulses to smooth the micropipette tip and in the same time remove the dental wax away from the tip.

3.11 Silver Wire Preparation and Chloriding Procedures

In order to coat the silver wire with a thin layer of silver chloride, one of the following three methods can be applied (*see* **Note 35**).

1. The simplest way is to soak the silver wire in bleach for ≈20 min or until a purple-gray color appears. This operation has to be done before each experiment and remember to clean the bleach off before use.
2. The second most common way is electroplating. Use a 9 V battery and two alligator clips to electroplate the silver wire with chloride by making the wire positive with respect to a chloride solution (3 M KCl) and passing a current through the electrode for 10–15 s (or until you see a change in color from bright silver to purple-gray). It is highly recommended to reverse the polarity while plating the electrode at least once. In fact it is good practice to remove any leftover chloride from the old wire by reversing the polarity before the first coating.
3. To obtain the most durable silver chloride wire, dip the end of the silver wire in pre-molten silver chloride for less than a second, which will leave a tiny deposit around the tip.

3.12 Getting Cells Ready for Patching

HEK-293 cells will be electroporated on the day before use and on the day of the recording the following procedure applies.

1. Remove the medium and add 1 ml of MM on the coverslips.
2. Add 1 ml of anti-CD8-coated beads (Dynabeads M-450).
3. To allow for a good dispersion of the micro beads, add another 1 ml of MM and shake very gently (left to right—forward and backward—no circular movement).
4. Allow at least 20–60 min for conjugation before starting your recording session.

3.13 Patch-Clamp General Procedure

On the day of recording from your cells, do the following:

1. Pull and store a dozen micropipettes (as described above).
2. Assure that your vibration-free table is floating.
3. Start your equipment (computer, amplifier, A/D box).
4. Start your acquisition software and the program you will use to monitor the cell resistance while approaching the cell (usually referred to as seal protocol (*see* **Note 36**)).
5. Check that membrane voltage holding V_m is at 0 mV.
6. Fill an ice bucket and place 1 ml syringe filled with some filtered micropipette medium. In case no pre-filtered internal medium is at hand, place a sterile disposable syringe filter (0.22 μm) between your syringe and the narrow needle (*see* **Note 37**).
7. Fill a Petri dish with some filtered bath solution (*see* **Note 38**).
8. Ground your bath using a microelectrode holder half cell and reference electrode.
9. Gently and carefully, place one of the coverslips containing the plated HEK-293 cells transfected with protein of interest and CD8 plasmid in your chamber.
10. Find a cell to patch (*see* **Note 39**).
11. Focus your microscope a bit above the cell.
12. Fill the recording pipette and gently tap out any entrapped air bubbles (*see* **Note 40**).
13. Fit the RP to the pipette holder, itself mounted on the amplifier headstage (*see* **Note 41**).
14. Apply positive air pressure in your pipette using 1 ml syringe.
15. Center micromanipulator "knobs."
16. Move micromanipulator towards the chamber and swivel and lock according to the instructions of the manufacturer.
17. Lower your recording pipette into the bath and monitor the approach to the cell by looking at both your microscope and your computer monitor while the seal test protocol is running. The moment the pipette touches the bathing medium, the current signal changes shape (the resistance should read 2–3 MΩ).
18. Correct the liquid junction potential (JP).
19. Bring your RP tip to the field of vision by making short moves with your micromanipulator (*see* **Note 42**).
20. Once you find the tip, bring it to the center of the field, change the objective back to high magnification, and focus on the tip. You will notice that your pipette is continuously leaking medium out (*see* **Note 43**).

21. Carefully bring the tip of the RP close to the cell. Move your manipulator down and correct from time to time by using the other directions. Focus back and forth on the RP and on the cell. Monitor and correct the JP as you go.
22. Once at close proximity of the cell (almost touching), position your RP tip in an area of the cell membrane that is free from the beads. You will notice that the membrane underneath the RP tip slightly changes shape forming a sort of "invagination" away from the pipette (this is a good sign).
23. Final check of the PJ—correct if needed.
24. Release the positive pressure and watch the current signal changing its shape as well as the membrane resistance getting high and reaching values of few Giga Ohms (*see* **Note 44**).
25. Set the holding voltage to −70 or −80 mV and ascertain that your seal is still good.
26. Minimize the capacitance spike transients by adjusting C-fast.
27. Switch from "bath" mode to "cell" mode using amplifier knob or do it through the acquisition software.
28. Break the underlying patch of membrane by applying firm negative pressure. Notice the capacitance getting bigger. The resistance has now dropped because of the access to the interior of the cell.
29. Reset the clock.
30. Wait for 3–5 min before starting the recording. This will allow dialysis and equilibration of the interior of the cell with the internal pipette recording solution.
31. Meanwhile, use the whole-cell parameter knobs on the amplifier (or the software) to compensate for the capacitive transients (*see* **Note 45**).
32. Start recording by choosing the voltage protocol saved earlier in the protocol folder (*see* **Note 46**).

3.14 The Yeast System

1. The organism is grown in liquid or solid medium (prepared by adding 2 % w/v Bacto agar) by inoculating from glycerol stock or lyophilized culture.
2. Incubate the inoculated medium at 28–30 °C for 48 h. The liquid medium is incubated in a shaking incubator at 250 rpm. Growth can be monitored by taking $O.D_{600\ nm}$ readings at different time intervals. An $O.D_{600\ nm} = 1.0$ is an indication that the exponential growth phase was reached.
3. To make spheroplasts, exponentially growing yeast cells were first incubated in solution "A" containing 50 mM KH_2PO_4 pH 7.2 and 0.2 % β-mercaptoethanol (β-ME) incubating at 25–30 °C for 30 min with occasional shaking.

4. After 30-min incubation, the cell suspension is diluted 1:1 with buffer containing 50 mM KH_2PO_4 pH 7.2 + 0.2 % β-ME and 2.4 M Sorbitol and then adding 0.01–0.03 mg/ml Zymolyase and 25 mg/ml BSA.
5. Incubate cell suspension with gentle shaking at 30 °C for 45 min. This treatment of yeast cell with Zymolyase 20T in the presence of BSA and β-ME results in stable protoplasts, which are harvested by centrifugation at 500 × *g* for 5 min.
6. The spheroplasts pellet is finally resuspended in buffer containing 200–250 mM KCl, 10 mM $CaCl_2$, 5 mM $MgCl_2$, 5 mM MES buffer pH 7.2 adjusted with Tris base, and 0.5–1 % (w/v) glucose (*see* **Note 47**).

3.15 Candidate Gene Cloning and Yeast Transformation

1. Routine molecular biology techniques can be followed for cloning and expression of plant transporters (including CNGCs) in yeast (*S. cerevisiae* or *P. pastoris*) which include the following:
2. Design sequence-specific primers (35–50 bp long) flanked by desired restriction enzyme sites.
3. Amplify the target gene using cDNA as template, high-fidelity DNA-polymerase, and dNTPs. This reaction mix is incubated in a thermal cycler for amplification under optimized parameters (*see* **Note 48**).
4. Digest PCR product and cloning vector with the restriction enzyme and setup a ligation reaction using T4 DNA ligase.
5. The ligated product is then first transformed into chemically competent *Escherichia coli* DH5a cells for amplification, followed by sequencing to validate the identity of the cloned sequence.
6. The constructed plasmid can then be transformed into yeast cells either using yeast transformation kit or by electroporation.
7. The transformants can be selected by plating the transformation mix on SC amino acid "drop-out" agar plates prepared to allow growth of yeast transformants with the desired auxotrophic marker.
8. For cloning and expression of heterologous protein in yeast, a number of 2 μ-based or yeast integrative plasmids containing constitutive or inducible promoter are available commercially or from various laboratories upon request. Some of these vectors that we have tested are *pYES2*, with galactose-inducible *GAL1* promoter, for cloning and expression in *S. cerevisiae*, pSM1052, with phosphoglycerate kinase (PGK) promoter for constitutive expression [28]; and pPICZ AB&C, with methanol-inducible *AOX1* promoter for high-level expression in *P. pastoris* (Invitrogen; Catalog # V190-20).

3.16 *Xenopus laevis* Oocytes

Besides the low cost and easy handling of the frogs, a female frog can be used over and over providing a time lapse to allow for recovery from the surgery required for harvesting the oocytes (*see* **Note 49**). A female can lay large eggs in sufficient numbers that are easily handled using only forceps and dissecting microscope or binocular. Electrophysiological assays for protein function could be done in the 48 h following the microinjection of the oocytes with mRNA or cDNA and the oocytes can survive up to a week after that depending on the quality and the type of the protein "channel" or "transporter" expressed. One of the advantages of using Xenopus oocytes is that it provides the ability to inject multiple species of mRNA simultaneously for co-expression, thus allowing the study of channels formed by multiple subunits possible [36].

1. For the purchase of Xenopus refer to your own local dealer(s). Some companies might accept international orders; check with these companies if they can deliver to your own country. Visit the following Web sites for more details: xenopus.com, enasco.com/xenopus/, or ecocyte-us.com.
2. Caring for the frogs is of utmost importance in order to get healthy oocytes (*see* **Note 19**).
3. For the surgical removal of Xenopus ovary tissue—partial oophorectomy—make sure to have the medium freshly prepared and the surgical tools disinfected (use 70 % ethanol) and/or sterilized.
4. Wash your hands before and after handling the frogs.
5. Wipe clean the surgical area with 70 % ethanol.
6. Use 1.5 l of water containing 2.25 g of MS-222 (tricaine mesylate or ethyl 3-aminobenzoate methanesulfonate) to anesthetize the animal (*see* **Note 50**).
7. Place the anesthetized frog on its back on a damp tissue.
8. Using the scalpel, make a shallow incision of <1 cm in the skin at the lower part of the abdomen lateral to the midline.
9. Now that the abdominal connective tissue and muscle are exposed, lift the muscles gently with the forceps and use the scissors to make a small cut.
10. Extend the cut in the muscle and connective tissue by about the same or even a bit less than the 1 cm skin incision length.
11. The ovary lobes are located just under the abdomen muscle. Use the forceps to grasp a lobe and pull it out very carefully and gently to get as much egg sacks exposed as possible.
12. Use scissors to trim the lobes off and place immediately in a Petri dish containing the calcium-free OMM buffer (*see* Subheading 2).

13. You might want to quickly assess the quality of the eggs under a binocular microscope (*see* **Note 51**).
14. If satisfied with the general health of the eggs, close the incision; if not, take few more ovary sacks and recheck the quality. If this is the first time you close an incision, we recommend you practice this step with someone who has some expertise in suturing (*see* **Note 52**).
15. At this stage, mark the Xenopus used with a small colored bead that you sow to the skin (*see* **Note 53**).
16. Wash your Xenopus with some distilled water and place her to recover in a container filled with cold water.

3.17 Enzymatic Digestion of the Follicular Cell Layer

1. Cut open the individual lobes, flatten them, and cut into 10–20 mm pieces (*see* **Note 54**).
2. Transfer them to another Petri dish containing zero calcium OMM buffer.
3. In a 50 ml Falcon tube, weigh 35 mg of collagenase and dissolve in 20 ml of Ca buffer.
4. Transfer the ovary pieces into the collagenase solution and place the tube (closed) on a shaker platform (set to 80–100 rpm). The time may vary but 1–2 h should lead to a complete separation of the oocytes from one another. Defoliculated oocytes are now ready to be microinjected (*see* **Note 55**).
5. Wash the oocytes at least three times using the normal OMM (i.e., containing calcium).
6. Under the binocular microscope, select only healthy looking oocytes that are in developmental stage 5 or 6 (animal and vegetal poles are well differentiated) and place them in a new Petri dish containing the normal OMM.
7. Store the washed oocytes at a temperature around 18 °C (*see* **Note 56**).

3.18 Gene Expression in Oocytes

The expression of heterologous genes can be achieved by microinjecting either cDNA or cRNA into oocytes and cRNA expression can be achieved by injecting nanogram of cRNA into the oocyte cytoplasm. In order to express cDNA, it needs to be injected into the oocyte nucleus; this method is less convenient since it requires visual localization of the nucleus. Whichever method is used, the ORF needs to be first cloned into a vector such as *pEXO* or *pGEMHE* [36, 37]. Both plasmids contain the 5′- and 3′-untranslated region of the *Xenopus* β-globin gene for enhanced expression in Xenopus oocytes. Below we describe how to prepare cRNA:

1. cRNA is obtained by transcribing the cloned cDNA in vitro using linearized cloned vector as template and an in vitro transcription kit.

2. Once the cRNA is transcribed, it is then cleaned by phenol:chloroform extraction followed by ethanol:ammonium acetate precipitation and 70 % ethanol (v/v) wash.
3. Finally, solubilize the pellet in UltraPure™ DNase/RNase-Free DI Water (*see* **Note 57**).

3.19 Microinjection of the Xenopus Oocytes

1. Place 3–5 μl drop of cRNA mixture of interest (enough to inject ~60–100 oocytes) on top of a small strip of parafilm [38].
2. Prepare an injection micropipette (one-step pull; *see* below in Subheading 3.21) using N-51 glass capillaries and fill it with the mixture from the drop (follow the instructions of your injection instrument). One could add a dye to the cRNA mixture, such as 1 % bromophenol blue (a pH indicator), to help with the visualization of the injection solution as suggested in [38] (*see* **Note 58**).
3. Place your filled micropipette on your injection device (*see* **Note 59**).
4. Place your oocytes under a dissecting microscope.
5. Bring your micropipette close to the oocyte and swiftly stab it with your sharp micropipette roughly at a 45° angle.
6. Microinject 50 nl of solution. The microinjection point should be right at the boundary between the poles.
7. Microinject a series of 30–50 oocytes and gently rinse them twice with OMM medium.
8. Transfer them to a new Petri dish or culture plate.
9. Repeat the same operation with water as a control. For this step fewer oocytes will suffice.
10. Incubate the injected oocytes at 18 °C (*see* **Note 60**).

3.20 TEVC

While patch clamping requires only one microelectrode, TEVC, as the name indicates, uses two microelectrodes: one records voltage continuously and the other injects current. Moreover, the micropipette tips are not used to patch a piece of membrane but instead they are simply inserted inside the oocytes. Consequently, no complicated pipette fabrication is required and the impediment of getting tight seals with GΩ membrane resistance is eliminated altogether. The use of good-quality micromanipulators (drift-free) is essential for mechanical stability and getting longer experimental recordings. You also do not really need a microscope; binocular microscope is all you need (oocytes in phase V–VI are >1 mm diameter). Protocols for signal acquisition are the same as for patch clamping.

3.21 Sharp Micropipette Fabrication

1. Use borosilicate glass (thin wall with filament—GC100TF-10/ OD: 1.0 mm and ID: 0.78 mm from WARNER INSTRUMENTS) and/or N-51 glass (from KIMAX).

2. Adjust "heat," "gravity," and "pull" force to make the micropipette in a one-step pull (follow Sutter Instruments' instructions manual).

3.22 General Procedure for TEVC

1. Oocytes have been injected a few days earlier and are ready for voltage clamp (see injection protocol above).
2. Backfill two pipettes with 3 M KCl solution using a fine syringe needle (that can fit into the 0.78 mm ID of your pipettes) and place them in a storage box (away from the dust).
3. Place one electrode into each of the two holders and insert the holders into the two headstages of your amplifier (*see* **Note 61**).
4. Using the coarse manipulators, immerse the tips of your electrodes into the recording bath solution. One should read about 1 MΩ for each microelectrode. Read the values following the procedure of the amplifier you are using (*see* **Note 62**).
5. Look through the compound binocular microscope to visualize the tips.
6. Zero the pipette offset for both electrodes (*see* **Note 63**).
7. Impale the oocytes successively with both microelectrodes. The DC meter of the amplifier will read the resting membrane potential for electrode P1 (V1) and P2 (V2) (*see* **Note 64**).
8. Reset your clock and enter all relevant parameters in your logbook.
9. You are now ready to record whole-cell currents from the injected oocytes. Follow the instructions for your particular amplifier (*see* **Note 65**) (Fig. 1).

4 Notes

1. All medium solutions are made with high-quality deionized (DI) water (≥18 MΩ resistance). Reverse osmosis (RO) water can be used for any of the other tasks.
2. All solutions are filtered through 0.22 mm Millipore filters except for solutions containing enzymes.
3. Antibiotics when needed are usually added to the solutions just a few minutes before their use.
4. For the calcium-free medium, simply omit calcium from the list.
5. You can make your own Faraday cage or buy one. Having said that, one could omit it completely from the list if the patch-clamp setup is to be dedicated to measuring high current densities as in the case of heterologously expressed ion channels. Indeed, all that is needed is a good grounding of the setup. For excised patches, a Faraday cage is advisable.

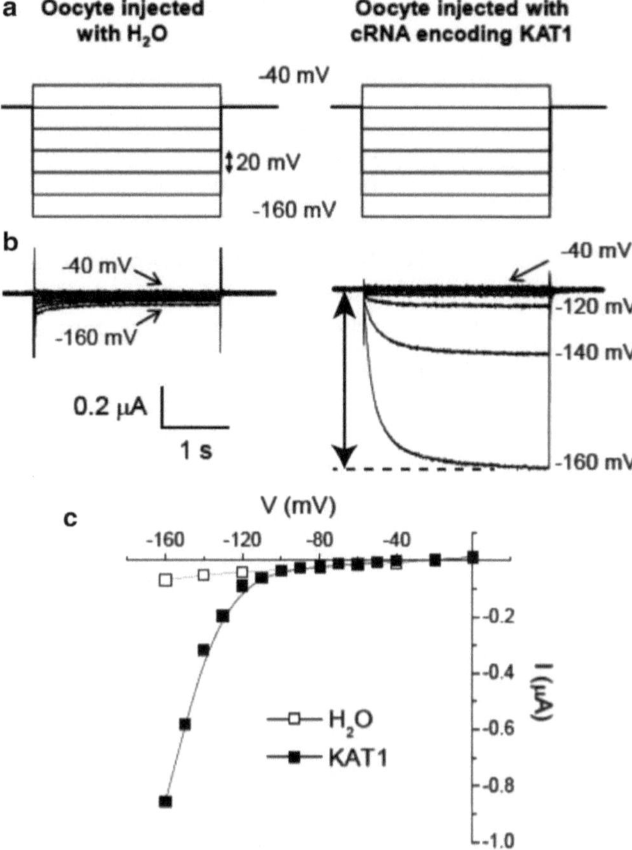

Fig. 1 Current–voltage (*I–V*) characteristic of KAT1-expressing oocytes. Example showing a set of voltage protocols (**a**) and the corresponding current traces (**b**) measured from two oocytes: one injected with water (50 nl) showing the control background currents (i.e., the amount of currents due to the endogenous expression of native channels and transporters; *left panel*) and the other one injected with cRNA (50 ng) encoding the *Arabidopsis* inward K^+-rectifier KAT1 (*right panel*). (**c**) Graphing the *I–V* relation shows that KAT1 is an inward rectifier that activates in response to hyperpolarizing voltages starting from ~–120 mV. The oocyte injected with water (*open squares*) shows a linear relation; sign of very little background current activation especially when compared to the one injected with KAT1 cRNA (*filled squares*). The *arrow* indicates how the currents were measured. External and internal solution compositions are given in Subheading 2. These oocytes were assayed 3 days following injection with either water or cRNA (unpublished data from Lemtiri-Chlieh F, Mercier R, Berkowitz GA)

6. Contrast enhancement is recommended. Opting for epifluorescence microscopy (or EFM) is a good idea especially for the detection of transfected cells using EGFP technology.
7. A motorized micromanipulator is preferable but not necessary. Choose either right- or left-handed micromanipulators depending on your preference.

8. Or any other means to position and hold the micromanipulators allowing easy access of the patch micropipette to the sample located in the recording chamber.
9. A perfusion chamber and a perfusion system is required if the bath composition needs changing (i.e., adding ligands, and blockers, or modifying the ion composition). In case a perfusion system is used, suction and waste is also required.
10. Nowadays, they come in two variants: the classic ones with the usual manual knobs on the front or the new knob-free ones where the settings are controlled by software.
11. To apply the experimental voltage commands as well as to set the amplifier parameters: These programs come with a dongle key that can be hooked to a USB connection. Most analysis modules run without a dongle.
12. Clampex (pClamp) program is still running on a Windows XP operating system.
13. Will provide line isolation, better and consistent ground, as well as surge protection.
14. For manufacturing all micropipettes including sharp microelectrodes for injecting oocytes and micropipettes to use with TEVC.
15. This is optional but can come handy to make good blunt patch pipettes and, therefore, increase the success rate of getting high seal resistances.
16. Once the medium is prepared from either individual ingredients or premixed powder, dissolve in high-quality double-distilled DI water and autoclave at 121 °C for 15 min.
17. If the medium is prepared from individual ingredients, it is prepared from mixing 10× concentrated solution A and B where solution A contains 17 g/l of YNB and 50 g/l of ammonium sulfate and solution B contains 20 g/l Glc. Both solutions are filter sterilized and diluted to formulate 1× working solution in sterilized DI water.
18. Add solution A and B (as in **Note 17**) to a 1× concentration under aseptic conditions. The CSM is available as complete, individual, or in combination with amino acid "drop-out" (www.mpbio.com/US/Pages/Search.aspx?k=CSM). The SC medium is used for screening and selection.
19. For keeping the frogs, a standing water tank can be used. The plastic water tank is covered with a transparent and perforated heavy lid. The wholes on the lid should be small (about 30 mm diameter) to stop the frogs from jumping out and keep the water well oxygenated and low in toxic compounds. Up to four females could share a 10 l tank and the water should preferably be produced by reverse osmosis (RO) system. Add

the following mineral salts to the bathing frog water tank: 700 mg/l NaCl, 10 mg/l KCl, and 20 mg/l $CaCl_2$. Water should be changed two to three times a week depending on how many frogs are there in the water tank and this should be done in a day when no feeding is taking place. Natural lighting would be preferable but a good alternative is to use a full-spectrum fluorescent bulb situated about one to two meters from the surface of the water on a light/dark cycle of 14-h light and 10-h dark. The room temperature containing the water tanks could be set up to 18–21 °C. Frog food is commercially available and feeding should be done twice or thrice a week.

20. Most patch equipment such as vibration isolation table, computer, A/D interface, and software can be used for TEVC too; therefore only some of the required items to run TEVC recording experiments are highlighted.
21. Maintaining a healthy cell population is particularly important for electrophysiological assay of heterologously expressing cells and requires some careful considerations. We described how to thaw and plate (subculture) cells from the cryovial and also how to cryopreserve cells in order to use them for further future work. It is assumed that the reader is aware of the aseptic (sterile) techniques required to successfully prevent microbial contamination.
22. Thawing should be done quickly (1 or 2 min at the most); hence everything for this operation needs to be ready ahead of time in tissue culture hood. There are two ways of thawing cells, either direct plating or centrifuge first (to get rid of the DMSO contained in the freezing medium) followed by plating.
23. Have some disinfectant (e.g., 10 % sodium hypochlorite) in your waste container.
24. Pipette up and down gently in order not to damage the cells but enough to disaggregate them.
25. For full detail on how to perform a viable cell count, go to the following Web site: http://www.abcam.com/index.html?pageconfig=resource&rid=11454.
26. FBS contained in the MM inactivates trypsin, so one needs to remove it before applying trypsin.
27. This step is critical as too much time spent in trypsin can damage the cells but not enough and detachment is not achieved.
28. Aim for 1×10^6 cells/ml.
29. A 1:5 split requires 2 ml of cell suspension into 8 ml of pre-warmed culture medium and these cells will be ready for transfection of the protein of interest in about a week.

30. Freezing the cells will insure that one has a renewable source of cells in case of unforeseen problems such as contamination. This step is best done with cells that have not been subcultured more than 6×.
31. Successful recording of ion transporters of any kind requires that the cells express these proteins at sufficient levels at the cell membrane to generate 1–2 μA of current and at the same time cells should be plated at low enough density to allow isolation and easy access by the patch pipette.
32. Trypsin hinders electroporation, so remove it. Also note that at this stage, one could, if needed, split some cells for future experiments in the pre-warmed medium.
33. If cells do not adhere easily, one can use coverslips that have been cleaned and also coated with a film of poly-l-lysine.
34. Sweep duration will depend on the sampling rate. Note that the higher the sampling rate, the better, but the files created will be much bigger as well. Higher sampling rate is required for fast activating channels (such as neuronal Nav channels). In this case, up to two sampling intervals per episode can be used (high sampling rate used initially during the fast activation of Nav and then the program switches to low sampling rate for the lower inactivation kinetics of Nav).
35. Remember to do the following before attempting any of the chloriding methods described in the methods: (1) clean your newly acquired batch of silver wire with ethanol to remove any finger residuals/oils and (2) before re-coating a wire that has been chlorided previously, remove any trace of the old layer of AgCl with some fine steel wool or pass the wire quickly through a flame. The color should look bright silver again.
36. The seal protocol is basically a small square voltage signal: 2–5 mV and 20–50 ms long and given at a high frequency.
37. Use a slightly hypo-osmotic solution for the recording pipette compared to the bath. This will improve seal resistance in the "whole-cell" configuration (and also excised outside-out). An osmometer is therefore an important instrument to have in an electrophysiology lab.
38. For the perfusion system: Fill your external 60 ml syringes with the desired filtered bath solution and initiate the perfusion system. Make sure not to have any bubbles and that the perfusion is running smoothly at about 1–2 ml/min and avoid flooding the chamber as the salt in the medium can damage the microscope optics.
39. The cell should look as healthy as possible given that they have been shocked with high voltage the day before and they should be bound to few micro beads, an indication that they have

been transfected with the CD8 plasmid. Note that it is not guaranteed that they have been transfected with the protein of interest but there is higher probability that they might have.

40. Air bubbles prevent closing of the electronic circuit (<=>high resistance reading).
41. Make sure that Ag/AgCl wire has been chlorided (*see* Subheading 3 above on how to chloride) and is in contact with the pipette solution.
42. Use right and left or forward and backward moves—do not move your manipulator up and down as this might break the tip of the pipette and basically ruin a set of cells. A good tip here is to start with the low magnification objective to locate the shadow of your pipette.
43. This is normal because of the positive pressure applied earlier. This is key in keeping the tip of your RP clean since a dirty tip means that no high resistance sealing will be formed between the pipette and the cell membrane.
44. The higher the seal value, the better. 1 GΩ seal is an acceptable value for the cell-attached configuration. If an instantaneous change in resistance and shape is not achieved, one can establish the seal by aspiring gently with a syringe and block while monitoring the resistance. Note that if the seal takes too long to establish, do not waste your time. Get a new RP, look for another cell to patch, and repeat the procedure to get a seal.
45. Enter all relevant cell parameter values into your experiment logbook (holding voltage, whole-cell capacitance, series resistance). Programs will also save these parameters automatically.
46. One important rule to remember: When recording digitized signals at 2 kHz sampling frequency, filter at 10 kHz using LowPass Bessel filter knob on your amplifier.
47. Follow the manufacturer's protocol provided with the DNA polymerase for gene amplification, since this depends on the type of high-fidelity DNA polymerase you are going to use.
48. It has been observed that leaving the protoplast suspension in the buffer containing glucose for up to 7 days will result in approximately fourfold increase in diameter. However, stable Gigaseals are best obtained using freshly prepared protoplasts, 1–4 h after final re-suspension [39].
49. The frequency of this operation can be done monthly if the surgical incision is made on alternative sides. Check with your institution however to see how many times you can operate on the same female. Note that many institutions allow five survival surgeries on a single female, with the sixth being a terminal operation.

50. It is highly recommended to use a non-recently fed animal. Put the frog in the bath for ~30 min or until it falls into a deep sleep. One could use a quick toe pinch to evaluate the responsiveness of the animal.
51. Few oocytes could be defoliculated with the forceps and if they look and feel generally healthy (i.e., firm consistence and uniform size and coloration). Pay attention to the interface between the animal (dark color) and vegetative (creamy white color) poles, which should be distinctly visible.
52. Take the suturing needle and insert it ~1 mm lateral to the incision (making sure to take muscle and fascia). Use the forceps to grasp a bit of the muscle on the other side of the incision (also ~1 mm lateral) and then pass the needle through it. Close the incision with the simple uninterrupted pattern of suturing and surgeon's square knot. Repeat the suturing with the skin layer and cut the excess thread [40].
53. This is especially useful in case one has more than one female Xenopus in the same water tank. This way one can use another female for the next experiment and leave enough time for this one to recover from the operation.
54. Corresponding to ~50–100 eggs: Wash them with calcium-free buffer in order to reduce murkiness.
55. Collagenase will only digest the follicular cell layer but it will not remove the vitteline membrane. Stripping the oocytes from their vitteline membrane is only required for patch clamping. Sometimes, visual checking that the follicular layer has been digested is not easy. To make sure of this step, one could prepare a sharp micropipette and see if it can be inserted easily into oocytes; the sharp tip of the micropipette should slide through with ease.
56. Change the OMM medium initially after 6 h and later on change the medium daily making sure of the quality of the oocytes. Remove unhealthy and/or dead-looking eggs.
57. Alternatively the cRNA can be purified using Qiagen nucleic acid purification kit.
58. Injecting the oocytes is usually done 1 day following the enzymatic digestion. Make two separate batches: one batch will be microinjected with 50 nl of the solution containing cRNA (or cDNA) encoding your protein of interest and the second batch will be microinjected with 50 nl of DI water. Initially, one could use different concentrations of cRNA and later on maybe stick to one or two concentrations that give a good amount of recorded current (between 0.5 and 2 μA). As mentioned earlier, cRNA is easier to inject as you only inject the cytoplasm. Injecting cDNA is comparatively harder as it has to be injected into the nucleus.

59. Make sure that droplet of solution at the tip of your micropipette is seen each time before pushing the injection button.
60. These oocytes could last for up to a week.
61. Make sure that Ag/AgCl wire has been chlorided and is in contact with the KCl solution.
62. For the GeneClamp 500 (GC500): Set the MODE to "SETUP"; DC METER on "V1" and "I2/V2"; SCALED OUTPUT on "I2"; Set the FREQUENCY (LOW-PASS FILTER) to 500; Set the GAIN to "x1"; to read resistance values, depress "R1" for "P1" and "R2" for "P2."
63. Using GC500 amplifier, simply push "Zero 1" and "Zero 2"—the panel meter will read 0 mV for P1 and P2.
64. Acceptable values range from –40 to –70 mV.
65. For the GC500: Switch the MODE to "VOLTAGE CLAMP"; set the GAIN as needed; set the stability to "200 μs"; set the dial for the holding potential in "OFF" position; the DC meter displays the holding potential in mV (V1) and the current required to clamp the membrane at a constant potential in μA (I2) (in this particular amplifier, the third headstage is a virtual ground and is used to actively clamp the bath potential to zero).

References

1. Gurdon JB, Lane CD, Woodland HR, Marbaix G (1971) Use of frog eggs and oocytes for the study of messenger RNA and its translation in living cells. Nature 233:177–182
2. Cao YW, Anderova M, Crawford NM, Schroeder JI (1992) Expression of an outward-rectifying potassium channel from maize messenger-RNA and complementary RNA in Xenopus Oocytes. Plant Cell 4:961–969
3. Cao YW, Ward JM, Kelly WB, Ichida AM, Gaber RF et al (1995) Multiple genes, tissue-specificity, and expression-dependent modulation contribute to the functional diversity of potassium channels in *Arabidopsis thaliana*. Plant Physiol 109:1093–1106
4. Véry AA, Gaymard F, Bosseux C, Sentenac H, Thibaud JB (1995) Expression of a cloned plant K^+ channel in Xenopus Oocytes—analysis of macroscopic currents. Plant J 7:321–332
5. Schroeder JI (1995) Heterologous expression of higher plant transport proteins and repression of endogenous ion currents in Xenopus oocytes. Methods Cell Biol 50:519–533
6. Tapper AR, George AL Jr (2003) Heterologous expression of ion channels. Methods Mol Biol 217:285–294
7. Terhag J, Cavara NA, Hollmann M (2010) Cave Canalem: how endogenous ion channels may interfere with heterologous expression in Xenopus oocytes. Methods 51:66–74
8. Bernaudat F, Frelet-Barrand A, Pochon N, Dementin S, Hivin P et al (2011) Heterologous expression of membrane proteins: choosing the appropriate host. PLoS One 6:e29191
9. Yesilirmak F, Sayers Z (2009) Heterelogous expression of plant genes. Int J Plant Genomics 2009:96482
10. Davenport R (2002) Glutamate receptors in plants. Ann Bot 90:549–557
11. Leng Q, Mercier RW, Yao W, Berkowitz GA (1999) Cloning and first functional characterization of a plant cyclic nucleotide-gated cation channel. Plant Physiol 121:753–761
12. Leng Q, Mercier RW, Hua BG, Fromm H, Berkowitz GA (2002) Electrophysiological analysis of cloned cyclic nucleotide-gated ion channels. Plant Physiol 128:400–410
13. Hua BG, Mercier RW, Leng Q, Berkowitz GA (2003) Plants do it differently. A new basis for potassium/sodium selectivity in the pore of an ion channel. Plant Physiol 132: 1353–1361

14. Hua BG, Mercier RW, Zielinski RE, Berkowitz GA (2003) Functional interaction of calmodulin with a plant cyclic nucleotide gated cation channel. Plant Physiol Biochem 41:945–954
15. Urquhart W, Gunawardena A, Moeder W, Ali R, Berkowitz GA, Yoshioka K (2007) The chimeric cyclic nucleotide-gated ion channel ATCNGC11/12 constitutively induces programmed cell death in a Ca^{2+} dependent manner. Plant Mol Biol 65:747–761
16. Yoshioka K, Moeder W, Kang HG, Kachroo P, Masmoudi K et al (2006) The chimeric Arabidopsis cyclic nucleotide-gated ion channel 11/12 activates multiple pathogen resistance responses. Plant Cell 18:747–763
17. Mercier RW, Rabinowitz NM, Ali R, Gaxiola RA, Berkowitz GA (2004) Yeast hygromycin sensitivity as a functional assay of cyclic nucleotide gated cation channels. Plant Physiol Biochem 42:529–536
18. Balagué C, Lin BQ, Alcon C, Flottes G, Malmstrom S et al (2003) HLM1, an essential signaling component in the hypersensitive response, is a member of the cyclic nucleotide-gated channel ion channel family. Plant Cell 15:365–379
19. Christopher DA, Borsics T, Yuen CYL, Ullmer W, Andeme-Ondzighi C et al (2007) The cyclic nucleotide gated cation channel AtCNGC10 traffics from the ER via Golgi vesicles to the plasma membrane of Arabidopsis root and leaf cells. BMC Plant Biol 7:48–51
20. Gobert A, Park G, Amtmann A, Sanders D, Maathuis FJM (2006) Arabidopsis thaliana cyclic nucleotide gated channel 3 forms a non-selective ion transporter involved in germination and cation transport. J Exp Bot 57: 791–800
21. Sherman T, Fromm H (2009) Physiological roles of cyclic nucleotide gated channels in plants. In: Baluška F, Mancuso S (eds) Signaling in plants, signaling and communication in plants. Springer, Berlin, pp 91–106
22. Maser P, Thomine S, Schroeder JI, Ward JM, Hirschi K et al (2001) Phylogenetic relationships within cation transporter families of Arabidopsis. Plant Physiol 126:1646–1667
23. Zelman AK, Dawe A, Gehring C, Berkowitz GA (2012) Evolutionary and structural perspectives of plant cyclic nucleotide-gated cation channels. Front Plant Sci 3:95
24. Ma W, Yoshioka K, Gehring C, Berkowitz G (2010) The function of cyclic nucleotide-gated channels in biotic stress. In: Demidchik V, Maathuis F (eds) Ion channels and plant stress responses, signaling and communication in plants. Springer, Berlin, pp 159–174
25. Yuen C, Christopher D (2010) The role of cyclic nucleotide-gated channels in cation nutrition and abiotic stress. In: Demidchik V, Maathuis F (eds) Ion channels and plant stress responses, signaling and communication in plants. Springer, Berlin, pp 137–157
26. Lemtiri-Chlieh F, Thomas L, Marondedze C, Irving H, Gehring C (2011) Cyclic nucleotides and nucleotide cyclases in plant stress responses. In: Shanker AK, Venkateswarlu B (eds) Abiotic stress response in plants—physiological, biochemical and genetic perspectives. InTech, Croatia, pp 137–182
27. Lemtiri-Chlieh F, Ali R, Berkowitz G (2004) Use of yeast as a heterologous expression system for electrophysiological analysis of plant CNGC channels. ASPB meeting, Lake Buena Vista, FL USA. http://abstracts.aspb.org/pb2004/public/P36/7652.html. Accessed 27 Sep 2012
28. Ali R, Zielinski RE, Berkowitz GA (2006) Expression of plant cyclic nucleotide-gated cation channels in yeast. J Exp Bot 57: 125–138
29. Sunkar R, Kaplan B, Bouche N, Arazi T, Dolev D et al (2000) Expression of a truncated tobacco NtCBP4 channel in transgenic plants and disruption of the homologous Arabidopsis CNGC1 gene confer Pb^{2+} tolerance. Plant J 24:533–542
30. Ali R, Ma W, Lemtiri-Chlieh F, Tsaltas D, Leng Q et al (2007) Death don't have no mercy and neither does calcium: Arabidopsis cyclic nucleotide gated channel2 and innate immunity. Plant Cell 19:1081–1095
31. Ton VK, Rao R (2004) Functional expression of heterologous proteins in yeast: insights into Ca^{2+} signaling and Ca^{2+}-transporting ATPases. Am J Physiol Cell Physiol 287:C580–C589
32. Terpitz U, Raimunda D, Westhoff M, Sukhorukov VL, Beauge L et al (2008) Electrofused giant protoplasts of *Saccharomyces cerevisiae* as a novel system for electrophysiological studies on membrane proteins. Biochim Biophys Acta 1778:1493–1500
33. Li P, Anumanthan A, Gao XG, Ilangovan K, Suzara VV et al (2007) Expression of recombinant proteins in Pichia pastoris. Appl Biochem Biotechnol 142:105–124
34. Senatore A, Boone AN, Spafford JD (2011) Optimized transfection strategy for expression and electrophysiological recording of recombinant voltage-gated ion channels in HEK-293T cells. J Vis Exp 47:e2314
35. Brown AL, Johnson BE, Goodman MB (2008) Making patch-pipettes and sharp electrodes with a programmable puller. J Vis Exp e939

36. Liman ER, Tytgat J, Hess P (1992) Subunit stoichiometry of a mammalian K^+ channel determined by construction of multimeric cDNAs. Neuron 9:861–871
37. Duprat F, Lesage F, Fink M, Reyes R, Heurteaux C et al (1997) TASK, a human background K^+ channel to sense external pH variations near physiological pH. EMBO J 16:5464–5471
38. Cohen S, Au S, Panté N (2009) Microinjection of Xenopus leavis oocytes. J Vis EXp e1106
39. Bertl A, Bihler H, Kettner C, Slayman CL (1998) Electrophysiology in the eukaryotic model cell Saccharomyces cerevisiae. Pflugers Arch 436:999–1013
40. Schneider PN, Hulstrand AM, Houston DW (2010) Fertilization of Xenopus oocytes using the host transfer method. J Vis Exp e1864

Chapter 7

Noninvasive Microelectrode Ion Flux Estimation Technique (MIFE) for the Study of the Regulation of Root Membrane Transport by Cyclic Nucleotides

Natalia Maria Ordoñez, Lana Shabala, Chris Gehring, and Sergey Shabala

Abstract

Changes in ion permeability and subsequently intracellular ion concentrations play a crucial role in intracellular and intercellular communication and, as such, confer a broad array of developmental and adaptive responses in plants. These changes are mediated by the activity of plasma-membrane based transport proteins many of which are controlled by cyclic nucleotides and/or other signaling molecules. The MIFE technique for noninvasive microelectrode ion flux measuring allows concurrent quantification of net fluxes of several ions with high spatial (μm range) and temporal (ca. 5 s) resolution, making it a powerful tool to study various aspects of downstream signaling events in plant cells. This chapter details basic protocols enabling the application of the MIFE technique to study regulation of root membrane transport in general and cyclic nucleotide mediated transport in particular.

Key words Membrane transport, MIFE, Cell signaling, Cyclic nucleotides, Microelectrodes, Ion fluxes, Stress responses, Channels, Membrane transporters

1 Introduction

Signaling mechanisms enable living organisms to adjust cell behavior to developmental needs and environmental constraints. Such mechanisms sense external information, transduce the messages into internal (chemical or electrical) signals, and trigger highly specific responses. At the cellular level the signal transduction machinery involves multiple components such as receptors, various second messengers (either protein or nonprotein molecules), transcription factors, and downstream targets, e.g., membrane ion transporters [1–6]. In plants an increasing number of molecules have been shown to function as second messengers that mediate signals and these include Ca^{2+} [1], cyclic nucleotides [7], cytosolic pH [8], ROS [9], nitric oxide [10], and lipids [11].

Chris Gehring (ed.), *Cyclic Nucleotide Signaling in Plants: Methods and Protocols*, Methods in Molecular Biology, vol. 1016,
DOI 10.1007/978-1-62703-441-8_7,

Changes in ion permeability and subsequently intracellular ion concentrations play a role in intracellular and intercellular communication, both as signal transducers and elicitors of physiological responses. In plant roots, critical physiological and developmental processes depend on transport processes: most nutrients are taken from the soil in the ionic form [12], water transport is partially mediated by proteins [13], root hair growth and nodule formation by N-fixing rhizobia in legumes are controlled by ion gradients [14], finally adaptive responses to soil adverse conditions, such as drought, salinity, and mechanical perturbations, rely on ion channels [15–17].

In light of the above, elucidating regulatory mechanisms of ion transport is essential for plant biologists to understand both adaptive and developmental processes in living organisms. Importantly, plant membranes host hundreds of transport proteins. For cations, 46 unique families are known, including >800 members in *Arabidopsis thaliana* [18], i.e., 5 % of the entire *Arabidopsis thaliana* genome. For K^+ transport alone, 75 genes from 7 different families have been identified [19, 20]. The activity of these transporters is controlled by numerous factors such as the membrane potential (voltage) [21], pH, ROS [22], stretching [15], phosphorylation status of the transporters [23, 24], and various ligands. Ligand binding transporters undergo conformational changes that lead to its activation, inhibition, or changes in voltage sensitivity. Such regulatory modifications have been reported to occur in response to hormones (e.g., auxin), ions (Ca^{2+} and H^+), trinucleotides (ADP and ATP) [21], amino acids [25, 26], phosphoinositides (e.g., PtdInsP2 and PtdInsP3 [27]), and cyclic nucleotides (cAMP and cGMP) [28]. Moreover, transporter activities may also be controlled at the transcriptional level that in turn affects the actual number of transporters localized at the membrane [29, 30].

Noninvasive microelectrode ion flux measuring (the MIFE system) allows concurrent quantification of net fluxes of several ions with high spatial (several μm) and temporal (ca. 5 s) resolution [31] and has been successfully used over the last decade to elucidate many different aspects of signal transduction pathways in living organisms. Selected examples include receptor-like activity in Arabidopsis root plasma membrane evoked by ATP [32] and ADP [33], cross-talks between polyamines and ROS signal transduction pathways [34], SOS signaling pathway [5], control of channel activity by plant natriuretic peptides [35, 36] and cyclic nucleotides [37], early signaling events associated with pathogen elicitors [38], regulation of transporters activity by MAP kinase cascade in fungi [39], and various aspects of light perception and signaling [40–42]. Here we detail specific protocols for the application of the MIFE technique for the identification and quantification of ion flux responses to signaling molecules in general and to the cyclic nucleotide cGMP in particular.

2 Materials

2.1 Plant Sample Preparation and Treatment

1. Plant growth medium: 0.43 % w/v Murashige and Skoog Basalt salt mixture (MS), 1 % sucrose, 0.35 % w/v Phytagel (*see* **Note 1**).
2. 90 × 16 mm Petri dishes.
3. Autoclave.
4. *Arabidopsis thaliana* seeds.
5. Autoclaving flask.
6. Parafilm.
7. Cyclic GMP stock solutions (*see* **Note 2**): 100 μM 8-bromoguanosine 3′5′-cyclic monophosphate sodium salt (Br-cGMP) dissolved in deionized water. The concentrations of stock solutions should be 50–100 higher than the working solutions.
8. 100 μM 3′5′-cyclic monophosphate sodium salt (cGMP) dissolved in deionized water.
9. Basic measuring solution (BSM): 0.5 mM KCl, 0.1 mM $CaCl_2$ (*see* **Note 3**), Tris-base and 2-(*N*-morpholino) ethanesulfonic acid (MES) (*see* **Note 4**), pH 6.
10. 55 × 15 mm Petri dishes.
11. Plastic mesh with an exclusion size of 1–2 mm.
12. Custom-made measuring chamber (*see* **Note 5**).
13. Fine tweezers.
14. Custom-made specimen holders (*see* **Note 6**).
15. Micropipettes: 0.5–10, 10–100, and 200–1,000 μL.

2.2 Electrode Fabrication

1. One-stage vertical electrode puller.
2. Nonfilamentous borosilicate glass capillaries 1.5 mm OD × 0.86 mm.
3. Electrode holders (e.g., E45W-F15PH; Warner Instruments, USA).
4. 10 mL syringe with a nonmetallic needle to fill electrodes (e.g., MicroFil MF34G-5, 0.1 mm ID; WPI, Sarasota, FL, USA).
5. Small oven that can heat to 250 °C, gloves, and metal electrode racks with metal covers.
6. Silanising agent (e.g., tributylchlorosilane; Fluka Chemicals 90796) (*see* **Note 7**).
7. Liquid ion exchangers (LIXs) (Fluka of Sigma-Aldrich, USA) (*see* **Note 8**).
8. Back-filling solutions are specific for each LIX. For the most common ions these are: H^+ (15 mM NaCl, 40 mM KH_2PO_4),

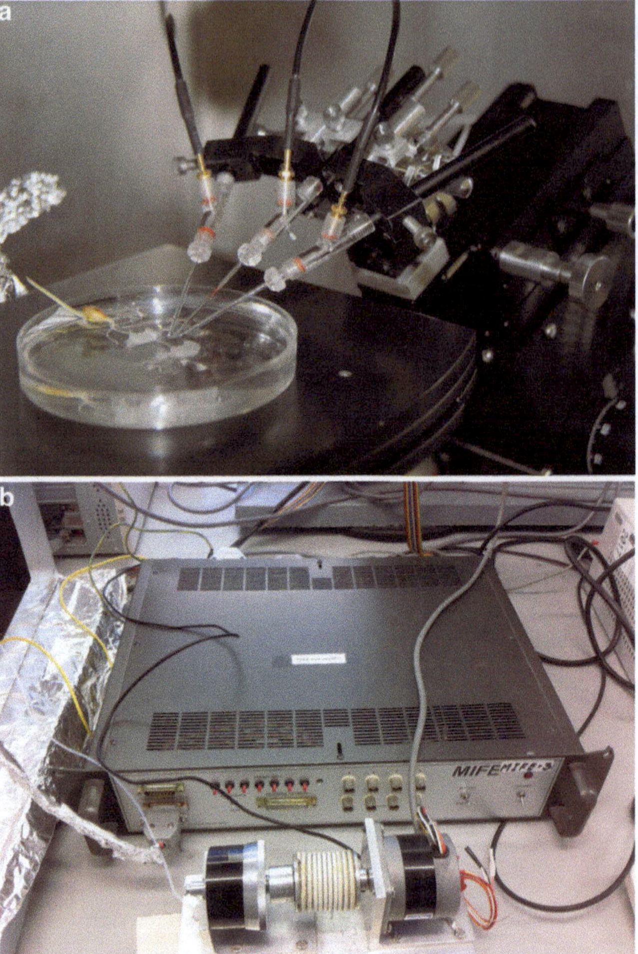

Fig. 1 Close-up of MIFE setup using inverted microscope configuration. (**a**) Plant specimen immobilized in a holder at the bottom of a standard Petri dish. Three MIFE microelectrodes come down from the right; these are assembled on the cartridge of a hydraulic manipulator and positioned close to the specimen using fine and coarse three-dimensional mechanical manipulators. (**b**) MIFE electronics and a one-dimensional hydraulic manipulator connected to the stepping motor to control electrode movement

K^+ (200 mM KCl), Na^+ (500 mM NaCl), Ca^{2+} (500 mM $CaCl_2$), and Cl^- (500 mM $MgCl_2$).

9. Electrode filling station consisting of two simple micromanipulators and a stereomicroscope.
10. Reference electrode: Chloride silver wire, glass capillary (50 μm diameter), 1 M KCl in 2 % w/v agar, and 0.25 N HCl.

2.3 MIFE Equipment Components

1. Microscope with long-working distance objectives (100× or 200×) as shown in Fig. 1.
2. MIFE™ system including main amplifier, controller, preamplifier, multi-manipulator providing 3-axis positioning and a stepper motor (available for purchase from Research Office Commercialization Unit, University of Tasmania, Australia) (Fig. 1).
3. PC for system control and data acquisition, with CIO-DAS08 card for analogue to digital conversion and installed MIFE CHART and MIFEFLUX software.
4. Anti-vibration table.
5. Faraday cage.

2.4 MIFE Calibration

The calibration solution is identical to the back-filling solutions (*see* **Note 9**) and deionized water must be used to prepare calibration solutions.

3 Methods

3.1 Plants Growth and Treatment

1. Dissolve MS medium and sucrose in distilled water, and stir with a magnetic stir to ensure proper mixing, adjust the pH to 6.0 using KOH and HCl, transfer to a volumetric flask, and add water to reach the final volume. Transfer the MS solution to an autoclaving flask; add the Phytagel and autoclave for at least 15 min at 121 °C.
2. Pour the autoclaved medium into Petri dishes and wait until it solidifies. Take sterilized *Arabidopsis thaliana* seeds and distribute over the medium surface. Seal Petri dishes with Parafilm.
3. Keep Petri dishes at 4 °C for 2 days to stratify the seeds. Afterwards, place the Petri dishes into a growth chamber providing the biological conditions required for the experiment. Position them vertically and keep them there for 5–8 days (*see* **Note 10**).
4. To pretreat the seedlings, prepare 4 mL of BSM containing the second messenger and pour into a 55 × 15 mm Petri dish. Then, insert the 55 mm plastic mesh and use the tweezers to gently transfer the seedlings into the mesh holes and ensure that the roots are completely immersed (*see* **Note 11**).
5. Take the small Petri dishes into the growth room for 24 h (*see* **Note 12**).

3.2 Electrode Fabrication

1. Insert nonfilamentous borosilicate glass capillaries into a vertical pipette puller.
2. Adjust the puller settings to produce electrodes with tip diameter of ca. 2 μm and switch on the electrode puller.
3. Store pulled electrodes in a stainless steel or aluminum covered rack in a vertical position before silanising.
4. For silanising, place electrode blanks uncovered in a rack with tips upright and base down and keep in the oven set at 250 °C overnight.
5. 10–15 min before silanization, place a steel cover over the electrode blanks.
6. Add 65 μL of tributylchlorosilane on the rack under the cover using a micropipette.
7. After 10 min, remove the lid and bake electrodes for a further 30 min.
8. Turn the oven off and let the electrodes cool down.
9. Position silanized and cooled blanks at the electrode filling station.
10. Back-fill the electrodes with appropriate back-filling solution using the syringe and nylon needle to fill three-fourth of the barrel's length under the stereomicroscope (*see* **Note 13**).
11. Front-fill the blank tip with LIX by putting it briefly into contact with the LIX-containing tube to achieve the column length of ~100 μm.
12. Label the electrodes and place them into BSM.
13. Fabricate the reference electrode by galvanizing a silver wire in a 0.25 N HCl solution for 15 min. Fill a glass capillary with 1 M KCl in 2 % agar. Place the wire into the capillary and seal with Parafilm.

3.3 Calibration of the Electrodes

1. Mount three electrodes in the MIFE holder and connect the reference electrode. Immerse electrodes in appropriate standard solution.
2. Open the MIFE CHART software.
3. Using the MIFECHART routine, set channel offsets.
4. Open the MIFE chart function and define the ions that are going to be measured.
5. With electrodes immersed into the calibration solution, record the voltage outputs for 15–20 s. Press F7 to store the readings.
6. Repeat the procedure for at least two more standards for the same ion.
7. Run the averaging routine. For each ion the program will calculate the equation that relates electric potential (mV) and ion concentration (mM) (*see* **Note 14**). The values are stored automatically in an AVC file.

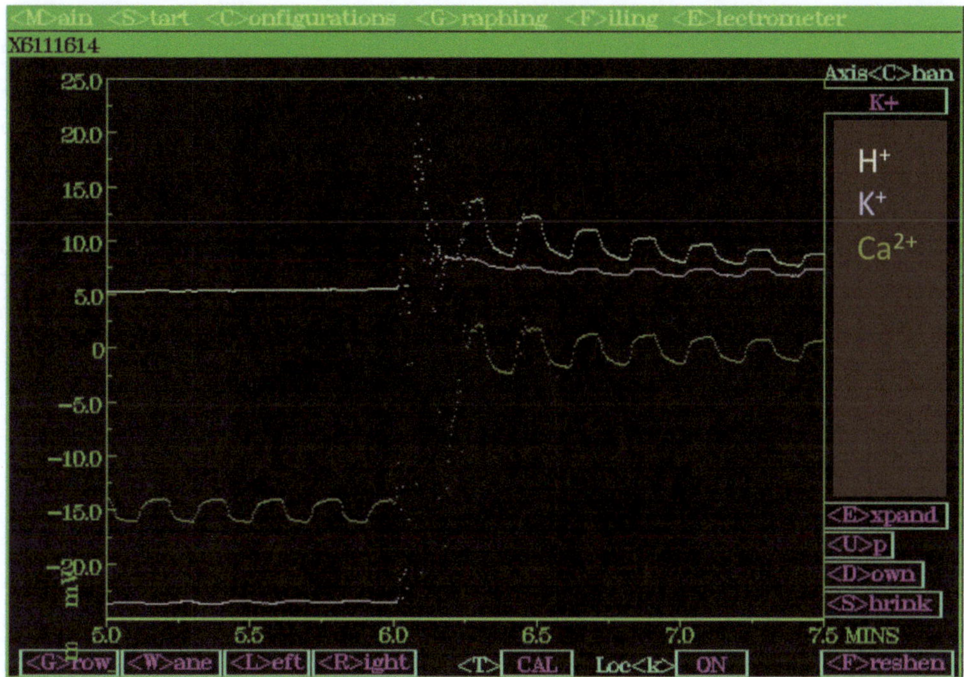

Fig. 2 Screenshot of output traces of MIFE measurements of H^+, K^+, and Ca^{2+}

3.4 Sample Preparation for MIFE

1. To immobilize the sample, take a specimen holder and humidify it with drops of BSM, use the tweezers to place the seedling root in the middle of the holder and wrap it with Parafilm (*see* **Note 15**). Place the holder with immobilized seedling attached into the measuring chamber. Fill the chamber with BSM solution containing the desired concentration of cGMP and ensure that the plant specimen is completely submersed.
2. Put the chamber under dim green light for 1 h.

3.5 Measuring Net Ion Fluxes

1. Mount the measuring chamber in the multi-holder. Turn on the microscope light.
2. Position the microelectrodes using the microscope. Use the fine mechanical micromanipulator to position the electrodes in the same plane (*see* **Note 16**).
3. Position the ion selective microelectrodes with their tips close together, 20–40 μm above the root surface (*see* **Note 17**).
4. Put the reference electrode inside the chamber.
5. Start a new file and give it a name.
6. Turn on the stepper motor and start recording (*see* **Note 18**).
7. Measure ion fluxes for 5–10 min to ensure that they reach steady state. Typical output traces for H^+, K^+, and Ca^{2+} are shown in Fig. 2.

8. Add appropriate chemical compound (treatment) to the chamber. Mix it gently but thoroughly with 5 mL pipette (*see* **Note 19**).
9. Record ion fluxes for as long as required to obtain a biological response (*see* **Note 20**).
10. Stop data acquisition and turn off the stepper motor.
11. Create an AVM file by using MIFEFLUX routine. Enter values for the following variables: Root radius (μm), the distance between electrode and the root surface (μm), buffer concentration (mM), and the Valid Data Interval (s).
12. Remove the measuring chamber and the electrodes.

3.6 Data Analysis

1. Start the MIFEFLUX program.
2. Type the name of the AVC and AVM files for which you want to calculate the fluxes.
3. Select cylindrical geometry.
4. The program creates a FLX file (ASCII format) that can be opened by a spreadsheet program (e.g., Excel).

4 Notes

1. Here we describe plant growth in solid media, however, they can also be grown hydroponically or on solid agar.
2. Experiments can be performed using other signaling molecules (e.g., phosphoinositides, cAMP, sugars, and signal peptides).
3. The specific ionic composition of measuring solutions depends on the specific purpose of the experiment. A few general principles, however, have to be observed: (1) solution ionic composition should be as simple as possible to avoid potential confounding effects of interfering ions or LIX poisons and (2) concentrations of ions whose fluxes are to be measured should be kept as low as practically (physiologically) possible to maximize signal-to-noise ratio of the measured signal. In this context, we have found that BSM solution (0.5 mM KCl + 0.1 mM $CaCl_2$) is a good compromise in most cases.
4. The need for pH buffer is determined by the attempt to reduce the confounding effects of the Donnan exchange in the cell wall [43] on Ca^{2+} flux measurements. However, if one attempts to measure net H^+ fluxes, the amount of TRIS and MES used should be reduced to only 1–2 mM, or not be used at all.
5. The simplest type of measuring chamber is a Petri dish; this will be suitable for MIFE configuration with an inverted microscope.
6. Various types of custom-made holders may be used; their shape and design depends strongly on the geometry of the measured

specimen, e.g., root or leaf segment; epidermal peel, etc. The basic principle is to ensure complete immobilization of the specimen without much mechanical trauma/stimulation.

7. When handling tributylchlorosilane use gloves and work under the fume hood in order to avoid skin contact and inhalation of this corrosive compound.
8. LIXs are cocktails containing selective ionophores for a range of specific ions. The list of commercial LIXs can be found at http://www.sigmaaldrich.com. Of these, some have rather poor selectivity (e.g., Na^+ LIX discriminates poorly between K^+ and Ca^{2+} [44], while Mg^{2+} LIX is highly sensitive to Ca^{2+} [45]). This should be taken into account when planning experiments. The problem of poor selectivity may be also partially overcome by either mathematical procedures [45] or using specific treatment protocols [46].
9. A three-point calibration is used routinely and works well; the range of standards should be broad enough to cover the concentrations of nutrients used in experiments.
10. The seedlings grow vertically so that the roots do not penetrate the solid medium.
11. The treatment protocols depend on the research questions and some examples include: (1) pretreatment of the roots with the signaling molecules and then study how this modifies ion flux responses to one environmental stress (e.g., salinity, hypoxia, oxidative stress), (2) comparison of the responses of wild type and mutant lines to reveal the functional role of a specific gene in response to signaling molecules, (3) the use of inhibitors to characterize the type of channels that are investigated, and (4) root treatments during ion fluxes to test the effects of channel gating molecules.
12. A 24 h treatment allows the signaling molecules to penetrate the cells and to elicit changes at the protein expression level. Functionally relevant induced events that change the number of channels expressed at the plasma membrane can be detected by this approach (e.g., transcriptional and posttranscriptional regulation of ion transporters).
13. The LIX needs time to acclimate to the electrode. Put the new filled electrodes in BSM, let them acclimate for 1 h, or several hours for Cl^- electrodes before using them for the first time [47].
14. The values for the slope, intercept, and coefficient of correlation for each calibrated ion are displayed. It is critical that the degree of correlation is high ($r^2=0.999$), the slope for monovalent ions should be greater than 50 mV and for divalent ions more than 25 mV. The electrodes that do not fulfill either of these criteria have to be replaced.

15. This step must be performed quickly, and it is important to ensure that the root is always in contact with solution. Also, do not touch the part of the root that will then be measured, while immobilizing it in the chamber. Inappropriate handling with tweezers may either cause direct damage to the root or activate mechano-sensitive channels.
16. Use high magnification under the microscope; when the three electrodes are in the same plane they all look focused.
17. The choice of specific root zone (e.g., elongation, meristem, mature, root hair) depends on the research questions asked. These zones differ in gene expression patterns [48], ion flux profiles, and intrinsic sensitivity to the treatments [49]. In our experience, the elongation zone is most responsive to treatments.
18. The motor moves the electrodes in a square-wave manner between two positions. Make sure the distances remain unchanged throughout the experiment, and adjust it otherwise.
19. After the addition of a solution, it usually takes up to 1 min for the solution to mix and reach unstirred layer condition. This period of recording must be discarded from the data analysis [49]. Some molecules interfere with the performance of the LIX [31], perform blank experiments to test if that is the case.
20. The default option is 1 h; by that time most ion fluxes reach the steady-state level. In some species, this process may be much quicker.

References

1. Sanders D, Pelloux J, Brownlee C, Harper JF (2002) Calcium at the crossroads of signaling. Plant Cell 14:S401–S417
2. Thiel G, Weise R (1999) Auxin augments conductance of K^+ inward rectifier in maize coleoptile protoplasts. Planta 208:38–45
3. Hager A (2003) Role of the plasma membrane H^+-ATPase in auxin-induced elongation growth: historical and new aspects. J Plant Res 116:483–505
4. Batelli G, Verslues PE, Agius F, Qiu Q, Fujii H, Pan S et al (2007) SOS2 promotes salt tolerance in part by interacting with the vacuolar H^+-ATPase and upregulating its transport activity. Mol Cell Biol 27:7781–7790
5. Shabala L, Cuin TA, Newman IA, Shabala S (2005) Salinity-induced ion flux patterns from the excised roots of Arabidopsis sos mutants. Planta 222:1041–1050
6. Kim TH, Bohmer M, Hu HH, Nishimura N, Schroeder JI (2010) Guard cell signal transduction network: advances in understanding Abscisic acid, Co_2, and Ca^{2+} signaling. Annu Rev Plant Biol 61:561–591
7. Newton RP, Smith CJ (2004) Cyclic nucleotides. Phytochemistry 65:2423–2437
8. Felle HH (2001) pH: Signal and messenger in plant cells. Plant Biol 3:577–591
9. Apel K, Hirt H (2004) Reactive oxygen species: metabolism, oxidative stress, and signal transduction. Annu Rev Plant Biol 55: 373–399
10. Besson-Bard A, Pugin A, Wendehenne D (2008) New insights into nitric oxide signaling in plants. Annu Rev Plant Biol 59:21–39
11. Meijer HJG, Munnik T (2003) Phospholipid-based signaling in plants. Annu Rev Plant Biol 54:265–306
12. Maathuis FJM (2009) Physiological functions of mineral macronutrients. Curr Opin Plant Biol 12:250–258
13. Chaumont F, Moshelion M, Daniels MJ (2005) Regulation of plant aquaporin activity. Biol Cell 97:749–764

14. Laohavisit A, Davies JM (2010) The role of ion channels in plant salt tolerance. In: Demidchik V, Maathuis F (eds) Ion channels and plant stress responses, signaling and communication in plants. Springer, Berlin, pp 69–86
15. Monshausen GB, Gilroy S (2009) Feeling green: mechanosensing in plants. Trends Cell Biol 19:228–235
16. Shabala S, Cuin TA (2008) Potassium transport and plant salt tolerance. Physiol Plant 133:651–669
17. Amtmann A, Beilby MJ (2010) The role of ion channels in plant salt tolerance. In: Demidchik V, Maathuis F (eds) Ion channels and plant stress responses, signaling and communication in plants. Springer, Berlin, pp 23–46
18. Maser P, Thomine S, Schroeder JI, Ward JM, Hirschi K, Sze H et al (2001) Phylogenetic relationships within cation transporter families of Arabidopsis. Plant Physiol 126:1646–1667
19. Shabala S (2003) Regulation of potassium transport in leaves: from molecular to tissue level. Ann Bot 92:627–634
20. Very AA, Sentenac H (2002) Cation channels in the Arabidopsis plasma membrane. Trends Plant Sci 7:168–175
21. Krol E, Trebacz K (2000) Ways of ion channel gating in plant cells. Ann Bot 86:449–469
22. Demidchik V, Shabala SN, Coutts KB, Tester MA, Davies JM (2003) Free oxygen radicals regulate plasma membrane Ca^{2+} and K^+-permeable channels in plant root cells. J Cell Sci 116:81–88
23. Bunney TD, van den Wijngaard PWJ, de Boer AH (2002) 14-3-3 protein regulation of proton pumps and ion channels. Plant Mol Biol 50:1041–1051
24. Li LG, Kim BG, Cheong YH, Pandey GK, Luan S (2006) A Ca^{2+} signaling pathway regulates a K+channel for low-K response in Arabidopsis. Proc Natl Acad Sci U S A 103:12625–12630
25. Demidchik V, Essah PA, Tester M (2004) Glutamate activates cation currents in the plasma membrane of Arabidopsis root cells. Planta 219:167–175
26. Michard E, Lima PT, Borges F, Silva AC, Portes MT, Carvalho JE et al (2011) Glutamate receptor-like genes form Ca^{2+} channels in pollen tubes and are regulated by pistil D-serine. Science 332:434–437
27. Suh BC, Hille B (2005) Regulation of ion channels by phosphatidylinositol 4,5-bisphosphate. Curr Opin Neurobiol 15:370–378
28. Talke IN, Blaudez D, Maathuis FJM, Sanders D (2003) CNGCs: prime targets of plant cyclic nucleotide signalling? Trends Plant Sci 8:286–293
29. Maathuis FJM, Filatov V, Herzyk P, Krijger GC, Axelsen KB, Chen SX et al (2003) Transcriptome analysis of root transporters reveals participation of multiple gene families in the response to cation stress. Plant J 35:675–692
30. Gojon A, Nacry P, Davidian JC (2009) Root uptake regulation: a central process for NPS homeostasis in plants. Curr Opin Plant Biol 12:328–338
31. Shabala L, Ross T, McMeekin T, Shabala S (2006) Non-invasive microelectrode ion flux measurements to study adaptive responses of microorganisms to the environment. FEMS Microbiol Rev 30:472–486
32. Demidchik V, Shang ZL, Shin R, Thompson E, Rubio L, Laohavisit A et al (2009) Plant extracellular ATP signalling by plasma membrane NADPH oxidase and Ca^{2+} channels. Plant J 58:903–913
33. Demidchik V, Shang ZL, Shin R, Colaco R, Laohavisit A, Shabala S et al (2011) Receptor-like activity evoked by extracellular ADP in Arabidopsis root epidermal plasma membrane. Plant Physiol 156:1375–1385
34. Zepeda-Jazo I, Velarde-Buendia AM, Enriquez-Figueroa R, Bose J, Shabala S, Muniz-Murguia J et al (2011) Polyamines interact with hydroxyl radicals in activating Ca^{2+} and K^+ transport across the root epidermal plasma membranes. Plant Physiol 157:2167–2180
35. Maryani MM, Shabala SN, Gehring CA (2000) Plant natriuretic peptide immunoreactants modulate plasma-membrane H^+ gradients in *Solanum tuberosum* L. leaf tissue vesicles. Arch Biochem Biophys 376:456–458
36. Ludidi N, Morse M, Sayed M, Wherrett T, Shabala S, Gehring C (2004) A recombinant plant natriuretic peptide causes rapid and spatially differentiated K^+, Na^+ and H^+ flux changes in *Arabidopsis thaliana* roots. Plant Cell Physiol 45:1093–1098
37. Pharmawati M, Shabala SN, Newman IA, Gehring CA (1999) Natriuretic peptides and cGMP modulate K^+, Na^+, and H^+ fluxes in *Zea mays* roots. Mol Cell Biol Res Commun 2:53–57
38. Tegg RS, Melian L, Wilson CR, Shabala S (2005) Plant cell growth and ion flux responses to the streptomycete phytotoxin thaxtomin A: calcium and hydrogen flux patterns revealed by the non-invasive MIFE technique. Plant Cell Physiol 46:638–648
39. Lew RR, Levina NN, Shabala L, Anderca MI, Shabala SN (2006) Role of a mitogen-activated protein kinase cascade in ion flux-mediated turgor regulation in fungi. Eukaryot Cell 5: 480–487

40. Babourina O, Newman I, Shabala S (2002) Blue light-induced kinetics of H^+ and Ca^{2+} fluxes in etiolated wild-type and phototropin-mutant Arabidopsis seedlings. Proc Natl Acad Sci U S A 99:2433–2438
41. Zivanovic BD, Cuin TA, Shabala S (2007) Spectral and dose dependence of light-induced ion flux responses from maize leaves and their involvement in leaf expansion growth. Plant Cell Physiol 48:598–605
42. Levina NN, Dunina-Barkovskaya AY, Shabala S, Lew RR (2002) Blue light modulation of ion transport in the slime mutant of *Neurospora crassa*. J Membr Biol 188:213–226
43. Shabala S, Newman I (2000) Salinity effects on the activity of plasma membrane H^+ and Ca^{2+} transporters in bean leaf mesophyll: masking role of the cell wall. Ann Bot 85:681–686
44. Chen Z, Newman I, Zhou M, Mendham N, Zhang G, Shabala S (2005) Screening plants for salt tolerance by measuring K^+ flux: a case study for barley. Plant Cell Environ 28:1230–1246
45. Knowles A, Shabala S (2004) Overcoming the problem of non-ideal liquid ion exchanger selectivity in microelectrode ion flux measurements. J Membr Biol 202:51–59
46. Cuin TA, Bose J, Stefano G, Jha D, Tester M, Mancuso S et al (2011) Assessing the role of root plasma membrane and tonoplast Na^+/H^+ exchangers in salinity tolerance in wheat: in planta quantification methods. Plant Cell Environ 34:947–961
47. Shabala SN, Newman IA, Morris J (1997) Oscillations in H^+ and Ca^{2+} ion fluxes around the elongation region of corn roots and effects of external pH. Plant Physiol 113:111–118
48. Birnbaum K, Shasha DE, Wang JY, Jung JW, Lambert GM, Galbraith DW et al (2003) A gene expression map of the Arabidopsis root. Science 302:1956–1960
49. Newman IA (2001) Ion transport in roots: measurement of fluxes using ion-selective microelectrodes to characterize transporter function. Plant Cell Environ 24:1–14

Chapter 8

Calcium Imaging of the Cyclic Nucleotide Response

Martin R. McAinsh, Stephen K. Roberts, and Lyudmila V. Dubovskaya

Abstract

Calcium (Ca^{2+}) is a key component of the signalling network by which plant cells respond to developmental and environmental signals. A change in guard cell cytosolic free Ca^{2+}($[Ca^{2+}]_{cyt}$) is an early event in the response of stomata to both opening and closing stimuli, and cyclic nucleotide-mediated Ca^{2+} signalling has been implicated in the regulation of stomatal aperture. A range of techniques have been used to measure $[Ca^{2+}]_{cyt}$ in plant cells. Here we describe a potential method for imaging cyclic nucleotide-induced changes in $[Ca^{2+}]_{cyt}$ in guard cells using the cameleon ratiometric Ca^{2+} reporter protein.

Key words Ca^{2+}, Cyclic nucleotide, Arabidopsis, Guard cells, Cameleon, Imaging

1 Introduction

Calcium (Ca^{2+}) is probably the most important second messenger in plants and is a key component in the signalling network by which plant cells respond to a diverse range of developmental and environmental signals [1, 2]. Changes in cytosolic free calcium concentration ($[Ca^{2+}]_{cyt}$) are observed in many different cell types in response to a diverse range of abiotic and biotic stimuli, examples of which include osmotic, salt, and drought signals [3, 4], oxidative stress [5, 6], cold [7, 8], gaseous pollutants [9, 10], light [11], plant hormones [12, 13], pathogens (elicitors) [7], and symbiotic signals [14, 15].

Cyclic nucleotides were first discovered in the 1950s through studies of the effects of the hormones epinephrine and glucagon upon glycogen phosphorylase activity in dog liver [16], and their actions in complex mammalian eukaryotic cell are well characterized. They are also present in the simplest prokaryotes, the *Eubacteria* and *Archae* [17], and have been shown to be present in various tissues of plants [18–20], where they are implicated in the regulation of important plant processes such as stomatal opening and closure [21, 22], cation flux regulation [23–25], chloroplast development [26], gibberellic acid functioning [18], and pathogen response [19, 27].

Chris Gehring (ed.), *Cyclic Nucleotide Signaling in Plants: Methods and Protocols*, Methods in Molecular Biology, vol. 1016, DOI 10.1007/978-1-62703-441-8_8, © Springer Science+Business Media New York 2013

Several reports suggest a role of cyclic nucleotides in plant Ca^{2+} signalling possibly through the activation of Ca^{2+}-permeable channels. cAMP stimulates Ca^{2+} influx in cultured carrot cells [28] whilst cAMP and cGMP both induce increases in $[Ca^{2+}]_{cyt}$ in protoplasts of aequorin-expressing tobacco [29]. In addition, dibutyryl cAMP (a membrane-permeable analogue of cAMP) has been shown to increase $[Ca^{2+}]_{cyt}$ in pollen tubes [30]. Ca^{2+}-permeable cation channels activated by the cyclic nucleotides cAMP and cGMP (cyclic nucleotide-gated cation channels — CNGCs) have been shown to be involved in Ca^{2+} signalling and the transduction of sensory stimuli in animals [31]. Plant CNGCs were first identified in barley [32] but are also present in many other plant species, including Arabidopsis, rice, and tobacco [32–34] — the Arabidopsis genome includes 20 full-length CNGC genes [35]. In addition, cAMP has been shown to stimulate hyperpolarization-activated Ca^{2+} channel (HACC) activity in guard cells [36].

Studies of the role of CNGCs in established Ca^{2+}-dependent processes provide further evidence of a role of cyclic nucleotides in plant Ca^{2+} signalling. Mutants of the Arabidopsis CNGC2 and CNGC4 genes exhibit altered patterns of pathogen-induced cell death during attack by *Pseudomonas syringae*, suggesting that cyclic nucleotide-induced influx in Ca^{2+} through these ion channels is a component of the pathway(s) by which pathogen-mediated responses are modulated [37, 38]. Similarly, a mutation that generates a chimeric CNGC-encoding gene, CNGC11/12, constitutively activates Arabidopsis defense responses and produces stunted plants which exhibit enhanced resistance to the virulent pathogen *Hyaloperonospora parasitica* Emco5 [39]. The Arabidopsis CNGC18 gene, which encodes a Ca^{2+}-permeable channel in pollen tubes [40], has been implicated in the generation of the tip-focused Ca^{2+} gradient in pollen tubes and the regulation of pollen tube growth. Overexpression of CNGC18 results in the formation of short, wide pollen tubes which exhibit depolarized growth which is enhanced at high external $[Ca^{2+}]$ and suppressed at low external $[Ca^{2+}]$ [41] whilst the cngc18 knockout mutants produce short, thin pollen tubes which exhibit non-directional growth before prematurely bursting [40]. However the mechanisms by which cyclic nucleotides activate these Ca^{2+}-permeable channels remain unknown.

Several techniques have been used to measure $[Ca^{2+}]_{cyt}$ in plants including Ca^{2+}-sensitive microelectrodes, fluorescent Ca^{2+}-indicators, Ca^{2+}-sensitive photoproteins, and fluorescence resonance energy transfer (FRET)-based cameleon reporters [42, 43]. Ca^{2+}-sensitive microelectrodes can be used only in cells that are able to withstand impalement with two electrodes or a double-barrelled electrode. In addition, microelectrodes suffer from slow response times and difficulties with calibration. These problems are particularly acute in plant cells in which the high turgor often results in partial displacement of the sensor, and the subsequent loss of sensitivity, following impalement. Consequently, the use of Ca^{2+}-sensitive electrodes has been limited to only a few studies in plants and algae, for example, *see* refs. 44, 45. In contrast, fluorescent Ca^{2+}-sensitive indicators (e.g., Quin-2, Fura-2, and Indo-1) have been used extensively to measure plant $[Ca^{2+}]_{cyt}$ [43]. These allow investigations of $[Ca^{2+}]_{cyt}$ and their dynamics via their changes in their excitation and emission spectra. However, although the use of fluorescent indicators has enabled significant advances in our understanding of Ca^{2+} signalling in plants [1, 46, 47], there are inherent limitations to use of such indicators (including loading difficulties, indicator loss/sequestration, high buffer capacity). Ca^{2+}-sensitive photoproteins, such as aequorin [43], emit light on binding $Ca^{2+}$with the luminescence being directly proportional to $[Ca^{2+}]_{cyt}$. Initially, measurements of plant $[Ca^{2+}]_{cyt}$ using aequorin were restricted to a limited number of cell types due to the need to microinject this high molecular weight protein into cells [48]. Aequorin has subsequently been introduced into plants by stable transformation techniques providing a noninvasive method for monitoring $[Ca^{2+}]_{cyt}$ [7]. The aequorin protein has been successfully targeted to the cytosol [7, 8], to specific cell types [49], including guard cells [50], and to organelles [8, 49]. However, due to the very low fluence rate of aequorin in most cases, the measurements had to be performed using whole seedlings or larger parts (whole tissues) of a limited number of species tractable to transformation techniques. In addition, differences in the stability, distribution, or localization of aequorin in cells and differences in the permeability of cells to the luminophore coelentrazine also have to be taken into account.

Cameleons are green fluorescent protein (GFP)-based ratiometric Ca^{2+}reporters [43, 51] consisting of a fusion protein comprising a cyan-emitting version of GFP (CFP) linked to calmodulin and a calmodulin-binding peptide (M13) and an enhanced yellow-emitting GFP (YFP). Binding of Ca^{2+} to the calmodulin domain induces a conformational change that can be detected by FRET between the component cyan and yellow fluorescent protein domains. $[Ca^{2+}]_{cyt}$ can therefore be measured by determining the efficiency of FRET (Fig. 1). Cameleons, like aequorin, can be introduced into plants by stable transformation techniques and can be targeted to specific cell types [51, 52] and organelles enabling the contribution of the major plant Ca^{2+} stores (the apoplast and the vacuole) to Ca^{2+} signalling to be studied [53]. The Ca^{2+} binding properties of the indicator can also be modified to enable investigations of Ca^{2+} dynamics in plant organelles which contain much higher Ca^{2+} concentrations, like within peroxisomes [54] and the endoplasmic reticulum [55]. In addition, the simultaneous expression of differentially targeted cameleon reporters has the potential to allow simultaneous recordings of mitochondrial and nuclear

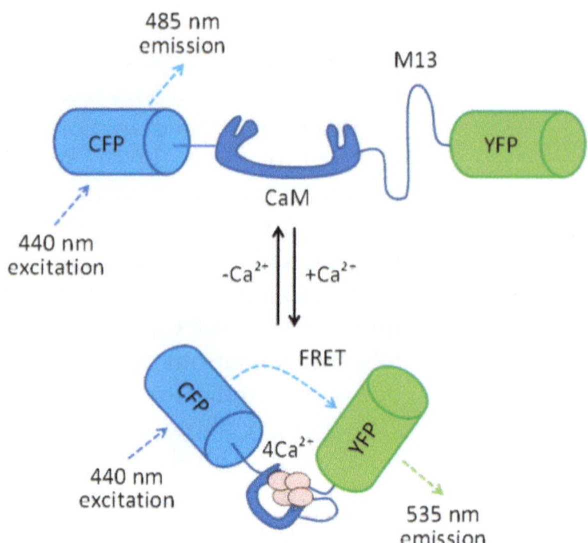

Fig. 1 Fluorescence characteristics of the cameleon Ca^{2+} reporter. Ca^{2+} binds to the calmodulin (CaM) domain between the CFP (cyan fluorescent protein) and YFP (yellow fluorescent protein) causing it to bind to the M13 peptide. The resultant conformational change causes CFP to come into close proximity to YFP, allowing FRET (fluorescence resonance energy transfer) to occur and YFP fluorescence to be produced from CFP excitation. As Ca^{2+} levels increase, the 485-nm emission from CFP (FRET donor) decreases, whereas the 535-nm emission from YFP (FRET acceptor) increases. The ratio of the 535-nm/485-nm (440-nm excitation) can be calibrated to give values to $[Ca^{2+}]_{cyt}$.

Ca^{2+} dynamics [47]. As with aequorin the use of cameleon Ca^{2+} reporters has been restricted to a limited number of plant species including *Arabidopsis thailana* [13, 56] and *Medicago truncatula* [57, 58]. Nevertheless, Ca^{2+} cameleon-expressing plants have proved instrumental in dissecting the role of Ca^{2+} oscillations in the response of stomatal guard cells to abiotics stress (e.g., ABA, cold stimulus, oxidative stress) [13, 59] and symbiosis signalling [14, 15]. Furthermore, plants expressing the improved version of the Ca^{2+} cameleon reporter [60] have been used to study cell type-specific Ca^{2+} oscillations evoked by extracellular nucleoside-triphosphates [61], toxic metals [62], or in growing root hair tips [63] and mechanically stimulated roots [64].

It is well established that changes in $[Ca^{2+}]_{cyt}$ are an early event in the response of stomata to opening and closing stimuli [65]. There are several reports of a role for cyclic AMP and/or cyclic GMP in stomatal opening. For example, cyclic AMP has been shown to reverse abscisic acid (ABA)- or Ca^{2+}-induced inhibition of both stomatal opening and whole-cell inward K^+ currents, an effect that was mimicked by activators of adenylatecyclase or inhibitors of phosphodiesterase [66]. A cGMP-mediated Ca^{2+} pathway has also been implicated in both auxin- [67, 68] and kinetin- [69, 70] induced stomatal opening and ABA-induced stomatal closure [22].

Therefore, in this chapter, the authors will describe a potential method for imaging cyclic nucleotide-induced changes in guard cell $[Ca^{2+}]_{cyt}$ in Arabidopsis stably expressing the cameleon ratiometric Ca^{2+} reporter protein.

2 Materials

1. Plant material: *Arabidopsis thaliana* (ecotype *Landsberg erecta*) seedlings stably expressing the calcium cameleon reporter, YC3.60, were grown from seed (*see* **Note 1**). Seeds were surface sterilized by agitation in 70 % (v/v) ethanol for 5 min and then sown on growth medium (*see* **Note 2**) with approximately 20–25 seeds per plate, using aseptic technique in a laminar flow hood. Plates were sealed using microporous tape and given at least a 2-day stratification treatment in the dark at 4 °C. Plates were subsequently transferred for 10–14 days to an Arabidopsis growth cabinet (*see* **Note 3**) with a 16/8-h light/dark regime and photosynthetic photon flux density (PPFD) of 130 μmol/m^2/s at day/night temperatures of 20 ± 2 °C/18 ± 2 °C (*see* **Note 4**). At this stage seedlings usually had two cotyledons and two true leaves. Seedlings were then transferred into 2.5 cm^2 plug trays filled with a sterile soil mixture composed of a peat-based compost and washed horticultural sand with a 3:1 ratio sieved through a 0.4 × 0.4 cm mesh tray. After 4–5 weeks plants were transplanted into 5 cm^2 pots (*see* **Note 5**). Plants in plug trays and pots were maintained in a growth room with a 16/8-h light/dark regime and PPFD ranging between 130 and 160 μmol/m^2/s at a temperature of 20 ± 2 °C and watered once a day from the bottom.
2. Isolation buffer: 10 mM 2-(*N*-morpholino)ethanesulfonic acid (MES) in distilled H_2O, pH 6.2 (KOH) (*see* **Notes 6** and **7**).
3. Perfusion buffer: 10 mM MES and 50 mM KCl in distilled H_2O, pH 6.2 (KOH). Store at 4 °C (*see* **Notes 6** and **7**).
4. Growth media: Half-strength Murashige and Skoog basal medium salts (2.2 g/l) [71], 1 g/l sucrose, 6 g/l plant tissue culture grade agar, pH 5.8 (KOH) (*see* **Note 2**).
5. Perfusion system: Perfusion can routinely be provided under gravity from a temperature controlled reservoir mounted approximately 50 cm above the specimen (*see* **Note 8**). The reservoir consists of a small (6-l) heated water bath and a purpose-built cooling coil. Perfusion media are delivered to the specimen along an insulated pipe. Excess media are removed from the specimen under vacuum [12, 72] (*see* **Note 9**).
6. Fluorescence microscopy: Specimens are viewed using an inverted epifluorescence microscope (*see* **Note 10**).

Fluorescence excitation is provided by a xenon light source (*see* **Note 11**). Excitation and emission wavelengths are specified using a "cameleon filter set" excitation: 440-nm, 20-nm bandwidth; emission 1 [CFP], 485-nm, 40-nm bandwidth; emission 2 [YFP]: 535-nm, 30-nm bandwidth; dichroic mirror: 455-nm dichroic long pass (*see* **Notes 12** and **13**). Typically, a 60× oil objective with a numerical aperture of 1.4 and non-fluorescent immersion oil are used for all measurements. Fluorescence emissions are quantified using a charge-coupled device (CCD) camera (*see* **Notes 14** and **15**).

3 Methods

3.1 Preparation of Leaf Epidermis

1. Apply a very thin coating of Hollister 7730 medical adhesive (Hollister Inc., Libertyville, Illinois, USA) to the surface of a cover slip.
2. Detach the leaf to be peeled immediately prior to each experiment (*see* **Note 16**).
3. Gently press the leaf abaxial side down onto the medical adhesive-coated surface of the cover slip (*see* **Note 17**).
4. Carefully use a razor blade to remove the cuticle and mesophyll layers of the leaf leaving the lower leaf epidermal layer containing stomatal complexes intact.
5. Maintain the prepared leaf epidermis in isolation buffer which has been aerated with CO_2-free air for at least 1 h, i.e., air that has been passed through soda lime, prior to use.

3.2 Perfusion System

1. Pipe a ring of petroleum jelly around the epidermal strip using a 1-ml syringe.
2. Use shards of broken cover slips attached to the bottom cover slip using low melting point wax (e.g., from Agar Scientific) to create a small open perfusion chamber approximately 0.5 × 1.0 cm and one cover slip deep (*see* **Notes 18** and **19**).
3. Place a drop of CO_2-free perfusion buffer at 20 °C in the perfusion chamber to prevent the epidermis from drying out.
4. Mount the perfusion system on the microscope stage, with the exposed epidermis upward, as if it were a standard microscope slide.
5. Place the inlet and outlet of the perfusion system at the front and rear of the perfusion chamber, respectively.
6. Perfuse the specimen continuously (6 ml/min) with CO_2-free perfusion buffer at 20 °C in the dark (*see* **Note 20**).

3.3 Calcium Imaging

1. Use ratio imaging to monitor spatially localized changes in $[Ca^{2+}]_{cyt}$ (*see* **Note 21**).
2. Measure the background autofluorescence signal at both emission wavelengths prior to each experiment (*see* **Note 22**).
3. Select guard cells with a high level of YFP fluorescence at the start of the experiment for analysis (*see* **Note 23**).
4. Record alternate 485- and 535-nm emission (440-nm excitation) images (*see* **Note 24**).
5. Integrate both signals over individual frames (*see* **Note 25**).
6. Subtract the autofluorescence signals from each pair of averaged 485- and 535-nm images, pixel-by-pixel, at the end of the experiment (*see* **Notes 26–28**).
7. Divide the autofluorescence-subtracted 535-nm (YFP) images by the corresponding autofluorescence-subtracted 485-nm (CFP) image, on a pixel-by-pixel basis, to produce a series of YFP/CFP-nm ratio images [15, 56, 63, 73, 74].

3.4 Calibration

1. Perform an in situ calibration of the YFP/CFP ratio to $[Ca^{2+}]_{cyt}$.
2. Record the maximum YFP/CFP ratio (R_{max}) using treatment with 1 M $CaCl_2$ or 50 % ethanol to raise Ca^{2+} to saturating levels [63] (*see* **Note 29**).
3. Record the minimum YFP/CFP ratio (R_{min}) following treatment with 1 mM 1,2-bis(*o*-aminophenoxy)ethane-*N*,*N*,*N*′,*N*′-tetraacetic acid (BAPTA)-AM (Molecular Probes).
4. Calculate $[Ca^{2+}]_{cyt}$ according to the equation:

$$\left[Ca^{2+}\right]_{cyt} = K_d \left(R - R_{min}\right) / \left(R_{max} - R\right)^{1/n},$$

where R represents the YFP/CFP ratio measured during the experiment [51], n represents the Hill coefficient that has been determined as 1 for YC3.6, and the K_d for Ca^{2+} = 250 nM [60] (*see* **Note 30**).

4 Notes

1. Transgenic plants stably expressing the calcium cameleon reporter, YC3.60, can be generated by *Agrobacterium*-meditated transformation, *see* refs. 56, 59.
2. Growth media should be sterilized by autoclaving at 120 °C for 30 min and allowed to cool to approximately 50 °C. It can then be poured into sterile non-vented 90 mm petri dishes in a laminar flow hood (approximately 25 ml of media per dish). Prepared plates can be sealed in a bag and stored at 4 °C until required.

3. A number of manufactures produce growth chambers that are specifically designed for the growth of Arabidopsis plants. These provide optimized controlled temperature, light, and humidity conditions for growing Arabidopsis plants from seed to maturity.
4. The illumination provided by Philips (UK) 740 TL warm white lamps, with supplementary lighting provided by 25 W incandescent lamps, is ideal for Arabidopsis growth.
5. The 5 cm^2 Arabasket pots that form part of the Arabidopsis Arasystem growing kit (Beta Tech, Gent, Belgium) are ideal for growing 4- to 5-week-old Arabidopsis plants.
6. Always use tissue culture grade MES (e.g., from Sigma-Aldrich). MES buffers should be aerated with CO_2-free air for 1 h before use and during experiments. CO_2-free air can be obtained by pumping air through a 15-cm column of soda lime [12].
7. MES buffers tend to become contaminated even when stored at 4 °C. This can cause problems with perfusion systems and the microinjection of cells. Therefore, unused buffers should be discarded regularly.
8. The perfusion rate can be adjusted by altering the height of the reservoir and/or the diameter of the perfusion tubing although alterations in the rate of delivery of plant hormones such as abscisic acid have been shown to affect stomatal responses [75].
9. Alternative perfusion systems used in studies of guard cell $[Ca^{2+}]_{cyt}$ include the exchange of perfusion media using low-noise peristaltic pumps [76, 77].
10. A range of inverted epifluorescence microscopes have been used in cameleon imaging systems [15, 56, 63, 73, 74].
11. Excitation using a 75 W xenon lamp (Osram, Germany) is frequently used in cameleon imaging systems although mercury lamps have also been used [56]. Neutral density filters are used to attenuate the light by 97 % (3 % light transmission) to reduce the exposure of the fluorescent reporters and cells to epifluorescence excitation.
12. Typically, emission filters are positioned in front of the CCD camera using a wheel and shutter which is software controlled, e.g., using Metafluor software (MDS, Inc., Toronto, Canada) [73, 74]. However, image splitters such as the Optosplit II emission image splitter from Cairn Research (UK) enable the images at both emission wavelengths to be captured simultaneously, eliminating the lag in image acquisition inherent in the use of filter wheels.
13. A number of manufactures produce specific "cameleon filter sets" (e.g., Chroma Technology, USA) although these can also

be assembled using an appropriate combination of excitation (440-nm) and emission (CFP, 485-nm; YFP, 535-nm) filters and dichroic mirror (455-nm long pass).

14. Hamamatsu ORCA (Hamamatsu Photonics Hamamatsu City, Japan) [78] and CoolSNAP (Photometrics, AZ, USA) cameras [73, 74] have both been used successfully to image guard cell Ca^{2+} dynamics using cameleon Ca^{2+} reporters. However, sensitivity, stability, and noise levels of ultra-low-light CCD cameras are constantly improving.
15. Confocal scanning laser microscopy (CLSM) can also be used for cameleon reporter-based measurements of plant $[Ca^{2+}]_{cyt}$ [15, 56, 63].
16. All experiments are conducted during the middle of the photoperiod, between 10 a.m. and 6 p.m., to minimize the effects of diurnal changes in stomatal responses.
17. Due to the short working distance of the epiflourescence lenses typically used in fluorescence imaging systems it is essential that the epidermal strip is completely flat on the cover slip.
18. The perfusion chamber is fabricated by using 2-mm strips of low melting point wax and a low power soldering iron to solder the cover slips together.
19. Large volume perfusion chambers introduce a lag period in the changeover of perfusion media during which mixing occurs. This can affect the kinetics of stimulus-induced changes in $[Ca^{2+}]_{cyt}$. The small volume of the perfusion chamber allows rapid, almost instantaneous changeover of the perfusion media.
20. Specimens should be perfused with CO_2-free perfusion buffer at 25 °C in the dark for at least 10 min before use to determine whether they will retain focus.
21. Conventional imaging [73, 74] and CLSM [56, 63] have both been used successfully to image guard cell Ca^{2+} dynamics using cameleon Ca^{2+} reporters.
22. Plant tissues frequently autofluoresce at the excitation/emission wavelengths used for fluorescence-based measurements of $[Ca^{2+}]_{cyt}$. In cameleon reporter-based measurements, if the autofluorescence signal is significant relative to the FRET signal, it has a damping effect on the dynamic range of the YFP/CFP ratio. Imaging of untransformed plants/tissues under conditions identical to those used in Ca^{2+} imaging experiments is therefore an essential control to characterize any potential problems from contaminating autofluorescence.
23. YFP bleaches faster than CFP upon excitation at 440 nm [77]. Therefore, select guard cells that exhibit high initial YFP fluorescence for analysis.

24. Limit the area from which the fluorescence is recorded to a single guard cell using the emission diaphragm of the microscope.
25. The signal from cells is often quite low making measurements extremely noisy. The noise can be reduced by integrating both the 485- and 535-nm emission (440-nm excitation) over a number of measurements, increasing the signal-to-noise ratio.
26. When imagining guard cell $[Ca^{2+}]_{cyt}$, autofluorescence subtraction is complicated by cell movements owing to stomatal closure. If the autofluorescence is relatively low (e.g., <10 % of the FRET signal), uniform across the sample, and stable, a mean autofluorescence calculated from the cytoplasmic region of an untransformed control not expressing the cameleon Ca^{2+} reporter may be subtracted from emission images, pixel by pixel, online prior to the YFP/CFP ratio calculation [56].
27. Care needs to be exercised when performing such background subtractions, as autofluorescence can change in intensity and spectrum with time and so could vary from time point to time point within each pixel of an image. Such changes could be interpreted incorrectly as changes in $[Ca^{2+}]_{cyt}$. Furthermore, different cells/tissues can develop intense autofluorescence that can be mistaken for dynamic Ca^{2+} changes (e.g., dead and dying cells), reinforcing the importance in monitoring the autofluorescence of untransformed controls over time prior to experiments.
28. It may be necessary to subtract any baseline drift due to faster bleaching of YFP by linear subtraction.
29. Using ionophores to determine minimum and maximum ratio values in vitro has previously failed to consistently raise the $[Ca^{2+}]_{cyt}$ as commonly observed for plant cells [56]. The values of R_{min} and R_{max} derived using purified cameleon protein have therefore also been used to convert YFP/CFP ratio values to $[Ca^{2+}]_{cyt}$ by fitting to the previously derived in vitro calibration curve.
30. Due to the inherent uncertainties of the precise in vivo K_d in such in situ calibrations, the raw YFP/CFP ratio data is frequently presented in addition to, or instead of, a calibrate $[Ca^{2+}]_{cyt}$ value.

Acknowledgments

The authors thank the European Community, The Royal Society, the Natural Environment Research Council (UK), the Biotechnology and Biological Sciences Research Council (UK), INTAS, and Belarusian Republican Foundation for Fundamental Research for funding.

References

1. Dodd AN, Kudla J, Sanders D (2010) The language of calcium signaling. Annu Rev Plant Biol 61:593–620
2. Kudla J, Batistic O, Hashimoto K (2010) Calcium signals: the lead currency of plant information processing. Plant Cell 22: 541–563
3. Knight H, Trewavas AJ, Knight MR (1997) Calcium signalling in *Arabidopsis thaliana* responding to drought and salinity. Plant J 12: 1067–1078
4. Ranf S, Wunnenberg P, Lee J et al (2008) Loss of the vacuolar cation channel, AtTPC1, does not impair Ca^{2+} signals induced by abiotic and biotic stresses. Plant J 53:287–299
5. McAinsh MR, Clayton H, Mansfield TA et al (1996) Changes in stomatal behavior and guard cell cytosolic free calcium in response to oxidative stress. Plant Physiol 111:1031–1042
6. Evans NH, McAinsh MR, Hetherington AM et al (2005) ROS perception in *Arabidopsis thaliana*, the ozone-induced calcium response. Plant J 41:615–626
7. Knight MR, Campbell AK, Smith SM et al (1991) Transgenic plant aequorin reports the effects of touch and cold-shock and elicitors on cytoplasmic calcium. Nature 352:524–526
8. Knight H, Trewavas AJ, Knight MR (1996) Cold calcium signaling in Arabidopsis involves two cellular pools and a change in calcium signature after acclimation. Plant Cell 8:489–503
9. Clayton H, Knight MR, Knight H et al (1999) Dissection of the ozone-induced calcium signature. Plant J 17:575–579
10. Short EF, North KA, Roberts MR et al (2012) A stress-specific calcium signature regulating an ozone-responsive gene expression network in Arabidopsis. Plant J 71:948–961
11. Shacklock PS, Read ND, Trewavas AJ (1992) Cytosolic free calcium mediates red light-induced photomorphogenesis. Nature 358: 753–755
12. McAinsh MR, Brownlee C, Hetherington AM (1990) Abscisic acid-induced elevation of guard cell cytosolic Ca^{2+} precedes stomatal closure. Nature 343:186–188
13. Allen GJ, Chu SP, Harrington CL et al (2001) A defined range of guard cell calcium oscillation parameters encodes stomatal movements. Nature 411:1053–1057
14. Ehrhardt DW, Wais R, Long SR (1996) Calcium spiking in plant root hairs responding to Rhizobium nodulation signals. Cell 85: 673–681
15. Capoen W, Sun J, Wysham D et al (2011) Nuclear membranes control symbiotic calcium signaling of legumes. Proc Natl Acad Sci U S A 108:14348–14353
16. Rall TW, Sutherland EW, Berthet J (1957) The relation of epinephrine and glucagon to liver phosphorylase. J Biol Chem 224: 1987–1995
17. Botsford JL, Harman JH (1998) cAMP in prokaryotes. Microbiol Rev 56:100–132
18. Penson SP, Schuurink RC, Fath A et al (1996) cGMP is required for gibberellic acid-induced gene expression in barley aleurone. Plant Cell 8:2325–2333
19. Durner J, Wendehenne D, Klessig D (1998) Defence gene induction in tobacco by nitric oxide, cyclic GMP and cyclic ADP-ribose. Proc Natl Acad Sci U S A 95:10328–19333
20. Donaldson L, Ludidi N, Knight MR et al (2004) Salt and osmotic stress cause rapid increases in *Arabidopsis thaliana* cGMP levels. FEBS Lett 569:317–320
21. Newton RP, Smith CJ (2004) Cyclic nucleotides. Phytochemistry 65:2423–2437
22. Dubovaskaya LV, Bakakina YS, Kolesneva EV et al (2011) cGMP-dependent ABA-induced stomatal closure in the ABA-insensitive Arabidopsis mutant abi1-1. New Phytol 191: 57–69
23. Maathuis FJM, Sanders D (2001) Sodium uptake in Arabidopsis roots is regulated by cyclic nucleotides. Plant Physiol 127: 1617–1625
24. Essah PA, Davenport R, Tester M (2003) Sodium influx and accumulation in Arabidopsis. Plant Physiol 133:307–318
25. Maathuis FJM (2006) cGMP modulates gene transcription and cation transport in Arabidopsis roots. Plant J 45:700–711
26. Bowler C, Neuhaus G, Yamagata H et al (1994) Cyclic GMP and calcium mediate phytochrome phototransduction. Cell 77:73–81
27. Ma Y, Szostkiewicz I, Korte A, Moes D et al (2009) Regulators of PP2C phosphatase activity function as abscisic acid sensors. Science 324:1064–1068
28. Kurosaki F, Kaburakim H, Nishi A (1994) Involvement of plasma membrane-located calmodulin in the response decay of cyclic nucleotide-gated cation channel in cultured carrot cells. FEBS Lett 340:193–196
29. Volotovski ID, Sokolovsky SG, Molchan OV et al (1998) Second messengers mediate increases in cytosolic calcium in tobacco protoplasts. Plant Physiol 117:1023–1030
30. Malho R, Camacho L, Moutinho A (2000) Signalling pathways in pollen tube growth and reorientation. Ann Bot 85:59–68

31. Kaupp UB, Seifert R (2002) Cyclic nucleotide-gated ion channels. Physiol Rev 82:769–824
32. Schuurink RC, Shartzer SF, Fath A et al (1998) Characterization of a calmodulin-binding transporter from the plasma membrane of barley aleurone. Proc Natl Acad Sci U S A 95: 1944–1949
33. Kohler C, Merkle T, Neuhaus G (1999) Characterisation of a novel gene family of putative cyclic nucleotide- and calmodulin-regulated ion channels in Arabidopsis thaliana. Plant J 18:97–104
34. Arzai T, Kaplan B, Fromm H (2000) A high-affinity calmodulin-binding site in a tobacco plasma-membrane channel protein coincides with a characteristic element of cyclic nucleotide-binding domains. Plant Mol Biol 42: 591–601
35. Talke IN, Blaudez D, Maathuis FJM et al (2003) CNGCs: prime targets of plant cyclic nucleotide signalling? Trends Plant Sci 8: 286–293
36. Lemtiri-Chlieh F, Berkowitz GA (2004) Cyclic adenosine monophosphate regulates calcium channels in the plasma membrane of *Arabidopsis* leaf guard and mesophyll cells. J Biol Chem 279:35306–35312
37. Clough SJ, Fengler KA, Yu IC et al (2000) The *Arabidopsisdnd1* "defense, no death" gene encodes a mutated cyclic nucleotide-gated ion channel. Proc Natl Acad Sci U S A 97: 9323–9328
38. Balagué C, Lin BQ, Alcon C et al (2003) HLM1, an essential signaling component in the hypersensitive response, is a member of the cyclic nucleotide-gated channel ion channel family. Plant Cell 15:365–379
39. Yoshioka K, Moeder W, Kang HG et al (2006) The chimeric Arabidopsis CYCLIC NUCLEOTIDE-GATED ION CHANNEL11/12 activates multiple pathogen resistance responses. Plant Cell 18: 747–763
40. Frietsch S, Wang YF, Sladek C et al (2007) A cyclic nucleotide-gated channel is essential for polarized tip growth of pollen. Proc Natl Acad Sci U S A 104:14531–14536
41. Chang F, Yan A, Zhao LN et al (2007) A putative calcium-permeable cyclic nucleotide-gated channel, CNGC18, regulates polarized pollen tube growth. J Integr Plant Biol 49: 1261–1270
42. Rudd JJ, Franklin-Tong V (2001) Unravelling response-specificity in Ca^{2+} signalling pathways in plant cells. New Phytol 151:7–33
43. Swanson SS, Choi WG, Chanoca A et al (2011) *In vivo* imaging of Ca^{2+}, pH, and reactive oxygen species using fluorescent probes in plants. Annu Rev Plant Biol 62:273–297
44. Miller AJ, Sanders D (1987) Depletion of cytosolic free calcium induced by photosynthesis. Nature 326:397–400
45. Felle H (1988) Auxin causes oscillations of cytosolic free calcium and pH in Zea mays coleoptiles. Planta 174:495–499
46. McAinsh MR, Pittman JK (2009) Shaping the calcium signature. New Phytol 181:275–294
47. Batistič O, Kudla J (2012) Analysis of calcium signaling pathways in plants. Biochim Biophys Acta 1820:1283–1293
48. Williamson RE, Ashley CC (1982) Free $Ca^{2+}$and cytoplasmic streaming in the alga Chara. Nature 296:647–651
49. Kiegle E, Moore CA, Haseloff J et al (2000) Cell-type-specific calcium responses to drought, salt and cold in the *Arabidopsis* root. Plant J 23:267–278
50. Dodd AN, Jakobsen MK, Baker AJ et al (2006) Time of day modulates low-temperature Ca^{2+} signals in Arabidopsis. Plant J 48:962–973
51. Miyawaki A, Llopis J, Heim R et al (1997) Fluorescent indicators for Ca^{2+} based on green fluorescent proteins and calmodulin. Nature 388:882–887
52. Emmanouilidou E, Teschemacher AG, Pouli AE et al (1999) Imaging Ca^{2+} concentration changes at the secretory vesicle surface with a recombinant targeted cameleon. Curr Biol 9: 915–918
53. Krebs M, Held K, Binder A et al (2012) FRET-based genetically encoded sensors allow high-resolution live cell imaging of Ca^{2+} dynamics. Plant J 69:181–192
54. Costa A, Drago I, Behera S et al (2010) H_2O_2 in plant peroxisomes: an in vivo analysis uncovers a Ca^{2+}-dependent scavenging system. Plant J 62:760–772
55. Iwano M, Entani T, Shiba H et al (2009) Fine-tuning of the cytoplasmic Ca^{2+} concentration is essential for pollen tube growth. Plant Physiol 150:1322–1334
56. Allen GJ, Kwak JM, Chu SP et al (1999) Cameleon calcium indicator reports cytoplasmic calcium dynamics in *Arabidopsis* guard cells. Plant J 19:735–747
57. Kosuta S, Hazledine S, Sun J et al (2008) Differential and chaotic calcium signatures in the symbiosis signaling pathway of legumes. Proc Natl Acad Sci U S A 105:9823–9828
58. Chabaud M, Genre A, Sieberer BJ et al (2011) Arbuscular mycorrhizal hyphopodia and germinated spore exudates trigger Ca^{2+} spiking in the legume and nonlegume root epidermis. New Phytol 189:347–355

59. Allen GJ, Chu SP, Schumacher K et al (2000) Alteration of stimulus-specific guard cell calcium oscillations and stomatal closing in Arabidopsis det3 mutant. Science 289:2338–2342
60. Nagai T, Yamada S, Tominaga T et al (2004) Expanded dynamic range of fluorescent indicators for Ca^{2+} by circularly permuted yellow fluorescent proteins. Proc Natl Acad Sci U S A 101:10554–10559
61. Tanaka K, Swanson SJ, Gilroy S et al (2010) Extracellular nucleotides elicit cytosolic free calcium oscillations in Arabidopsis. Plant Physiol 154:705–719
62. Rincon-Zachary M, Teaster ND, Sparks JA et al (2010) Fluorescence resonance energy transfer-sensitized emission of yellow cameleon 3.60 reveals root zone-specific calcium signatures in Arabidopsis in response to aluminum and other trivalent cations. Plant Physiol 152:1442–1458
63. Monshausen GB, Messerli MA, Gilroy S (2008) Imaging of the Yellow Cameleon 3.6 indicator reveals that elevations in cytosolic Ca^{2+} follow oscillating increases in growth in root hairs of Arabidopsis. Plant Physiol 147: 1690–1698
64. Monshausen GB, Bibikova TN, Weisenseel MH et al (2009) Ca^{2+} regulates reactive oxygen species production and pH during mechanosensing in Arabidopsis roots. Plant Cell 21: 2341–2356
65. Jin XC, Wu WH (1999) Involvement of cyclic AMP in ABA- and Ca^{2+}-mediated signal transduction of stomatal regulation in *Vicia faba*. Plant Cell Physiol 40:1127–1133
66. Cousson A (2001) Pharmacological evidence for the implication of both cyclic GMP-dependent and -independent transduction pathways within auxin-induced stomatal opening in *Commelina communis*(L.). Plant Sci 161:249–258
67. Cousson A (2003) Pharmacological evidence for a positive influence of the cyclic GMP-independent transduction on the cyclic GMP-mediated Ca^{2+}-dependent pathway within Arabidopsis stomatal opening in response to auxin. Plant Sci 164:759–767
68. Pharmawati M, Billington T, Gehring CA (1998) Stomatal guard cell responses to kinetin and natriuretic peptides are cGMP-dependent. Cell Mol Life Sci 54:272–276
69. Pharmawati M, Maryani MM, Nikolakopoulos T et al (2001) Cyclic GMP modulates stomatal opening induced by natriuretic peptides and immunoreactive analogues. Plant Physiol Biochem 39:385–394
70. Kim TH, Bohmer M, Hu HH et al (2010) Guard cell signal transduction network: advances in understanding abscisic acid, CO_2, and Ca^{2+}signaling. Annu Rev Plant Biol 61: 561–591
71. Murashige T, Skoog F (1962) A revised medium for rapid growth and bio assays with tobacco tissue cultures. Physiol Plant 15: 473–497
72. Ng CKY, Carr K, McAinsh MR et al (2001) Drought-induced guard cell signal transduction involves sphingosine-1-phosphate. Nature 410:596–599
73. Mori C, Murata Y, Yang Y et al (2006) CDPKs CPK6 and CPK3 function in ABA regulation of guard cell S-type anion- and Ca^{2+}-permeable channels and stomatal closure. PLoS Biol 4:10
74. Yang Y, Costa A, Leonhardt N et al (2008) Isolation of a strong Arabidopsis guard cell promoter and its potential as a research tool. Plant Methods 4:6
75. Trejo CL, Clephan AL, Davies WJ (1995) How do stomata read abscisic acid-signals. Plant Physiol 109:803–811
76. Schroeder JI, Hagiwara S (1990) Repetitive increases in cytosolic Ca^{2+} of guard cells by abscisic acid activation of nonselective Ca^{2+} permeable channels. Proc Natl Acad Sci U S A 87:9305–9309
77. Lemtiri-Chlieh F, MacRobbie EAC, Webb AAR et al (2003) Inositol hexakisphosphate mobilizes an endomembrane store of calcium in guard cells. Proc Natl Acad Sci U S A 100: 10091–10095
78. Young JJ, Mehta S, Israelsson M et al (2006) CO_2 signaling in guard cells: calcium sensitivity response modulation, a Ca^{2+}-independent phase, and CO_2 insensitivity of the *gca2* mutant. Proc Natl Acad Sci U S A 103: 7506–7511

Chapter 9

Identification and Quantitation of Signal Molecule-Dependent Protein Phosphorylation

Arnoud Groen, Ludivine Thomas, Kathryn Lilley, and Claudius Marondedze

Abstract

Phosphoproteomics is a fast-growing field that aims at characterizing phosphorylated proteins in a cell or a tissue at a given time. Phosphorylation of proteins is an important regulatory mechanism in many cellular processes. Gel-free phosphoproteome technique involving enrichment of phosphopeptide coupled with mass spectrometry has proven to be invaluable to detect and characterize phosphorylated proteins. In this chapter, a gel-free quantitative approach involving ^{15}N metabolic labelling in combination with phosphopeptide enrichment by titanium dioxide (TiO_2) and their identification by MS is described. This workflow can be used to gain insights into the role of signalling molecules such as cyclic nucleotides on regulatory networks through the identification and quantification of responsive phospho(proteins).

Key words Gel-free proteomics, Mass spectrometry, Phosphoproteomics, Quantitation, TiO_2 enrichment

1 Introduction

Phosphoproteomics is the subdivision of proteomics focusing on the identification, classification, and characterization of proteins subjected to phosphorylation. Phosphorylation is a reversible post-translational modification (PTM), where typically the gamma phosphate group from ATP is transferred to serine, threonine, tyrosine, or histidine residues by the enzymatic action of kinases [1–4]. These phosphate groups can then be removed from proteins by hydrolysis catalyzed by phosphatases, in the event of dephosphorylation. In addition, phosphorylation on arginine, lysine, aspartate, glutamate, and cysteine has been reported (*see* review ref. 5). Phosphorylation can alter protein activity level, protein function, interactions with binding partners, subcellular localization, and tertiary and quaternary structure, and influences key cellular processes including cell growth, metabolism, degradation of

Chris Gehring (ed.), *Cyclic Nucleotide Signaling in Plants: Methods and Protocols*, Methods in Molecular Biology, vol. 1016, DOI 10.1007/978-1-62703-441-8_9,

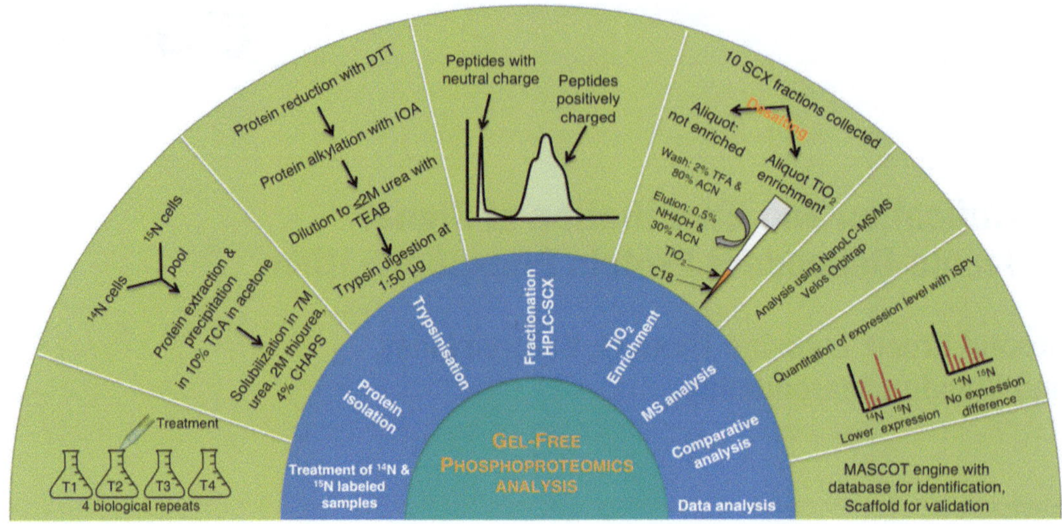

Fig. 1 Approach for the study of responsive phosphoproteome and total proteome using gel-free techniques. Immediately after treatment, an equal quantity of cells from all treatment-points grown in ^{15}N-enriched media is mixed together and is used as the control. Each treatment-point grown in ^{14}N-enriched media is mixed with an equal quantity of the control grown in ^{15}N-enriched media. The cells from each combined ^{14}N/^{15}N sample are homogenized to a fine powder in liquid nitrogen with mortar and pestle and proteins are precipitated in 10 % (w/v) TCA in acetone, washed in 80 % (v/v) acetone, and resolubilized in 7 M urea, 2 M thiourea, and 4 % (w/v) CHAPS. Proteins are reduced, alkylated, and proteolyzed with trypsin. Following detergent removal, sample fractionation is performed by strong cation exchange (SCX) and collected fractions are then desalted. An aliquot of each fraction is processed directly by mass spectrometry (MS), while the rest is enriched for phosphopeptides using TiO_2 prior to MS. The MS data is then processed with iSPY for quantitative analyses, where the peptide mass shift induced by the "heavy" ^{15}N-enriched media is used to differentiate the control from the sample of interest. The MS data is then processed with the MASCOT search engine using the appropriate database for the organism of interest for protein identification. Data are further processed with Scaffold and Scaffold PTM for validation of protein identification and posttranslational modification (PTM) assignment, respectively

proteins, apoptosis, and intracellular signalling [4, 5]. Knowledge of changes in the phosphorylation status of the proteome upon physiological response provides essential insights into signal transduction networks activated in cells in response to different stimuli, like biotic and abiotic stresses, which in turn lead to the discovery of targets for drug discovery and/or genetic modification [6].

In a typical proteomics workflow, proteins are digested to peptides prior to downstream mass spectrometric analysis. However, phosphopeptides are greatly outnumbered by non-phosphorylated peptides in a typical cellular protein digest and typically ionize less efficiently than their non-phosphorylated counterparts. Hence, it is necessary to enrich for phosphopeptides to ensure their efficient analysis [7]. Several enrichment techniques have been optimized over the years such as the immunoprecipitation with phospho-antibodies [8], cation and anion exchange chromatography [9], and use of titanium dioxide (TiO_2) [10, 11] or immobilized metal affinity chromatography (IMAC) [12, 13] for the separation of negatively charged (phospho)peptides. In the present experiment, phosphopeptides are enriched using titanium dioxide (TiO_2) following pre-fractionation by strong cation exchange (SCX).

Quantitative phosphoproteomics methods can be used to measure the dynamics of phosphorylation events. Relative quantitation, which involves the measurement of the relative amounts of proteins among samples, as opposed to absolute quantitation, which determines the actual concentration, is based on label-free techniques [14, 15], peptide labelling techniques such as iTRAQ or TMT [16, 17], or the in vivo addition of stable isotopes like ^{15}N [18, 19] or SILAC [20]. In this chapter, $^{14}N/^{15}N$ labelling was used as quantitation method (Fig. 1). *Arabidopsis thaliana* cell suspension cultures were grown in media containing either light ^{14}N or heavy ^{15}N, in the form of ammonium nitrate and potassium nitrate, as the sole nitrogen source, where ^{15}N-labelled samples were used as control. The use of ^{15}N as the control, which consists of all the treatments of a given experiment combined together prior to protein extraction, allows for a direct comparison between a given biological sample against the pool control. The methods used for fractionation and enrichment of phosphopeptides from complex protein extracts are in the context of a workflow designed to enable both the identification and relative quantitation of phosphorylation events. Although *Arabidopsis thaliana* was used in these experiments, a similar workflow can easily be adapted with other organisms, whereby cell culture can be adapted.

2 Materials

All solutions are prepared using analytical grade reagents and Milli-Q® "ultrapure" (Millipore) water (dH_2O), purified in successive steps of filtration and deionization, to attain a purity of 18.2 MΩ cm, except stated otherwise. All reagents and solutions were stored at room temperature, unless stated otherwise. It is important to follow safety regulations for all chemicals and biological waste handling and disposal. The material safety data sheets (MSDS) should be read prior to using any chemical.

2.1 Protein Extraction and Protein Concentration Estimation

1. Ten percent (w/v) trichloroacetic acid (TCA) in ice-cold acetone. Weigh 5 g of TCA and dissolve in 50 mL of 100 % (v/v) of acetone. Allow to cool down at −20 °C prior to use. Prepare fresh.
2. Eighty percent (v/v) acetone. Measure 80 mL of 100 % (v/v) acetone and mix with 20 mL of dH_2O.
3. 1 M Tris–HCl, pH 8.0: Add 800 mL of dH_2O to a 1 L glass beaker on a stirring plate. Add a clean rinsed stirring bar to the

beaker (*see* **Note 1**). Weigh 121.14 g of Tris and transfer to the glass beaker. Allow to dissolve completely and adjust to pH 8.0 with HCl. Transfer the solution to a 1 L measuring cylinder and adjust the final volume to 1 L with dH_2O.

4. Urea lysis buffer (ULB): 7 M urea, 2 M thiourea, 4 % (w/v) 3-[(3-cholamidopropyl)dimethylammonio]-1-propanesulfonate (CHAPS). Add about 4 mL of dH_2O to a 25 mL glass beaker on a stirring plate. Add a clean rinsed stirring bar to the beaker (*see* **Note 1**). Weigh 4.2 g of urea, 1.52 g of thiourea, and 0.4 g of CHAPS and transfer all components to the beaker. Add 0.5 mL of 1 M Tris–HCl (pH 8.0), 200 μL of phosphatase inhibitors, cocktail set V (Calbiochem, Merck Millipore, Darmstadt, Germany) (*see* **Note 2**). Dissolve one tablet of complete protease inhibitor cocktail tablet, EDTA free (Roche, Pennzberg, Germany) (*see* **Note 3**). Mix well and adjust the final volume to 10 mL with dH_2O (*see* **Note 4**).
5. Quick Start™ Bradford protein assay (Bio-Rad, Hercules, USA) (*see* **Note 5**). Allow the Quick Start 1× dye reagent to settle to room temperature prior to use.
6. Quick Start™ bovine serum albumin (BSA) standard: 2 mg/mL (Bio-Rad) (*see* **Note 6**).

2.2 Protein Digestion

1. 0.5 M dithiothreitol (DTT) stock: Weigh 77.1 mg of DTT and dissolve in 1 mL of high-pressure liquid chromatography (HPLC)-grade dH_2O.
2. 70 mM iodoacetamide (IOA): Weigh 12.95 mg of IOA and dissolve in 1 mL of HPLC-grade dH_2O. Wrap the tube with aluminum foil to protect tube from light.
3. 50 mM tetraethylammonium bicarbonate (TEAB): Dilute 100 μL of 1 M stock into 1.9 mL of HPLC-grade dH_2O (*see* **Note 7**).
4. Trypsin (sequencing grade modified trypsin, Promega, Madison, USA): Resuspend trypsin in 100 μL of trypsin resuspension buffer provided. Vortex well. Heat at 30 °C for 15 min.

2.3 Detergent Removal

1. Detergent removal spin columns (Pierce, Thermo Fisher Scientific, Rockford, USA) (*see* **Note 8**).
2. Wash/equilibration buffer: 50 mM Tris–HCl, pH 8.0. Measure 50 mL of 1 M Tris–HCl, pH 8.0 and dilute to 1 L with dH_2O.

2.4 Protein Separation by Strong Cation Exchange

1. HPLC system: Acquity UPLC® system (Waters, Milford, MA, USA), connected with a 250 μL injection loop, a 500 μL sample syringe, and a 30 μL volume needle, operating at 0.5 mL/min.

2. Column: PolySulfoethyl A™, 200×4.6 mm, 5 μm, 200 Å column packing (PolyLC, Columbia, MD, USA).
3. UV cell: Photodiode array (PDA, Waters) set up to read wavelengths between 210 and 400 nm.
4. SCX stock solution: 70 mM KH_2PO_4. Add about 800 mL of HPLC-grade dH_2O to a 1 L glass beaker on a stirring plate (*see* **Note 1**). Add a cleaned rinsed stirring bar to the beaker. Weigh 9.5 g of KH_2PO_4. Adjust to pH 2.7 with HCl. Adjust to 1 L with HPLC-grade dH_2O.
5. SCX buffer A: 7 mM KH_2PO_4, 30 % (v/v) ACN, pH 2.7. Dilute 100 mL of SCX stock solution with 600 mL of HPLC-grade dH_2O. Ensure that pH is 2.7. Add 300 mL of acetonitrile (ACN).
6. SCX buffer B: 7 mM KH_2PO_4, 350 mM KCl, 30 % (v/v) ACN, pH 2.7. Add 100 mL of SCX stock solution to 500 mL of HPLC-grade dH_2O in a 1 L glass beaker on a stirring plate. Weigh 26.1 g of KCl and transfer to the beaker. Adjust pH to 2.7. Add 300 mL ACN. Make up to 1 L with HPLC-grade dH_2O.
7. SCX buffer C: 50 mM KH_2PO_4, 500 mM NaCl, pH 7.5. Pour into a glass beaker 500 mL of SCX stock solution and place on a stirring plate. Weigh 29.22 g of NaCl and add to the beaker. Adjust to pH 7.5 with NaOH and add HPLC-grade dH_2O to 700 mL total volume.
8. LoBind® 2 mL microcentrifuge tubes (Eppendorf, Hamburg, Germany).
9. Software: Masslynx (Waters).

2.5 Desalting of Peptide Samples

1. C18 cartridges (Sep-Pak® C18 Vac, 3 cc, 500 mg, Waters).
2. 10 mL syringe.
3. Conditioning buffer: 100 % (v/v) ACN.
4. Equilibration buffer: 50 % (v/v) ACN, 0.5 % (v/v) acetic acid. Measure 50 mL of ACN and add 500 μL of acetic acid. Adjust volume to 100 mL.
5. Washing buffer: 0.1 % (v/v) trifluoroacetic acid (TFA). Add 100 μL of TFA to 100 mL of HPLC-grade dH_2O (*see* **Note 9**).
6. 0.5 % (v/v) acetic acid: Add 50 μL of acetic acid to 10 mL of HPLC-grade dH_2O.
7. Elution buffer: 50 % (v/v) ACN, 0.5 % (v/v) acetic acid. Measure 5 mL of ACN and add 50 μL of acetic acid. Adjust volume to 10 mL with HPLC-grade dH_2O.

2.6 Phosphopeptide Enrichment Using Titanium Dioxide

All buffers should be prepared prior to enrichment (*see* **Note 10**).

1. Washing buffer: 80 % (v/v) ACN, 2 % (v/v) TFA. Add 4 mL of 100 % (v/v) ACN to a 10 mL glass measuring cylinder. Add 100 μL of TFA and adjust volume to 5 mL with HPLC-grade dH_2O. Mix well and store at 4 °C.
2. Loading buffer: 1.3 M 2,5-dihydroxybenzoic acid (DHB), 80 % (v/v) ACN, and 2 % (v/v) TFA. Weigh 200 mg of DHB and transfer to a 2 mL microcentrifuge tube. Add 1 mL of washing buffer and mix well.
3. Elution buffer 1: 2.5 % (v/v) ammonia solution, pH≥10.5. Add 5 mL of HPLC-grade dH_2O to a 10 mL glass measuring cylinder. Then add 1 mL of 25 % (v/v) ammonium solution and adjust the pH to greater or equal to 10.5 with ammonium solution. Mix and adjust volume to 10 mL with HPLC-grade dH_2O.
4. Elution buffer 2: 30 % (v/v) of ACN. Add 5 mL of HPLC-grade dH_2O to a 10 mL glass measuring cylinder. Add 3 mL of 100 % (v/v) ACN and adjust the volume to 10 mL with HPLC-grade dH_2O.
5. 100 % (v/v) ACN.
6. Titanium dioxide (TiO_2) beads: Weigh 2 mg of TiO_2 (Titansphere TiO_2 5 mM, GL Sciences Inc., Tokyo, Japan) beads into a microcentrifuge tube (*see* **Note 11**).
7. StageTips, C18 material, GELoader® pipette tip (Proxeon Biosystems, Odense, Denmark).

2.7 Identification of Phosphopeptides by LC-MS/MS

1. MS resuspension buffer: 0.1 % (v/v) formic acid (FA) and 5 % (v/v) ACN. Add 5 mL of HPLC-grade dH_2O to a 10 mL glass measuring cylinder. Add 10 μL of 99 % (v/v) FA and 500 μL of 100 % (v/v) ACN. Mix and adjust the volume to 10 mL with HPLC-grade dH_2O.
2. LC-MS buffer A: 0.1 % (v/v) FA and 5 % (v/v) ACN. Add 50 mL of HPLC-grade dH_2O to a 100 mL glass measuring cylinder. Add 100 μL of 99 % (v/v) FA and 5 mL of 100 % (v/v) ACN. Mix and adjust the volume to 10 mL with HPLC-grade dH_2O. Sonicate the buffer for 15 min in a bath sonicator (*see* **Note 12**).
3. LC-MS buffer B: 0.1 % (v/v) FA and 90 % (v/v) ACN. Add 90 mL of 100 % (v/v) ACN. Add 100 μL of 99 % (v/v) FA and adjust the volume to 100 mL with HPLC-grade dH_2O. Mix well and sonicate the buffer for 15 min in a bath sonicator (*see* **Note 12**).
4. Autosampler vials and vial closures (National Scientific Company, Rockwood, TN).

5. Pre-column: 5 μm, 200 Å, 50 mm long × 0.2 mm Magic C18AQ (Michrom Bioresources, Auburn, California).
6. Separation column: 3 μm, 200 Å, 150 mm long × 0.075 mm Magic C18AQ (Michrom).
7. Mass spectrometer: LTQ Orbitrap Velos™ (Thermo Fisher Scientific, San José, USA) coupled to an EASY nLC system (Proxeon, Thermo Scientific).
8. Source: Advance CaptiveSpray source for Thermo (Michrom).

2.8 Data Processing

1. Software tools for raw data file conversion: iSPY (in house software, Cambridge Centre for Proteomics, University of Cambridge, UK), MSConvert (open-source software), and Proteome Discoverer (Thermo Fisher Scientific, Rockford, USA).
2. Software tools for peptide and protein identification: Mascot (Matrix Science, London, UK) and Scaffold (Proteome Software, Portland, USA).
3. Software tools for ^{15}N quantitation: iSPY (in house software, Cambridge Centre for Proteomics, University of Cambridge, UK), MSQuant (open-source software; http://msquant.alwaysdata.net/).
4. PTM detection: Scaffold PTM (Proteome Software, Portland, USA).

3 Methods

Perform all procedures at room temperature unless stated otherwise.

3.1 Protein Extraction and Protein Concentration Estimation

1. Mix the ^{14}N and ^{15}N protein samples in a one-to-one (w/w) ratio. Grind cells to a fine powder with mortar and pestle in liquid nitrogen and put in a 15 mL Falcon tube. Mix with 10 volumes of ice-cold 10 % (w/v) TCA in acetone, vortex well to ensure suspension of the ground sample, and incubate overnight at −20 °C.
2. On the following day, vortex well to suspend the material and centrifuge at 2,800 × *g* for 10 min. Decant the supernatant and resuspend the pellet in ice-cold 80 % (v/v) acetone with vigorous vortexing (*see* **Note 13**). Spin down at 2,800 × *g* for 10 min and decant the supernatant. This washing step with 80 % (v/v) acetone should be repeated two more times. At the end of the final wash, decant the supernatant and evaporate excess acetone under a fume hood for about 10–15 min (*see* **Note 14**).

Table 1
Gradient for the separation of tryptic peptides by SCX chromatography

Time (min)	Buffer A (%)	Buffer B (%)	Buffer C (%)
0	100	0	0
2	100	0	0
17	85	15	0
37	0	100	0
52	0	100	0
53	0	0	100
67	0	0	100
68	100	0	0
87	100	0	0

3. Resolubilize the protein pellet in 2–3 volumes of ULB overnight with vigorous vortexing (*see* **Note 15**). Spin down at 2,800 × *g* for 10 min and collect the supernatant to a fresh tube (*see* **Note 16**).
4. Once resolubilized, proteins can be stored at −80 °C until further use.
5. To estimate the concentration of proteins in the sample, a 5 μL aliquot of the protein extract was mixed with 15 μL of dH_2O and 1 mL of Quick Start™ Bradford protein assay and pipetted into a cuvette (*see* **Note 17**). In parallel, increasing concentrations of BSA from 0 to 1.5 mg/mL were prepared in a similar fashion (Table 1) and pipetted separately in a cuvette. Duplicates of every preparation were made and allowed to react for a minimum of 5 min to a maximum of 60 min, as recommended by the manufacturer. Absorbance of the sample and BSA standards was measured at 595 nm. Plot results from the BSA to build up a standard curve and deduce a trendline equation for the curve with Excel and use to calculate the estimated concentration of the protein sample.

3.2 Protein Digestion

To ensure good efficiency of trypsin digestion and identification of cysteine-containing peptides, proteins should first be reduced and alkylated, which are here described using DTT and IOA, respectively. The following protocol is described using 1 mg total proteins as the starting material. Prepare an estimated 1 mg total proteins in urea/thiourea-based buffer. The following protocol assumed an estimated concentration of 2 mg/mL.

1. Add 5 μL of DTT from the 0.5 M stock solution for 495 μL of total proteins.
2. Vortex well and incubate for 2 h at 37 °C.
3. Cool down the sample to room temperature.
4. Add 5 μL of IOA from the 70 mM stock solution to the reduced total proteins.
5. Vortex well and incubate for 30 min in the dark at room temperature.
6. Add another 5 μL of DTT from the 0.5 M stock solution to the reduced and alkylated total proteins (*see* **Note 18**).
7. Dilute the protein sample with 1.75 mL of 50 mM TEAB or 100 mM ammonium bicarbonate per 500 μL of protein sample to ensure that the final concentration of urea is ≤2 M (*see* **Note 19**).
8. Add 10–20 μg trypsin for 1 mg total proteins (*see* **Note 20**). In case of 1:50 dilution, add 100 μL of trypsin solution (20 μg), while for a dilution of 1:100, add 50 μL of trypsin solution (10 μg).
9. Vortex well and incubate for 4 h to overnight at 37 °C.

3.3 Detergent Removal

Use one fresh column for each sample. All centrifugation steps were carried out in a swing out rotor.

1. Remove the screw cap and the bottom plug and place the column in a 15 mL Falcon tube (collection tube).
2. Centrifuge the column in the collection tube at 1,000 × *g* for 2 min to remove the storage solution.
3. Add 2 mL of 50 mM of Tris–HCl (pH 8.0), centrifuge at 1,000 × *g* for 2 min, and discard the flow through. Repeat this step twice (*see* **Note 21**).
4. Place the column in a fresh 15 mL Falcon tube and pipette in the peptide sample.
5. Incubate for at least 2 min and then centrifuge at 1,000 × *g* for 2 min to collect the flow through, containing the detergent-free peptide sample.
6. Dry the desalted sample with tube lid opened using a Speed Vac concentrator set on low heat (*see* **Note 22**) until almost complete dehydration (*see* **Note 23**). Peptide samples can be immediately processed (*see* Subheading 3.4) or alternatively can be stored at −80 °C until further use.

3.4 Peptide Fractionation by Strong Cation Exchange

1. Resolubilize the dried peptides in an appropriate volume of buffer A. The volume should not exceed the LC loop volume. For high-resolution chromatography, it is important not to exceed the maximum column loading capacity. Pellet any

precipitate from the sample prior to loading on the HPLC at top speed for 5 min (*see* **Note 24**).

2. Purge the LC system with all three buffers for 20 min at flow rate of 8 mL/min. Reproducible SCX chromatography is obtained with long and constant equilibration times. Calibrate the LC system for needle, sample, and syringe volumes. Switch on the UV cell.
3. Equilibrate the Acquity HPLC® system by running buffer A for 30 min, then with buffer B for 30 min, and finally with buffer A for 30 min at a flow rate of 0.5 mL/min. Set up the gradient described in Table 1.
4. Inject the peptide sample, measuring wavelengths 210–400 nm (*see* **Note 25**).
5. Collect ten 4-min fractions from minute 0 to minute 40 in 2 mL Lobind® tubes (*see* **Note 26**).
6. After collection, desalt the fractions (*see* **Note 27**).

3.5 Peptide Desalting

1. Use one C18 cartridge for each sample and place it in a 15 mL Falcon tube. Condition the cartridge with 3 mL of conditioning buffer and discard the eluate (*see* **Note 28**).
2. Equilibrate the column with 3 mL of equilibration buffer and discard the eluate.
3. Wash the column with 3 mL of washing buffer.
4. Add TFA to the sample to 0.4 % (v/v) final concentration. Load the sample onto the column and discard the eluate (*see* **Note 29**).
5. Wash the column with 3 mL of washing buffer and discard the eluate.
6. Wash the column with 0.5 % (v/v) acetic acid and discard the eluate (*see* **Note 30**).
7. Elute the peptides from the column with 1 mL of elution buffer (*see* **Note 31**).
8. Dry the desalted sample with tube lid opened using a Speed Vac concentrator (*see* **Note 32**) set on low heat (*see* **Note 22**) until almost complete dehydration (*see* **Note 23**). Alternatively, the sample can be desiccated using a freeze drier on dry ice overnight. In case of using the latter technique, perforate the lids of the tubes with a needle and freeze-dry the fractions with the tube lid closed.
9. Peptide samples can be immediately processed (*see* Subheading 3.6) or alternatively can be stored at −80 °C or preferably −140 °C if available until further use.

3.6 Phosphopeptide Enrichment by TiO_2

1. Resuspend the peptide sample with 50 μL of loading buffer (*see* **Note 33**).
2. Condition 2 mg of TiO_2 beads per sample with 200 μL of 100 % (v/v) ACN and vortex well.
3. Centrifuge at 70 × *g* for 30 s and remove the supernatant. Add 200 μL of loading buffer, vortex well for at least 5 min, and centrifuge at 70 × *g* for 30 s.
4. Remove the supernatant and equilibrate the beads in 200 μL of washing buffer. Vortex well and pellet down the beads by centrifuging at 70 × *g* for 30 s.
5. Decant the supernatant and resuspend the beads in 10 μL of loading buffer (*see* **Note 34**).
6. Mix 10 μL of beads in loading buffer (*see* **Note 35**) with the peptide sample and vortex well (*see* **Note 36**). Incubate with shaking for a minimum of 5 min, but no longer than 15 min (*see* **Note 37**). Centrifuge at 70 × *g* for 30 s and pipette out about 30 μL of supernatant.
7. Resuspend the beads and pipette the entire volume (*see* **Note 36**) into the GELoader® pipette tip. Apply gentle air pressure using a 5 mL Combitip® pipette tip to elute the buffer (*see* **Note 38**).
8. Wash the beads three times with 45 μL of washing buffer each step.
9. Elute phosphopeptides with 30 μL of elution buffer 1 and collect to a fresh tube.
10. Perform a second elution with 10 μL of elution buffer 2 (*see* **Note 39**) and collect into the same tube as elution one.
11. Dry the collected phosphopeptides using a Speed Vac concentrator (*see* **Note 22**) and process by MS for identification immediately. Alternatively the phosphopeptides can be stored for a short period of time at −80 or −140 °C, if available, until further use.

3.7 Identification of Phosphopeptides by LC-MS/MS

1. Resuspend the phosphopeptides in 12 μL MS resuspension buffer. Vortex well to ensure complete resuspension and centrifuge at 20,800 × *g* for 5 min (*see* **Note 40**).
2. Pipette out 10 μL of the resolubilized phosphopeptides into an autosampler vial. Close the vial with a vial closure immediately (*see* **Note 41**). Centrifuge the vial at 145 × *g* for 30 s (*see* **Note 42**). Load the vials in the LC system connected with the MS and set up the run program.
3. Equilibrate the pre-column with a 12 μL volume of LC-MS buffer A at a flow rate of 8 μL/min and maximum pressure of 280 Bar.

4. Equilibrate the column with a 2.5 μL volume of LC-MS buffer A at a flow rate of 0.5 μL/min and maximum pressure of 280 Bar.
5. Set the volume for auto-sampler wash at 100 μL.
6. Pick up a volume of 5 μL of peptide mixtures at a flow rate of 10 μL/min and load 40 μL at a flow rate of 8 μL/min onto the pre-column and column connected to the nLC system, with the maximum pressure of 280 Bar to protect the column integrity. A three-step gradient of 0-40 % LC-MS buffer B in 40 min, then 40–90 % LC-MS buffer B in 5 min, and finally 90 % LC-MS buffer B for 15 min with a flow rate of 400 nL/min over 45 min can be applied for peptide elution.
7. The LTQ-Orbitrap is equipped with an Advance CaptiveSpray source and operated in positive mode. Data are acquired between m/z 350 and m/z 1,600 and the normalized collision-induced dissociation at 35.0 V. The top ten precursor ions are selected in the MS scan by Orbitrap with resolution r=60,000 for fragmentation in the linear ion trap. The spray voltage is 1.5 kV, the capillary voltage is 47.5 V, the capillary temperature is 250 °C, and a sheath and auxiliary gas flow of 35 and 15 arbitrary units are used, respectively. Data are recorded using Xcalibur™ software version 2.1 (Thermo Fisher Scientific).

3.8 Data Processing

1. Convert the files from the LTQ Orbitrap Velos from ".raw" to ".mzXML" using MSConvert with the following settings: Output file format ".mzXML"; write index; and use ".zlib" compression, and 32 bit binary encoding (*see* **Note 43**).
2. Convert the resulting ".mzXML" files to ".mgf" using Proteome Discoverer version 1.1.0 (Thermo Fisher Scientific).
3. Upload the ".mgf" to the Mascot tool. Run the search using the following settings: carbamidomethyl as a fixed modification (*see* **Note 44**); oxidation on methionine (M) residues and phosphorylation on serine (S), threonine (T), and tyrosine (Y) residues as variable modifications; 25 ppm of peptide tolerance, 0.8 Da of MS/MS tolerance; max of two missed cleavages, a peptide charge of +2, +3, or +4; and selection of decoy database.
4. Transfer the results from the Mascot search that are in the ".dat" format in the same folder as the ".mzXML," ".mgf," and ".raw" files.
5. Run the Mascot files obtained with Scaffold for improved identification.
6. Select the ^{15}N quantitation option in the iSPY (*see* **Note 45**) or MSQuant interface for ^{15}N quantitation. Other software such

as QuantiSpec [21], Prorata [22], and Census (http://fields.scripps.edu/census) can also be used.

7. Scaffold™ PTM software is used specifically to identify PTMs. The Mascot files in the ".dat" format are converted to ".mzident" using Scaffold™. Scaffold™ PTM uses the ".mzident" files with stringent set parameters including 95 % protein probability and 90 % unmodified peptide probability to identify the PTMs.

4 Notes

1. Pouring dH_2O into the beaker will help the stirring bar to start moving immediately and hasten the solubilization process.
2. The phosphatase inhibitor cocktail set V from Calbiochem-Millipore Merck contains four phosphatase inhibitors for the inhibition of both serine/threonine and protein tyrosine phosphatases. Each 50× concentrated vial includes 250 mM sodium fluoride, 50 mM β-glycerophosphate, 50 mM sodium pyrophosphate decahydrate, and 50 mM sodium orthovanadate (dilution factor 1:50). These inhibitors are highly toxic and should be used with extreme care and disposed of appropriately.
3. The complete protease inhibitor cocktail tablet, EDTA free from Roche, contains a mixture of several proteases that are potent for the inhibition of serine and cysteine protease but no metalloproteases. The tablet can be added directly into 10 mL of extraction media (dilution factor 1:10). These inhibitors are highly toxic and should be used with extreme care and disposed of appropriately.
4. Complete dissolution of high concentrations of urea and thiourea solution takes 2–4 h. Keep the solution below 30 °C to prevent carbamylation of primary amines.
5. The Quick Start™ Bradford protein assay is compatible with urea and thiourea present in ULB.
6. Alternatively a stock of 2 mg/mL of BSA can be prepared by dissolving 2 mg of BSA into 1 mL of ULB.
7. Alternative to TEAB, 100 mM ammonium bicarbonate can be used to dilute the protein sample to 2 M urea. To prepare 100 mM ammonium bicarbonate solution, weigh 0.79 g ammonium bicarbonate and transfer into a 50 mL beaker. Add 60 mL of HPLC-grade dH_2O and mix thoroughly until fully dissolved on a stirring block. Adjust the volume to 100 mL and store at room temperature.
8. In the present case, a 2 mL column volume was used, which can process a maximum sample volume of 500 μL. Several formats of columns are available from the same company.

9. The TFA is a strong acid. It is highly recommended to handle it in a fume hood. Additionally it is recommended to use only glass syringes to pipette with such high concentration of TFA to avoid polymer contamination in the mass spectrometer.
10. Prior preparation of buffers will ensure rapid process and thus will benefit the phosphopeptide enrichment.
11. Use 2 mg of TiO_2 beads per 100 μg sample.
12. Sonication of the buffers prior to LC-MS/MS is essential to limit the risk of pressure problems on the LC system as it will break down particles and degas the solution.
13. Ensure that the protein pellet is thoroughly washed with 80 % (v/v) acetone to remove the TCA, as it will interfere with downstream analysis.
14. Avoid letting the pellet to overdry, as this will render proteins more difficult to resolubilize.
15. The volume of ULB should be kept to a minimum of twice the volume of pellet to ensure buffering of the proteins and a pH around 7.5–8.0.
16. Centrifugation of the sample prior to use ensures that debris and putative insoluble materials are removed from the protein extracts.
17. The Quick Start™ Bradford protein assay is linear up to 1.5 mg/mL. If the protein sample is more concentrated it is important to dilute it. In the present experiment, it is diluted 4× in ULB (i.e., 5 μL of sample mixed with 15 μL of ULB).
18. Increasing the final concentration of DTT to 10 mM will quench unreacted IOA.
19. A concentration of urea >2 M will inhibit the activity of trypsin.
20. To ensure good rate of digestion, trypsin:proteins should be diluted at a 1:50 to 1:100 ratio.
21. Any buffer can be used instead of Tris–HCl provided the pH is within the range of 4–10. These include AMBIC, PBS, MES, MOPS, and bicarbonate. However the buffer should not contain any organic solvent.
22. The Speed Vac concentrator should be set on low heat to limit heating of the sample.
23. Desalted fractions should not be dried out completely (a few microliters should still be visible) as peptide resolubilization for downstream analysis will become difficult.
24. Any precipitate may block the column or increase the back pressure significantly. It is thus important to precipitate down any debris or particles by centrifugation at full speed for 5 min

at room temperature prior to loading onto the HPLC column.

25. The wavelength(s) chosen for measurement depends on the type of UV cell. In the present experiment, the cumulative wavelengths represent measurements from 210 to 400 nm.
26. The duration of fraction collection and volume of the fractions are dependent upon the desired resolution and the complexity of the starting sample as well as the downstream analysis by MS that will be chosen. Highly complex samples might need greater fractionation (i.e., through extended gradient duration or collection of smaller fractions).
27. Desalting is preferably done immediately after SCX fractionation. It is advisable to start early morning the SCX procedure so that the desalting can be carried out immediately after and the desiccation overnight. Alternatively, the fractions can be frozen at –20 °C for a few days prior to desalting.
28. Liquids used will flow through the column by gravity. However, to hasten the process the column in the 15 mL Falcon tubes can be centrifuged at low speed at 145 ×*g* for 30 s.
29. During this step, peptides will bind to the C18 material. It is recommended to allow the sample to flow through the column by gravitational force.
30. During this step, TFA is washed from the column.
31. The elution should be carried out using gravity just like for the sample loading step.
32. The elution buffer contains 50 % (v/v) ACN and it is thus difficult to keep sample frozen during the freeze-drying process. Therefore, the use of a Speed Vac concentrator in this particular case is preferred.
33. Use low-binding tubes to limit phosphopeptide loss due to sticking on the tube surface.
34. Cut the end of tips to widen the opening and avoid damaging the beads while pipetting them.
35. Ensure that the beads are in solution and pipetted into the tip.
36. It is vital to mix the sample thoroughly with DHB prior to loading the sample on the GELoader® tip to ensure bead specificity to phosphopeptides.
37. The incubation of the peptides with the beads should last a minimum of 5 min to ensure that phosphopeptides will be enriched. However, it is important to keep the incubation below 15 min to limit unspecific binding of non-phosphopeptides.

38. Completely elute out the buffer by applying pressure onto the tip using a 5 mL Combitip® (Eppendorf) pipette tip. The beads should at that point be well packed at the bottom of the tip, sitting on the C18 membrane.
39. The second elution step is carried out to ensure complete recovery of phosphopeptides from the beads.
40. The centrifugation step after resolubilization of the phosphopeptides ensures that any precipitate and bead particles are pelleted down.
41. The autosampler vial should be closed immediately to avoid evaporation of the sample that happens rapidly with ACN.
42. The centrifugation step just prior to loading onto the LC system ensures that the liquid sits at the bottom of the tube and that no air bubbles are trapped, which would result in high pressure of the LC system.
43. The binary encoding is dependent on the processing machine.
44. Since the proteins were alkylated prior to trypsin digestion the carbamidomethyl modification should be selected as fixed.
45. The iSPY tool utilizes the embedded Percolator software (open-source software) for improved peptide identifications and generates ".csv" report files on the peptide and protein identifications as well as provides quantitative data. The iSPY tool compares the ^{14}N peptide isotopic peak distribution with the ^{15}N ones from MS^1 dataset, through comparison of the theoretical mass difference between the heavy and light peptide and the typical isotopic distribution pattern. This means that the quantitation can be calculated although the incorporation rate of ^{15}N might not be 100 % and the isotopic peak pattern shifts towards a lower mass over charge range.

References

1. Besant PG, Tan E, Attwood PV (2003) Mammalian protein histidine kinases. Int J Biochem Cell Biol 35:297–309
2. Cohen P (2000) The regulation of protein function by multisite phosphorylation—a 25 year update. Trends Biochem Sci 25:596–601
3. Dhanasekaran N, Reddy EP (1998) Signaling by dual specificity kinases. Oncogene 17:1447–1455
4. Ubersax JA, Ferrell JE (2007) Mechanisms of specificity in protein phosphorylation. Nat Rev Mol Cell Biol 8:530–541
5. Cieśla J, Frączyk T, Rode W (2011) Phosphorylation of basic amino acid residues in proteins: important but easily missed. Acta Biochim Pol 58:137–148
6. Nakashima K, Ito Y, Yamaguchi-Shinozaki K (2009) Transcriptional regulatory networks in response to abiotic stresses in Arabidopsis and grasses. Plant Physiol 149:88–95
7. Heemskerk AAM, Busnel JM, Schoenmaker B, Derks RJE, Klychnikov O, Hensbergen PJ, Deelder AM, Mayboroda OA (2012) Ultra-low flow electrospray ionization-mass spectrometry for improved ionization efficiency in phosphoproteomics. Anal Chem 84:4552–4559
8. Thelemann A, Petti F, Griffin G, Iwata K, Hunt T et al (2005) Phosphotyrosine signaling networks in epidermal growth factor receptor overexpressing squamous carcinoma cells. Mol Cell Proteomics 4:356–376

9. Beausoleil SA, Jedrychowski M, Schwartz D, Elias JE, Villen J et al (2004) Large-scale characterization of HeLa cell nuclear phosphoproteins. Proc Natl Acad Sci U S A 101:12130–12135
10. Larsen MR, Thingholm TE, Jensen ON, Roepstorff P, Jorgensen TJD (2005) Highly selective enrichment of phosphorylated peptides from peptide mixtures using titanium dioxide microcolumns. Mol Cell Proteomics 4:873–886
11. Thingholm TE, Jorgensen TJ, Jensen ON, Larsen MR (2006) Highly selective enrichment of phosphorylated peptides using titanium dioxide. Nat Protoc 1:1929–1935
12. Ficarro SB, McCleland ML, Stukenberg PT, Burke DJ, Ross MM et al (2002) Phosphoproteome analysis by mass spectrometry and its application to Saccharomyces cerevisiae. Nat Biotechnol 20:301–305
13. Zheng HY, Hu P, Quinn DF, Wang YK (2005) Phosphotyrosine proteomic study of interferon alpha signaling pathway using a combination of immunoprecipitation and immobilized metal affinity chromatography. Mol Cell Proteomics 4:721–730
14. Old WM, Meyer-Arendt K, Aveline-Wolf L, Pierce KG, Mendoza A et al (2005) Comparison of label-free methods for quantifying human proteins by shotgun proteomics. Mol Cell Proteomics 4:1487–1502
15. Silva JC, Denny R, Dorschel C, Gorenstein MV, Li GZ et al (2006) Simultaneous qualitative and quantitative analysis of the Escherichia coli proteome: a sweet tale. Mol Cell Proteomics 5:589–607
16. Ross PL, Huang YN, Marchese JN, Williamson B, Parker K et al (2004) Multiplexed protein quantitation in Saccharomyces cerevisiae using amine-reactive isobaric tagging reagents. Mol Cell Proteomics 3:1154–1169
17. Thompson A, Schafer J, Kuhn K, Kienle S, Schwarz J et al (2003) Tandem mass tags: a novel quantification strategy for comparative analysis of complex protein mixtures by MS/MS. Anal Chem 75:1895–1904
18. Conrads TP, Alving K, Veenstra TD, Belov ME, Anderson GA et al (2001) Quantitative analysis of bacterial and mammalian proteomes using a combination of cysteine affinity tags and 15N-metabolic labeling. Anal Chem 73:2132–2139
19. Krijgsveld J, Ketting RF, Mahmoudi T, Johansen J, Artal-Sanz M et al (2003) Metabolic labeling of C. elegans and D. melanogaster for quantitative proteomics. Nat Biotechnol 21:927–931
20. Ong SE, Blagoev B, Kratchmarova I, Kristensen DB, Steen H et al (2002) Stable isotope labeling by amino acids in cell culture, SILAC, as a simple and accurate approach to expression proteomics. Mol Cell Proteomics 1:376–386
21. Haegler K, Mueller NS, Maccarrone G, Hunyadi-Gulyas E, Webhofer C et al (2009) QuantiSpec—quantitative mass spectrometry data analysis of N-15-metabolically labeled proteins. J Proteomics 71:601–608
22. Pan CL, Kora G, McDonald WH, Tabb DL, VerBerkmoes NC et al (2006) ProRata: a quantitative proteomics program for accurate protein abundance ratio estimation with confidence interval evaluation. Anal Chem 78:7121–7131

Chapter 10

Comparative Gel-Based Phosphoproteomics in Response to Signaling Molecules

Claudius Marondedze, Kathryn Lilley, and Ludivine Thomas

Abstract

The gel-based proteomics approach is a valuable technique for studying the characteristics of proteins. This technique has diverse applications ranging from analysis of a single protein to the study of the total cellular proteins. Further, protein quality and to some extent distribution can be first assessed by means of one-dimensional gel electrophoresis and then more informatively, for comparative analysis, using the two-dimensional gel electrophoresis technique. Here, we describe how to take advantage of the availability of fluorescent dyes to stain for a selective class of proteins on the same gel for the detection of both phospho- and total proteomes. This enables the co-detection of phosphoproteins as well as total proteins from the same gel and is accomplished by utilizing two different fluorescent stains, the ProQ-Diamond, which stains *only* phosphorylated proteins, and Sypro Ruby, which stains the entire subset of proteins. This workflow can be applied to gain insights into the regulatory mechanisms induced by signaling molecules such as cyclic nucleotides through the quantification and subsequent identification of responsive phospho- and total proteins.

Key words 2D gel electrophoresis, Comparative analysis, Fluorescent stain, Isoelectric focusing, Phosphoproteomics, SDS gel electrophoresis

1 Introduction

Two-dimensional gel electrophoresis (2-DE) is a relatively reproducible proteomics technique, widely used for fractionating and resolving proteins. The principle behind this technique is that proteins are separated by two different physicochemical properties, i.e., based on their net charges or isoelectric point (p*I*) by isoelectric focusing (IEF) and then their molecular mass by sodium dodecyl sulfate-polyacrylamide gel electrophoresis (SDS-PAGE) [1, 2]. Thus, a complex protein sample can be analyzed simultaneously with high resolution and invaluable information can be drawn [1, 3]. In recent years, several specific dye stains have been developed, to identify individual proteins that undergo posttranslational modifications such as phosphorylation [4, 5]

Chris Gehring (ed.), *Cyclic Nucleotide Signaling in Plants: Methods and Protocols*, Methods in Molecular Biology, vol. 1016, DOI 10.1007/978-1-62703-441-8_10, © Springer Science+Business Media New York 2013

and glycosylation [5]. The nonoverlapping spectral properties of available dyes enable co-detection of different subsets of the proteome by 2-DE. Current research utilizing multiple fluorescent stains used in conjunction with 2-DE is well placed to be directed at identifying novel biomarkers or even proteins involved in specific signaling pathways, in addition to quantitative and differential expression comparative analysis.

Posttranslational modifications such as phosphorylation and glycosylation are important in defining functional characteristics of a number of cellular proteins [6]. Phosphorylation *per se* involves the transfer of a phosphate group by kinases from ATP to target protein-specific amino acid residues, namely, serine, threonine, tyrosine, or histidine [7]. Phosphorylation and dephosphorylation (removal of phosphate group from amino acid residues by phosphatases) are principal signaling events that alter protein structure and function [8], which in turn modulates cellular processes including cell growth and apoptosis, and mediates intracellular signal transduction. As such, the field of phosphoproteomics, the study of phosphorylation events in a cell, organ, or tissue, is critical in providing insights into protein or pathway that may be deactivated as a result of a change in the phosphorylation status.

Here, we demonstrate that phosphorylated proteins can be efficiently resolved on a gel and visually observed by fluorescent scanning. A similar technique has been used to visualize total proteome by staining the gel with a fluorescent dye, Sypro Ruby stain [9]. Of importance to note is that a single gel can be used to extract data of the phosphoproteome as well as the total proteome. This is made possible by first staining the gel with a phosphostain, the ProQ-Diamond phosphoprotein stain, scanning the gel, and subsequently staining the same gel with the total proteome stain, the Sypro Ruby stain.

Two proteome profiles can therefore be obtained from the same gel, in which protein spots can be matched to one another for protein identification and phosphoproteome quantitation.

2 Materials

All solutions are prepared using analytical grade reagents and Milli-Q® "ultrapure" (Millipore) water (dH_2O), purified in successive steps of filtration and deionization, to attain a purity of 18.2 MΩ cm. All reagents and solutions were stored at room temperature unless stated otherwise. It is important to follow safety regulations for all chemicals and biological waste handling and disposal. The material safety data sheets (MSDS) should be read prior to using any chemical.

2.1 Protein Separation by 2-DE

1. Resolving gel buffer (4×): 1.5 M Tris–HCl (pH 8.8), 0.2 % (w/v) SDS (*see* **Note 1**). Add about 100 mL of dH_2O to a 1 L glass beaker on a stirring plate set to 30 °C. Add a clean rinsed stirring bar to the beaker (*see* **Note 2**). Weigh 181.72 g Tris and transfer to the beaker. Adjust dH_2O volume to 800 mL. Mix well and adjust pH to 8.8 with HCl (*see* **Note 3**). Weigh 2 g SDS (*see* **Note 1**), transfer to the beaker with Tris–HCl, and mix well until clear. Adjust to 1 L with dH_2O.
2. Forty percent (w/v) acrylamide:Bis solution (37.5:1 acrylamide:Bis, 2.6 % C). Store at 4 °C (*see* **Note 4**).
3. Ammonium persulfate (APS): 10 % (w/v) solution in dH_2O. Weigh 0.1 g APS into a microcentrifuge tube. Dilute in 1 mL dH_2O. Prepare fresh.
4. *N*,*N*,*N*′,*N*′-tetramethylethylenediamine (TEMED). Store at 4 °C (*see* **Note 5**).
5. Tris–glycine SDS running buffer (10×): 25 mM Tris, 1.92 M glycine, 0.1 % (w/v) SDS. Add about 100 mL of dH_2O to a 1 L glass beaker cylinder on a stirring plate set to 30 °C. Weigh 30.2 g Tris and transfer to the beaker. Adjust the volume to 600 mL with dH_2O and mix well. Weigh 144 g glycine and 10 g SDS (*see* **Note 1**), transfer to the beaker with Tris, and mix well until all components are dissolved. Make up to 1 L with dH_2O. To make 1× running buffer, dilute 100 mL of 10× running buffer stock with 900 mL dH_2O (*see* **Note 6**).
6. IPG buffer pH 3–10, linear.
7. Linear Immobiline™ DryStrip pH range 4–7, 7 cm (GE Healthcare Life Sciences) (*see* **Note 7**), generally referred to as immobilized pH gradient (IPG) strips.
8. Immobiline™ DryStrip cover fluid (GE Healthcare Life Sciences).
9. Paper electrode wicks.
10. One percent (w/v) bromophenol blue solution: Weigh 100 mg bromophenol blue and 60 mg Tris. Transfer to a 15 mL Falcon tube and add 10 mL of dH_2O. Vortex to dissolve.
11. IEF rehydration buffer: 7 M urea, 2 M thiourea, 4 % (w/v) CHAPS, and 0.002 % (w/v) bromophenol blue. Weigh 10.5 g urea, 3.8 g thiourea, and 1 g CHAPS (*see* **Note 8**) and transfer into a 50 mL beaker. Add 15 mL of dH_2O and 50 μL of 1 % (w/v) bromophenol blue solution (*see* **Note 9**) and mix thoroughly until fully dissolved on a stirring plate set to 30 °C (*see* **Note 10**). Dissolving the buffer components takes 2–4 h. Adjust the volume to 25 mL and aliquot 1.25 mL into microcentrifuge tubes. Store at −20 °C. At the time of use, add 6.25 μL IPG buffer and 3.5 mg dithiothreitol (DTT) per 1.25 mL aliquot of the thawed rehydration buffer and vortex well (*see* **Note 11**).

12. 1.5 M Tris base: Weigh 181.72 g Tris and transfer to a 1 L beaker. Add dH_2O to a volume of 800 mL. Mix well and adjust pH with HCl to 8.8 (*see* **Note 3**). Adjust volume to 1 L with dH_2O. Store at 4 °C.
13. Equilibration buffer: 50 mM Tris–HCl, pH 8.8, 6 M urea, 30 % (v/v) glycerol, 2 % (w/v) SDS, 0.002 % (w/v) bromophenol blue. Add 80 mL of dH_2O to a 250 mL glass beaker on a stirring plate. Weigh 72.07 g urea and 4.0 g SDS (*see* **Note 12**), and transfer to the beaker. Add 10 mL 1.5 M Tris base (pH 8.8), 69 mL glycerol, and 400 μL of 1 % (w/v) bromophenol blue solution (*see* **Note 9**). Adjust the final volume to 200 mL with dH_2O. Aliquot as 10 mL fractions in 15 mL Falcon tubes and store at −20 °C. Prior to strip equilibration, add 0.1 g DTT to a 10 mL aliquot and to a second 10 mL aliquot, add 0.25 g iodoacetamide (IOA) (*see* **Note 13**).
14. SDS reducing buffer (2×): 200 mM Tris, 40 % (v/v) glycerol, 10 % (w/v) SDS, 0.02 % (w/v) bromophenol blue, pH 6.8. Weigh 1.21 g Tris and 5 g SDS (*see* **Note 1**), and transfer into a 50 mL beaker. Add 15 mL of dH_2O, 20 mL glycerol, and 0.01 g bromophenol blue (*see* **Note 9**) and mix thoroughly until fully dissolved on a stirring plate set to 30 °C. Adjust the volume to 50 mL with dH_2O and aliquot 1 mL into microcentrifuge tubes. Store at −20 °C. At the time of use, add either 100 μL 2-mercaptoethanol (in fume hood) or 10 mg DTT per 1 mL aliquot of the thawed rehydration buffer to act as reductant.
15. PeppermintStick™ phosphoprotein molecular weight standards (Life Technologies, Grand Island, NY, USA). Mix 1 μL of the protein standards with 6 μL of 2× SDS reducing buffer and denature by heating for 5 min at 95 °C. Clarify by centrifugation for 2 min at 14,000 × *g* at room temperature.
16. Sealing agarose: 0.5 % (w/v) agarose and 0.002 % (w/v) bromophenol blue. Weigh 0.5 g low-melting agarose (*see* **Note 14**) and transfer to a 250 mL conical flask. Add 100 mL of 1× running buffer, swirl to disperse, and then add 200 μL of 1 % (w/v) bromophenol blue solution (*see* **Note 9**). Heat in a microwave on low heat or in a heating plate until the agarose is completely dissolved. Do not allow the solution to boil over. Aliquot into 1.5 mL microcentrifuge tubes (*see* **Note 15**). Cool and store at room temperature.
17. 2-Propanol (isopropanol) ACS reagent, ≥99.5 % purity.
18. Ettan™ IPGphor™ 3 isoelectric focusing system (GE Healthcare).
19. Mini-PROTEAN® Tetra cell (Bio-Rad).

2.2 Visualization of Phosphoproteins with the Pro-Q Diamond Phosphoprotein Stain

1. Fixing solution: 50 % (v/v) methanol and 10 % (v/v) acetic acid. Measure 500 mL methanol in a 1 L measuring cylinder. Add 100 mL acetic acid and 400 mL dH_2O (*see* **Note 16**). Store at room temperature.
2. Staining solution: 3× diluted Pro-Q Diamond phosphoprotein stain [10] (Life Technologies). Measure 33 mL Pro-Q Diamond Phosphoprotein stain in a 100 mL measuring cylinder. Add 66 mL dH_2O. Pour solution to a bottle covered with two layers of aluminium foil. Mix well before use. Store at 4 °C.
3. 1 M sodium acetate stock solution: Add 300 mL of dH_2O to a 500 mL glass beaker on a stirring block. Weigh 38.5 g sodium acetate and transfer to the beaker. Mix well and adjust pH with HCl to 4.0 (*see* **Note 2**). Adjust volume to 500 mL with dH_2O. Store at room temperature.
4. Pro-Q Diamond destaining solution: 50 mM sodium acetate (pH 4.0) and 20 % (v/v) acetonitrile. Measure 50 mL of 1 M sodium acetate stock solution in a 1 L measuring cylinder. Add 200 mL of acetonitrile and 750 mL of dH_2O. Pour solution into a clean bottle and store at room temperature. Mix well before usage.
5. SYPRO Ruby total protein stain.
6. SYPRO destaining solution: 10 % (v/v) methanol and 7 % (v/v) acetic acid. Measure 100 mL in a 1 L measuring cylinder. Add 70 mL of acetonitrile and 830 mL of dH_2O. Pour solution into a clean bottle and store at room temperature. Mix well before use (*see* **Note 17**).

3 Methods

Perform all procedures at room temperature unless specified otherwise. The technical approach detailed below is summarized in Fig. 1.

3.1 Rehydration/Reswelling of IPG Strips in Preparation of Protein Separation by 2D-PAGE

Hydrophobic proteins need high concentration of urea (7–8 M) to facilitate their solubility. The buffering capacity of urea causes a light pH increase in the acidic side of the gel. The solubility of highly hydrophobic proteins, such as membrane proteins, can be increased by the addition of nonionic detergents such as Triton X-100 or zwitterionic detergents like CHAPS or Zwittergent.

1. Aliquot 50 μg protein into a 1.5 mL microcentrifuge tube (*see* **Note 18**).
2. Adjust the sample volume to 125 μL with IEF rehydration buffer.

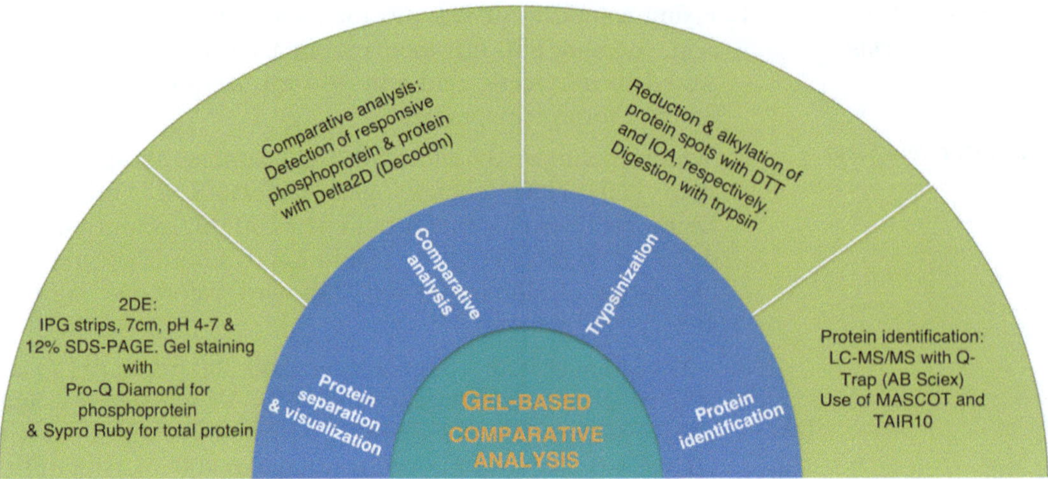

Fig. 1 Experimental approach for the study of comparative phosphoproteome and total proteome using gel-based techniques. Following sample preparation, proteins are separated by 2D gel electrophoresis using isoelectric focusing in the first dimension and SDS-PAGE in the second dimension. Phosphoproteins and total proteins are visualized by gel staining with the Pro-Q Diamond and SYPRO (Molecular Probes) stains, respectively, and imaged with the Typhoon (GE Healthcare) scanner. Gel images are used for comparative analysis with the Delta 2D (Decodon) software. Detected responsive phosphoproteins and protein spots (at p value ≤ 0.05) are reduced with DTT, alkylated with IOA, and in-gel digested with trypsin. The resulting peptides are analyzed by LC-MS/MS with the Q-Trap (AB Sciex) and the MASCOT search engine using the TAIR10 database for protein identification

3. Vortex to mix and centrifuge at 20,800×*g* for 2 min at room temperature (*see* **Note 19**).
4. Place the IPG box kit (GE Healthcare) on a flat level surface (*see* **Note 20**). Insert one disposable rehydration tray (also referred to as equilibration tray) into the IPG box. Carefully pipette out the entire volume of the sample rehydration buffer mix (125 μL), avoiding taking up any pelleted debris, and spread the solution in one of the channels of the rehydration tray over a region slightly smaller than the strip size (about 5 cm here).
5. Take a pack of IPG strips from −20 °C and place it on ice.
6. Take an IPG strip from the packaging using forceps; carefully, slowly, and swiftly remove the protective coverslip; and discard it (*see* **Note 21**).
7. Slowly place the IPG strip, gel side facing downwards (*see* **Note 22**), into the tray channel containing the sample by gently positioning the strip just on top of the mixture. Ensure that the sample has spread over the entire length of the strip to avoid unequal rehydration and dispersion of sample out of the strip borders. While placing the strip, ascertain no air bubbles are trapped under the strip. If any have been introduced, move the

strip up and down in an inclined motion until the bubble is out at the ends of the strip, by holding the strip with forceps.

8. When all samples and strips are placed, each in a separate tray channel, add on top of the strip some Immobiline™ DryStrip cover fluid to form a protective thin layer (*see* **Note 23**).
9. Close the IPG box and passively rehydrate the strip overnight. In case of passive rehydration, IPG strips are rehydrated without applying any voltage (this ensures that the sample will not aggregate as there is no field strength applied), while in active rehydration 50 V is applied to the IPG strips, and the latter requires the use of the IEF tray instead of the disposal rehydration tray. In active rehydration, incubation time varies from 1 to 99 h; however, for a 7 cm strip 12 h/overnight rehydration time is optimal.

3.2 Isoelectric Focusing of IPG Strips: The First Dimension of 2D-PAGE

IEF of IPG strips corresponds to the first dimension of the 2D electrophoresis technique [11]. Proteins migrate along the strip according to their charge until they reach their isoelectric point (p*I*), at which they carry no net charge. The focusing is performed on horizontal ceramics connected to a thermostatic circulator, as the process requires efficient cooling and controlled temperature. A high voltage is applied in order to attain completely focused proteins. Important to note is that the optimization of the running conditions is critical to prevent precipitation and aggregation of proteins as well as to accomplish reproducible resolution across replicates and treatments.

1. Following overnight passive rehydration, using forceps, gently take the strip out of the reswelling tray from the edges. Rinse the IPG strips by gently spraying dH_2O from a squeeze bottle onto the gel. Dry excess dH2O by applying the strip, gel facing up, on a lint-free wipe (*see* **Note 24**). Do not disturb the gel side.
2. Clean and dry the IPGphor instrument platform with lint-free paper wipe before placing the manifold. Position the manifold on the IPGphor and ensure that it is levelled using a spirit level at the center of the manifold tray.
3. Place the strip, gel facing up, onto the manifold/ceramic plate mounted on the IPGphor machine (*see* **Note 25**). The positive end of the strip should be placed on the anode (marked +) end of the IPGphor electrode plate.
4. Wet two paper wicks for each strip with dH_2O and allow for excess moisture to be absorbed on a dry lint-free wipe. Place the pre-damped wicks on both ends of the strip overlapping the end of the gel on the strip (about 2 mm on top of the gel end) (*see* **Note 26**). Cover each strip with Immobiline™ DryStrip cover fluid.

5. Carefully, place the electrode assembly on top of the wicks as close as possible to the wick end on top of the gel. Twirl the electrode cams into closed position of the exterior edge of the tray. Ensure that the electrode assembly is in contact with the wick by pressing the center part of the electrode while locking (*see* **Note 27**).
6. Close the safety lid. If you are working with fluorescent labelled samples, cover the safety lid with a light protective cover to limit signal quenching.
7. Switch on the IPGphor machine and start the IPGphor software on the PC. Set the communication between the software and the instrument. Load a program of choice. In the case of 7 cm long strips, a preset program is recommended as follows (*see* **Note 28**):

   Step 1: 500 V for 250 Vh; Step 2: 1,000 V for 500 Vh; Step 3: 10,000 V for 10,000 Vh. Maximum current and voltage allowed per strip are 10,000 V and 75 μA, respectively, when using the manifold, and 8,000 V and 50 μA, respectively, when using single-channel cup holder. The temperature is kept constant at 20 °C throughout the run.
8. Start the run and make sure that it is starting by checking the actual current. The initial current generally starts at 70–75 μA. This is due to the high traffic flow of proteins and should decrease after 10–15 min such as the 500 V set point is then reached (*see* **Note 29**). Such a run will take 2.5–3.0 h (Fig. 2) provided that the sample is relatively clean, i.e., with low amount of interfering substances such as salts and interfering substances.
9. When the run is complete, stop the IPGphor instrument and remove the electrode cams.
10. Remove the wicks and gently place the IPG strip into a 15 mL Falcon tube or an IPG equilibration tray using forceps on the edges of the gel strip (*see* **Note 30**).

3.3 Equilibration of Focused IPG Strips in Preparation for the Second Dimension of 2D-PAGE

1. Pipette 5 mL of equilibration buffer containing DTT into the Falcon tube for the first step of strip equilibration or 1 mL of the same buffer into the IPG equilibration tray. The 2D-PAGE workflow is visualized in Fig. 3.
2. Place the Falcon tube on a shaker and allow for gentle shaking to avoid damaging the strip. Incubate for 15 min.
3. After 15 min, gently aspirate out/decant the DTT containing buffer and replace with 15 mL of equilibration buffer containing IOA for the second step. Equilibrate by shaking for 15 min (*see* **Note 31**).
4. Remove the buffer immediately after the 15-min equilibration and rinse the strip in 1× SDS running buffer.

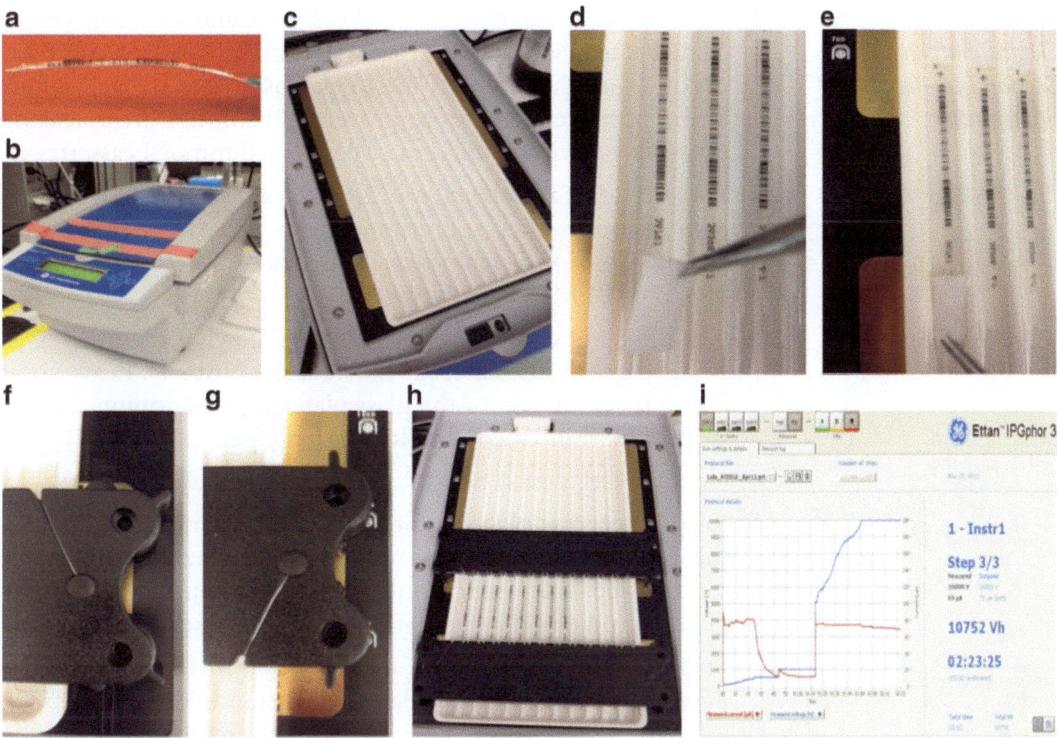

Fig. 2 Step-by-step description of the first dimension of 2D-PAGE: IPG strip focusing. (**a**) Reswelled IPG strip, 7 cm, after overnight passive rehydration; (**b**) Ettan™ IPGphor™ III (GE Healthcare); (**c**) positioning of the ceramic manifold IPG holder onto the Ettan™ IPGphor™ III; (**d**) positioning of the IPG strip and positioning of wet wicks; (**e**) outlook of strip position just prior to starting IEF; (**f**) electrode assembly, unlock; (**g**) electrode assembly, locked; (**h**) overview of the setup, ready for starting the IEF; (**i**) display of the reading of the IEF run

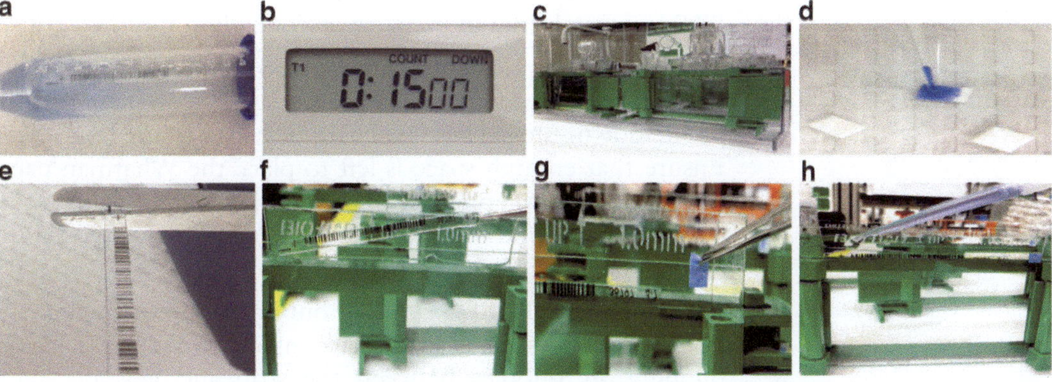

Fig. 3 Step-by-step description of the second dimension of 2D-PAGE: SDS-PAGE. (**a**) Strip in equilibration buffer; (**b**) equilibration for 15 min; (**c**) assembly of the Mini-PROTEAN® casting frame for SDS-PAGE preparation; (**d**) pipetting of the molecular marker on filter paper; (**e**) cutting of the end of the strip to fit in the SDS-PAGE casting plates; (**f**) positioning of the IEF strip after equilibration; (**g**) positioning of filter paper with molecular marker; (**h**) overlay with molten agarose solution

3.4 SDS Gel Electrophoresis: The Second Dimension of 2D-PAGE

1. Mount the gel cassette on a gel casting stand assembly (here a Mini-Protean® Tetra handcast system from Bio-Rad was used) making sure that the thin and thick glass plates are well aligned at the bottom and sides to prevent gel solution from leaking. Cast the gel with a 7.25 cm × 10 cm × 1.0 mm gel cassette.
2. For preparing a 12 % gel, mix 2.5 mL of resolving buffer, 3 mL of acrylamide stock solution, and 4.5 mL dH_2O in a 50 mL Falcon tube or Erlenmeyer flask. Add 100 μL of APS and 10 μL of TEMED (*see* **Note 5**).
3. Close the tube and mix by inverting slowly five times or by swirling gently and carefully the flask (*see* **Note 4**). Leave about 5 mm space on top of the glass plates for positioning the IPG strip and a thin layer of molten agarose gel.
4. Gently overlay with 500 μL of isopropanol (*see* **Note 32**).
5. Keep the excess of gel solution in the beaker/flask to control the polymerization process. Alternatively, when the gel is fully set, two separate layers should be visible through the thin glass plate on top of gel representing the gel and isopropanol layers.
6. Prepare the molecular marker by mixing 1 μL of PeppermintStick™ molecular weight standards with 6 μL of reducing buffer for each gel. Vortex well and incubate at 95 °C for 5 min. Centrifuge at 20,800 × *g* for 5 min at room temperature.
7. Prepare the molecular marker for loading onto the gel by pipetting 7 μL of the reduced PeppermintStick™ molecular weight standards onto a 4 × 4 mm piece of Whatman paper. Allow the paper to air-dry (*see* **Note 33**).
8. When the gel has polymerized, replace the isopropanol with 1× SDS running buffer in preparation to loading the IPG strip.
9. After rinsing the IPG strip with 1× SDS running buffer, load it on top of the casted gel. The gel faces the small plate and the plastic side of the gel strip will naturally stick to the large plate. Ensure that enough space is left to place the Whatman paper with the marker on one side of the gel. This can be achieved by cutting off one end of the plastic of the strip, avoiding excising the gel itself (*see* **Note 34**).
10. Remove excess 1× SDS running buffer on top of the gel and check that the strip is in direct contact with the gel. Avoid trapping air bubbles between the strip and gel. Place the marker on one end of the gel, as far away as possible from the strip and immediately overlay/seal with molten agarose gel (*see* **Note 35**).
11. Place the gel plates into the electrode assembly module and ascertain tight clamping to avoid buffer leakages. Place the electrode assembly into the electrophoresis tank and fill the

inner chamber with 1× SDS running buffer and the outer chamber to the recommended level, as indicated on the tank.

12. Carry out electrophoresis at 50 V for 10 min or until the bromophenol blue dye is about 2–5 mm into the resolving gel, then increase to 120 V, and run until the dye reaches the bottom of the gel (*see* **Note 36**).
13. After electrophoresis, disassemble the cell electrophoresis module and unseal the plates with a gel releaser. The gel remains on one side of the glass plates, generally the large plate. Gently transfer the gel to a clean gel tray with fixing solution using the gel releaser.

3.5 Phosphoprotein Visualization with the Pro-Q Diamond Phosphoprotein Stain

Only use powder-free gloves and high-purity solvents and dH_2O to limit dust particles and background staining on the gel. All gel incubation steps are carried out at room temperature with gentle shaking at 35–50 rpm and in the dark with the exception of the fixing. The Pro-Q Diamond phosphoprotein staining method described here is the optimized economic protocol [10].

1. Fix the gel with 100 mL fixing solution for at least 1 h with gentle shaking. Alternatively the gel can be fixed overnight. For safety reasons, the shaker can be placed under a fume hood.
2. After fixing, discard carefully the fixing solution and wash the gel in dH_2O twice for 15 min with gentle shaking.
3. Stain the gel with 60–70 mL of 3× diluted Pro-Q Diamond phosphoprotein stain solution. Cover the gel tray with two layers of aluminium foil and incubate for 2 h with gentle shaking.
4. Decant carefully the staining solution, avoiding touching the gel, as this will create fingerprints during imaging. Carefully pour about 100 mL of Pro-Q Diamond destaining solution, replace the foil cover on the gel tray, and incubate for 30 min with gentle shaking.
5. Decant carefully the staining solution and repeat the Pro-Q Diamond destaining step three more times. Start the imaging scanner at the beginning of the last Pro-Q Diamond destaining step to allow the laser to warm-up.
6. At the end of destaining, discard carefully the Pro-Q Diamond destaining solution and wash the gel in dH_2O twice for 5 min with gentle shaking.
7. Clean the scanner glass plate and lid with dH_2O and lint-free paper wipe. In case the glass plate was in direct contact with fluorescence material, 10 % (v/v) hydrogen peroxide or 75 % (v/v) ethanol and rinsing with dH_2O can be used to clean thoroughly.

8. Carefully place the gel on top of the glass plate. Only touch the sides and corners of the gel to avoid fingerprints. Ensure that no air bubbles are trapped between the gel and glass plate, as they will appear as black spots on the gel image.
9. Select the appropriate laser and emission filter for Pro-Q Diamond stain: excitation at 555 nm and emission at 580 nm. If using the Typhoon imaging system, Pro-Q Diamond stain is not one of the preset stains but is scanned with the green laser (excitation source at 532 nm) and emission filter at 560 nm. Only the two phosphoproteins, ovalbumin (45.0 kDa) and β-casein (23.6 kDa), from the PeppermintStick molecular marker should be visible. Background staining of the four non-phosphorylated proteins, β-galactosidase (116.25 kDa), bovine serum albumin (66.2 kDa), avidin (18.0 kDa), and lysozyme (14.4 kDa), can be seen occasionally, suggesting that further washes in dH_2O and/or reduction of laser intensity/contrast/light should be adjusted until they do not come into view anymore.
10. Following imaging of phosphoproteins, the gel is subsequently stained in Sypro Ruby total protein staining for quality control.
11. Stain the gel in 60 mL undiluted Sypro Ruby stain overnight.
12. Destain the gel in Sypro Ruby destaining solution for 30 min.
13. Rinse the gel in dH_2O twice for 5 min.
14. Clean the scanner glass plate and lid with dH_2O and lint-free paper wipe. In case the glass plate was in direct contact with fluorescence material, 10 % (v/v) hydrogen peroxide or 75 % (v/v) ethanol and further rinsing with dH_2O can be used to clean thoroughly.
15. Carefully place the gel on top of the glass plate. Only touch the sides and corners of the gel to avoid fingerprints. Ensure that no air bubbles are trapped between the gel and glass plate, as they will appear as black spots on the gel image.
16. Select the appropriate laser and emission filter for Sypro Ruby: excitation at 280 nm, and emission at 450/610 nm. If using the Typhoon imaging system, Sypro Ruby is one of the preset stains and is scanned with the green laser (excitation source at 532 nm) and emission filter at 610 nm. Here, all six proteins, β-galactosidase (116.25 kDa), bovine serum albumin (66.2 kDa), ovalbumin (45.0 kDa) and β-casein (23.6 kDa), avidin (18.0 kDa), and lysozyme (14.4 kDa), from the Peppermint stick molecular marker should be visible.

3.6 Comparative Analysis

After scanning, the gel images can be used for software-based comparative proteomics like the Delta2D v2 (Decodon, Greifswald, Germany), DeCyder (GE Healthcare Life Sciences), and Progenesis (Non-Linear Dynamics, Newcastle, UK). With regard to Delta2D, a demo version can be installed for learning purposes and even obtaining comparative statistics. However, the demo version does not allow for saving your data in a publishable format and a license might be necessary depending on your requirements.

4 Notes

1. Wear a protective mask while handling of SDS to prevent breathing in particles. SDS will not dissolve at room temperature; ensure warming up the buffer to about 35 °C for complete dissolution.
2. Pouring dH_2O to the beaker will help the stirring bar to start moving immediately and hasten the solubilization process.
3. The buffer needs to be brought to room temperature prior to adjusting the pH to 8.8.
4. Acrylamide is carcinogenic and should be handled with extreme care. Wear protective gloves, lab coat, safety goggles, and a facemask. Alternatively gels can be prepared under a fume hood.
5. TEMED is harmful by inhalation and can cause burns. Wear protective gloves, lab coat, safety goggles, and a facemask and handle in a fume hood. As soon as it is added, the acrylamide will start to polymerize and the solution should be poured in between the two plates immediately.
6. Pour and mix 10× running buffer stock and dH_2O gently to limit foaming from SDS.
7. If the protein of interest in a study falls within a narrow p*I* range, zoom strips can be used which cover a narrower p*I* range and will enable increased resolution of proteins within this p*I* range.
8. To minimize protein precipitation during IEF, proteins are resolubilized in a buffer composed of urea at high concentration, typically 7 M, along with a nonionic or zwitterionic detergent. The zwitterionic detergent, CHAPS, is the detergent of choice for protein separation by 2-DE. Other detergents can also be used such as Triton X-100 or ASB14, especially for membrane protein resolubilization. Conversely, SDS and other ionic detergents are not compatible to IEF [12].
9. The addition of bromophenol blue will not interfere or play a role in the migration of proteins during IEF. It is only added for helping tracking of the electrophoresis.

10. Once proteins are resolubilized in urea, care must be taken not to expose the sample to temperatures above 37 °C as this may lead to protein carbamylation [13].
11. Ensure that the urea is completely solubilized in the solution since the freezing may lead to precipitation of urea.
12. It is important to start with a minimal volume of dH_2O as the volume may rise beyond the desired final volume after addition of all the components.
13. The DTT and IOA should be added separately to the equilibration buffer aliquot just prior to use. The tube containing the IOA should be covered with foil to protect from light.
14. Use low-melting agarose to ensure rapid melting of the gel at the time of use at relatively low temperature (about 60 °C) to limit the risk of carbamylation when pouring on top of the strip for sealing.
15. Small aliquot volume is suitable for single use and will help limiting the risk of contamination with keratin, dust, and other contaminants due to unnecessary handling.
16. Prepare and handle the solution under a fume hood.
17. Following Pro-Q Diamond phosphoprotein stain, the gel should be stained with Sypro Ruby for total protein stain. This will allow visualization of the total proteome and serve as a control to the Pro-Q Diamond.
18. The maximum volume of sample should be kept to less than half of the total volume recommended for the strip reswelling. In the case of 7 cm long strip, the volume of sample should be at the maximum volume of 62 μL. If the sample exceeds this volume, then it can be precipitated using 100 % (v/v) acetone or the 2D clean-up kit (GE Healthcare Life Sciences). The sample should be solubilized in a buffer containing urea or a combination of urea and thiourea, nonionic or zwitterionic detergent, IPG buffer (carrier ampholytes), and a reducing agent, typically DTT.
19. The centrifugation step will pellet any salts or debris that may have remained in the sample and may later interfere with the strip reswelling and focusing.
20. A spirit level can be used to ensure that the surface on which the IPG box is placed is level. This is important to avoid the sample moving away from the IPG strip.
21. The gel sits on the plastic strip with writing on it. It is possible to handle the strip from both ends using forceps since the plastic is longer than the gel by a few mm on both sides.
22. For passive reswelling, the strip should be placed in such a way that writings are readable.

23. To avoid drying of the strip, especially if the IPG box is not fully closed, ensure that the strip is covered with a layer of mineral oil throughout the entire length of the equilibration tray well.
24. Rinsing of the gel strip will help removing any salts that may have crystallized as well as proteins that may have precipitated during the rehydration process.
25. In case of using a single-channel cup holder, the gel strip should be placed gel facing down as the electrodes are positioned on the bottom of the ceramic itself.
26. In case of using a single-channel cup holder, the wetted paper wicks need to be placed between the cup holder and the strip.
27. In case of using a single-channel cup holder, just place the plastic cover on top of the strip to ensure that the strip is in contact with the electrodes.
28. For other gel strip length and pH range, other recommended programs can be found on the manual "2D Electrophoresis, Principles and Methods" published by GE Healthcare Life Sciences.
29. If there is no current flow (i.e., reading indicates 0 μA), the focusing is not progressing. Restart the connection between the instrument and the software and ensure that the manifold is properly positioned. In addition, check if the electrode assembly is properly secured and is in full contact with the electrode area of the IPGphor and the gel strips.
30. It is convenient to start early in the morning so that the second dimension can be carried out on the same day. Conversely, IEF strips can be stored at −20 °C up to a few days until ready for the second dimension.
31. The duration of the equilibration can be extended to 20–30 min for each step according to the sample type. It is important that the duration of these two steps is kept exactly the same.
32. The overlay prevents inhibition of acrylamide polymerization as a result of chemical interaction when in contact with atmospheric oxygen, and in addition helps to level the resolving gel solution.
33. To facilitate downstream comparison of gels, ensure that all strips are positioned in a similar way, i.e., the positive side of the strip always at the edge of the plate, away from the molecular marker side.
34. The preparation of the molecular marker on a Whatman paper can be skipped if the gel prepared has a well for the standard. Check that the Whatman paper is not too thick and can fit in between the plates.

35. Avoid trapping air bubbles between the strip and the acrylamide gel, as this will prevent proteins from entering the second-dimension gel. Ensure that the agarose is molten and ready for use before placing the marker paper between the two glass plates to prevent proteins from the marker to start diffusing under the strip.
36. The low voltage applied at the beginning of the electrophoresis is to ensure that proteins enter well the gel and at the same time.

References

1. Anderson NG, Matheson A, Anderson NL (2001) Back to the future: the human protein index (HPI) and the agenda for post-proteomic biology. Proteomics 1:3–12
2. Wasinger VC, Cordwell SJ, Cerpa-Poljak A, Yan JX, Gooley AA et al (1995) Progress with gene-product mapping of the Mollicutes: *Mycoplasma genitalium*. Electrophoresis 16:1090–1094
3. Blackstock WP, Weir MP (1999) Proteomics: quantitative and physical mapping of cellular proteins. Trends Biotechnol 17:121–127
4. Schulenberg B, Aggeler R, Beechem JM, Capaldi RA, Patton WF (2003) Analysis of steady-state protein phosphorylation in mitochondria using a novel fluorescent phosphosensor dye. J Biol Chem 278:27251–27255
5. Wu J, Lenchik NJ, Pabst MJ, Solomon SS, Shull J, Gerling IC (2005) Functional characterization of two-dimensional gel-separated proteins using sequential staining. Electrophoresis 26:225–237
6. Cohen P (2000) The regulation of protein function by multisite phosphorylation—a 25 year update. Trends Biochem Sci 25:596–601
7. Dhanasekaran N, Reddy EP (1998) Signaling by dual specificity kinases. Oncogene 17:1447–1455
8. Seger R, Krebs EG (1995) Protein kinases. The MAPK signaling cascade. FASEB J 9: 726–735
9. Berggren K, Chernokalskaya E, Steinberg TH, Kemper C, Lopez MF et al (2000) Background-free, high sensitivity staining of proteins in one- and two-dimensioanl sodium dodecyl sulfate-polyacrylamide gels using a luminescent ruthenium complex. Electrophoresis 21:2509–2512
10. Agrawal GK, Thelen JJ (2005) Development of a simplified, economical polyacrylamide gel staining protocol for phosphoproteins. Proteomics 5:4684–4688
11. Righetti PG (1990) Immobilized pH Gradients: theory and methodology, vol 20, Laboratory techniques in biochemistry and molecular biology. Elsevier Science Publishers B.V, Amsterdam
12. Ames GF, Nikaido K (1976) Two-dimensional gel electrophoresis of membrane proteins. Biochemistry 15:616–623
13. McCarthy J, Hopwood F, Oxley D, Laver M, Castagna A et al (2003) Carbamylation of proteins in 2-D electrophoresis-Myth or reality? J Proteome Res 2:239–242

Chapter 11

An Affinity Pull-Down Approach to Identify the Plant Cyclic Nucleotide Interactome

Lara Donaldson and Stuart Meier

Abstract

Cyclic nucleotides (CNs) are intracellular second messengers that play an important role in mediating physiological responses to environmental and developmental signals, in species ranging from bacteria to humans. In response to these signals, CNs are synthesized by nucleotidyl cyclases and then act by binding to and altering the activity of downstream target proteins known as cyclic nucleotide-binding proteins (CNBPs). A number of CNBPs have been identified across kingdoms including transcription factors, protein kinases, phosphodiesterases, and channels, all of which harbor conserved CN-binding domains. In plants however, few CNBPs have been identified as homology searches fail to return plant sequences with significant matches to known CNBPs. Recently, affinity pull-down techniques have been successfully used to identify CNBPs in animals and have provided new insights into CN signaling. The application of these techniques to plants has not yet been extensively explored and offers an alternative approach toward the unbiased discovery of novel CNBP candidates in plants. Here, an affinity pull-down technique for the identification of the plant CN interactome is presented. In summary, the method involves an extraction of plant proteins which is incubated with a CN-bait, followed by a series of increasingly stringent elutions that eliminates proteins in a sequential manner according to their affinity to the bait. The eluted and bait-bound proteins are separated by one-dimensional gel electrophoresis, excised, and digested with trypsin after which the resultant peptides are identified by mass spectrometry—techniques that are commonplace in proteomics experiments. The discovery of plant CNBPs promises to provide valuable insight into the mechanism of CN signal transduction in plants.

Key words Cyclic nucleotide, Adenosine 3′, 5′-cyclic monophosphate (cAMP), Guanosine 3′5′-cyclic monophosphate (cGMP), Cyclic nucleotide-binding protein, Cyclic nucleotide-binding domain

1 Introduction

In plants, the presence of cyclic nucleotides (CNs), particularly adenosine 3′,5′-cyclic monophosphate (cAMP) and guanosine 3′,5′-cyclic monophosphate (cGMP), has been unequivocally established [1]. These CNs have been implicated to play an important role in a number of plant processes including chloroplast development, stomatal function, and responses to both abiotic and

Chris Gehring (ed.), *Cyclic Nucleotide Signaling in Plants: Methods and Protocols*, Methods in Molecular Biology, vol. 1016, DOI 10.1007/978-1-62703-441-8_11,

biotic stresses [2]. Despite this, very little is known about how the CN signal is decoded in the cell and specifically, the direct downstream protein targets of CNs that mediate their physiological effects [3]. BLAST searches using CN-binding proteins (CNBPs) that are found across kingdoms have failed to identify equivalent homologs in plants. It has therefore become necessary to adopt alternative approaches to the discovery of CNBPs in plants. Affinity pull-down is a technique used to purify a subpopulation of the proteome according to its affinity to a specific bait such as a drug or a second messenger [4]. In animals, affinity pull-down has been used to identify protein targets of the CN second messengers, cAMP and cGMP, through the use of CN-baits which are synthetic CNs attached to supports used for affinity purification [5, 6]. The proteins purified by CN-baits include both CNBPs that directly bind CNs as well as proteins that bind CNBPs and they are collectively referred to as the CN interactome. In animals, identification of the CN interactome has provided valuable insights into CN signaling mechanisms [7]. In contrast, in plants the CN interactome remains unresolved. Here, we describe a CN affinity pull-down technique that can be used to identify the plant CN interactome and thereby contribute to the discovery of downstream protein targets of plant CN signaling.

Cyclic nucleotide signaling pathways are well characterized in many diverse species ranging from bacteria to humans. In *Escherichia coli,* cAMP is produced in response to low glucose and directly binds the catabolite activator protein (CAP) transcription factor which in turn regulates the expression of many genes [8]. In animals, on the other hand, CN signaling is largely mediated through protein kinase A (PKA) and protein kinase G (PKG) that bind cAMP and cGMP, respectively [9]. The CNs activate their respective kinases by binding to the regulatory domain which releases the catalytic domain to initiate downstream phosphorylation cascades [10]. Soon after the discovery of CNs in animals, a family of phosphodiesterases (PDEs) was identified that are responsible for CN degradation. The PDEs bind CNs with varying affinity and specificity for cAMP and cGMP; and their activity ensures the transient nature of the CN signal—a key feature of any second messenger [11]. The dogma that CN-dependent kinases are the major intracellular receptors for CNs was challenged as more CNBPs were identified. The existence of other types of CNBPs was initially supported by the observation that some CN-mediated effects were resistant to protein kinase inhibitors. Subsequently, CNs were found to bind and regulate ion channels including cyclic nucleotide-gated channels (CNGCs) in rod photoreceptor cells [12] and olfactory sensory neurons [13] and hyperpolarization-activated cyclic nucleotide-modulated (HCN) channels in pacemaker cells of the heart [14].

Sequence and structural information suggests that, despite there being a number of different CNBPs across kingdoms, CN-binding domains (CNBDs) are well conserved with only two types present: (1) the cyclic nucleotide-binding (CNB) domain found in CAP, PKA, PKG, and CNGCs and (2) the GAF domain found in cGMP-binding PDEs, *Anabaena* adenylyl cyclase, and *E. coli* FhlA. More recent additions to the family of CNBPs have been identified through sequence-based searches for proteins that contain CNB domains. These include two exchange proteins directly activated by cAMP (Epacs) that are guanine-nucleotide-exchange factors (GEFs) [15] and four cGMP-binding proteins in the slime mold *Dictyostelium* (GbpA-D)—two of which contain GEF domains (one in combination with a protein kinase) while the other two are novel PDEs [16]. Indeed, CN signaling in *Dictyostelium* appears to be quite divergent as it also has unique extracellular cAMP receptors (cARs) [17]. Currently, almost 7,700 proteins have been identified that contain CNB domains. In prokaryotes these include transcription factors and channels while, in eukaryotes, these are protein kinases, channels, and GEFs [18]. In addition, more than 1,400 proteins have been identified that contain GAF domains including PDEs, adenylyl cyclases, and transcriptional regulators in prokaryotes, PDEs in eukaryotes, and photoreceptor proteins in cyanobacteria and plants [19].

In plants, a family of 20 CNGCs and a number of Shaker-type K^+ channels have been found to contain canonical signatures for CNB domains and a few of these have been demonstrated to be regulated by CNs [20, 21]. Additionally, GAF domains have been identified in phytochromes and ethylene receptors with no ascribed function [22]. Of note, no CN-dependent protein kinases or CN-specific PDEs have been identified in plants [22]. This has led to suggestions that (1) these signaling components were lost from the plant lineage together with the loss of cilia, particularly since these components are present in the motile algae *Chlamydomonas* [23], and (2) the expanded family of CNGCs is the primary target of CN signaling in plants [24]. In light of the unusual CNBPs in *Dictyostelium*, it is possible that plant CNBPs have also diverged significantly from those in bacteria and animals. Intriguingly, a plant-specific protein phosphatase 2C (PP2C) has been reported to contain a kinase, phosphatase, and CNB domain suggesting that it may be a novel CN-dependent kinase [18]. In support of this, there is evidence that CNs initiate phosphorylation in plants [25]. Alternatively, small plant proteins that appear to exclusively comprise CNB domains may form regulatory components of kinases [18]. Since the two known CNBDs have evolved independently [22], it is possible that other types of CNBDs have evolved that remain unidentified. Having exhaustively mined the available sequence data, it is now particularly pertinent to pursue experimental approaches to identify downstream components of plant

CN signaling. One such approach is to extract the plant CN interactome by CN affinity pull-down.

The CN affinity pull-down method entails (1) a non-denaturing protein extraction, (2) incubation of the proteins with a commercially available synthetic CN-bait, and (3) sequential elution to remove low-affinity and nonspecifically bound proteins and enrich for high-affinity proteins that are eluted in the final elution fractions or remain tightly bound to the bait. These proteins that bind the bait with high affinity form the CN interactome. The identity of the purified proteins is determined using standard proteomics approaches. In brief, the proteins are fractionated by one-dimensional sodium dodecyl sulfate-polyacrylamide gel electrophoresis (SDS-PAGE), the entire lane excised, and subjected to in-gel tryptic digest (IGTD) and the resultant peptides identified by mass spectrometry (MS)/MS analysis for de novo sequencing. To date there have been three studies that attempted to affinity purify CNBPs from plants. One failed to identify the resultant proteins while the other two identified nucleoside diphosphate kinase and glyceraldehyde 3-phosphate dehydrogenase—both of which have been identified in animal studies as low-affinity binding proteins due to their ability to bind other nucleotide-like compounds [26–28]. The method presented here differs from the previous unsuccessful attempts to affinity purify CNBPs from plants in that it (1) includes a sequential elution technique, (2) characterizes the entire complement of purified proteins, and (3) analyzes bead-bound proteins [27, 28]. Once identified, the plant CN interactome can be examined for the presence of known and modified CNBDs or conserved domains that may be novel CNBDs.

2 Materials

Most of the equipment required for this method will be present in a standard molecular biology lab. Additional equipment requirements worth noting include a rotator that can be placed in a cold room and a speedvac. Ultimately peptide samples must be submitted to a proteomics facility with a nanoflow high-performance liquid chromatography (HPLC) system coupled to a mass spectrometer capable of MS/MS analysis. Since the operation of such highly specialized equipment is normally performed by a dedicated expert, the technicalities of this procedure are not discussed in detail here.

2.1 Plant Tissue, Protein Extraction, and Protein Quantification

1. Plant material from a sequenced organism, in this case *Arabidopsis thaliana* leaf tissue (*see* **Note 1**).
2. Preprepared stock solutions for assay buffer (*see* **Note 2**): 1 M Tris–HCl pH 7.4; 1 M sucrose; 1 M $MgSO_4{\cdot}7H_2O$; 1 M KCl;

and 0.5 M ethylenediaminetetraacetic acid (EDTA) pH 8.0 made up with Milli-Q water and autoclaved. Ascorbic acid prepared as a 0.5 M stock with Milli-Q water, filter-sterilized, and stored at −20 °C.

3. Additional chemicals for assay buffer: Phenylmethanesulfonyl fluoride (PMSF) (*see* **Note 3**); isopropanol; and 100× plant-specific protease inhibitor cocktail (Sigma, catalog number P9599).
4. Chemicals for protein extraction: Poly(vinylpolypyrrolidone) (PVPP) (*see* **Note 4**) and liquid nitrogen.
5. Chemicals for protein quantification: Bio-Rad Protein Assay kit and 2 mg/mL Bovine Serum Albumin (BSA) standard (Bio-Rad).

2.2 Synthetic CN-Baits, Affinity Pull-Down, and Sequential Elution

1. Synthetic CN-agarose baits available from BioLog Life Science Institute (Bremen, Germany):

   8-(2-Aminoethylamino)cAMP agarose (8-AEA-cAMP-agarose, catalog number A 020)

   2-(6-Aminohexylamino)cAMP agarose (2-AHA-cAMP-agarose, catalog number A 054)

   8-(2-Aminoethylthio)cGMP agarose (8-AET-cGMP-agarose, catalog number A 019)

   N^2-(6-Aminohexyl)cGMP agarose (2-AH-cGMP-agarose, catalog number A 056) (*see* **Note 5**)
2. Negative control: Ethanolamine agarose (EtOH-NH-agarose, BioLog catalog number E 010) (*see* **Note 6**).
3. Chemicals for elution buffers: Sodium salts of adenosine diphosphate (ADP); adenosine monophosphate (AMP); cAMP; guanosine diphosphate (GDP); guanosine monophosphate (GMP); and cGMP (Sigma).

2.3 Protein Precipitation and Sample Preparation for SDS-PAGE

1. Chemicals for protein precipitation: HPLC-grade acetone.
2. Stock solution for sample application buffer (SAB): 0.5 M Tris–HCl pH 6.8 prepared with Milli-Q water and autoclaved.
3. Chemicals for SAB: 20 % (w/v) SDS (Bio-Rad); glycerol; bromophenol blue; and β-mercaptoethanol (β-ME).
4. 2× SAB: 125 mM Tris pH 6.8, 20 % (v/v) glycerol, 4 % (w/v) SDS, 0.005 % (w/v) bromophenol blue, 2 % (v/v) β-ME. For 10 mL 2× SAB add 3.3 mL Milli-Q water, 2.5 mL 0.5 M Tris–HCl pH 6.8, 2 mL glycerol, 2 mL 20 % (w/v) SDS, 5 mg bromophenol blue, and 200 μL β-ME (*see* **Note** 7). Dilute with Milli-Q water to make 1× SAB.

2.4 SDS-PAGE

1. Stock solutions for SDS-PAGE: 1.5 M Tris–HCl pH 8.8 prepared with Milli-Q water and autoclaved; 0.5 M Tris pH 6.8 (Subheading 2.3, **item 2**) and 10 % (w/v) ammonium persulfate (APS) prepared with Milli-Q water (*see* **Note 8**).
2. Additional chemicals for SDS-PAGE: 30 % (w/v) acrylamide/bis 29:1 and 20 % (w/v) SDS and *N*,*N*,*N*′,*N*′-tetramethylethylenediamine (TEMED).
3. 10× running buffer: 250 mM Tris, 1.92 M glycine, 1 % (w/v) SDS prepared in Milli-Q water. Dilute the stock solution with Milli-Q water to make 1× running buffer.
4. PageRuler prestained protein ladder (Fermentas, catalog number SM0671).
5. Coomassie stain and destain solutions.

2.5 In-Gel Tryptic Digest

High-quality HPLC-grade reagents and water must be used in all IGTD solutions.

1. Stock solution for IGTD buffers: 100 mM ammonium bicarbonate (NH_4HCO_3) prepared with HPLC water and filter-sterilized.
2. Additional chemicals for IGTD buffers: Acetonitrile (ACN); dithiotreitol (DTT); iodoacetamide (IOA); formic acid (FA); and sequencing-grade trypsin.
3. 50 mM NH_4HCO_3: Prepared by diluting the 100 mM NH_4HCO_3 stock with HPLC water.
4. 25 mM NH_4HCO_3, 50 % ACN: Prepared with HPLC water.
5. Digestion buffer: Reconstitute 10 ng/μL trypsin in 50 mM NH_4HCO_3.
6. 20 mM NH_4HCO_3: Prepared by diluting the 100 mM NH_4HCO_3 stock with HPLC water.
7. 5 % (v/v) FA, 50 % (v/v) ACN: Prepared with HPLC water.

2.6 Mass Spectrometry Protein Identification

1. Mascot search engine (currently version 2.3) for querying MS/MS data against the relevant protein database—for Arabidopsis it is currently the Arabidopsis_TAIR10 protein sequence database.
2. Scaffold (currently version 3.6) software for compiling and performing statistical analysis on the Mascot results.
3. Microsoft Excel is required for subsequent comparisons between the datasets.
4. Relevant database for evaluating nucleotide-binding sites; for Arabidopsis this is TAIR (www.arabidopsis.org).

3 Methods

The protein extraction, affinity pull-down, and sequential elution should be performed on the same day.

3.1 Protein Extraction

Perform all steps at 4 °C and minimize handling times. Conditions such as pH and temperature should be kept constant (*see* **Note 9**).

1. Prepare a stock solution of 0.1 M PMSF in isopropanol.
2. Prepare the assay buffer on the day of use. A total of 100 mL assay buffer is sufficient to extract proteins from approximately 2.5 g leaf tissue and perform pull-downs with the four different CN-baits and negative control (0.5 g tissue per pull-down) (*see* **Note 10**).
3. Assay buffer: 50 mM Tris–HCl pH 7.4, 0.25 M sucrose, 1 mM EDTA, 0.1 mM $MgSO_4{\cdot}7H_2O$, 10 mM KCl, 5 mM ascorbic acid, 1 mM PMSF, 1× protease inhibitor cocktail. To make 100 mL assay buffer add 65.79 mL Milli-Q water, 5 mL 1 M Tris pH 7.4, 25 mL 1 M sucrose, 200 μL 0.5 M EDTA, 10 μL 1 M $MgSO_4{\cdot}7H_2O$, 1 mL 1 M KCl, 1 mL 0.5 M ascorbic acid, 1 mL 0.1 M PMSF, and 1 mL 100× protease inhibitor cocktail (*see* **Note 11**). Incubate the assay buffer on ice to equilibrate to 4 °C.
4. Prepare the protein extraction buffer by aliquoting 6 mL assay buffer into a 50 mL tube and adding 30 mg (0.5 % w/v) PVPP.
5. In a precooled mortar and pestle, grind 2.5 g of tissue to a fine powder in liquid nitrogen, taking care not to allow any tissue to thaw.
6. Immediately add the frozen tissue to the 6 mL protein extraction buffer containing PVPP.
7. Aliquot the protein extraction into Eppendorf tubes.
8. Centrifuge in a microfuge at 12,000×*g* for 20 min at 4 °C. Carefully remove the supernatant into a fresh tube and discard the pellet (cell debris and insoluble PVPP). If visible particles remain, repeat the centrifugation for a further 10 min.

3.2 Protein Quantification

The protein concentration of the extraction is determined using the Bio-Rad Protein Assay kit that is designed on the Bradford method [29].

1. Prepare sufficient protein assay reagent by diluting the Bio-Rad Protein Assay kit fivefold with Milli-Q water and equilibrate to room temperature. For the quantification of a single protein extraction 10 mL is sufficient.
2. Prepare a 20 μL aliquot of Milli-Q water as a blank.

3. Prepare 20 μL of each BSA standard: 100, 250, 500, 750, 1,000, and 1,500 μg/mL BSA by diluting the 2 mg/mL BSA stock with the appropriate volume of Milli-Q water.
4. Prepare 20 μL of a fourfold dilution of the plant protein extraction with Milli-Q water.
5. Add 980 μL of protein assay reagent to the 20 μL blank, standards, and plant sample. Mix and allow the reaction to develop for at least 5 min, but not longer than 1 h.
6. Blank the spectrophotometer at OD_{595} and then measure the absorbance for all standards and the sample.
7. Quantify the concentration of the protein extraction by calculating the standard curve for BSA and then extrapolating the sample concentration and multiplying by the dilution factor (*see* **Note 12**).

3.3 Affinity Pull-Down

Perform all steps at 4 °C.

1. Resuspend the agarose beads and allow the resin to settle. For each of the CN-baits, add 200 μL beads to 1 mL assay buffer and pre-equilibrate by incubating on a rotator at 40 rpm at 4 °C for 30 min–2 h (*see* **Note 13**).
2. Remove the equilibration buffer from the beads by centrifuging at 100 × *g* for 30 s in a microfuge and then gently aspirating the supernatant, ensuring not to disturb the agarose bed (*see* **Note 14**).
3. Incubate 1–1.2 mL of the protein extraction solution (approximately 1.5 mg protein) with 200 μL pre-equilibrated beads on a rotator at 40 rpm at 4 °C for 1–4 h (*see* **Note 15**).
4. Centrifuge the beads at 100 × *g* for 30 s in a microfuge and carefully remove the supernatant, collecting this as the flow through.
5. Add 1 mL of wash buffer to the protein-bound beads, incubate on a rotator at 40 rpm at 4 °C for 5 min, then centrifuge as before, and collect the supernatant as the first wash.
6. Repeat **step 5** until the protein-bound beads have been washed a total of six times, collecting each wash fraction.

3.4 Sequential Elution

Perform all steps at 4 °C.

1. For cAMP-agarose baits the sequential elution strategy is as follows: 100 mM GDP, 100 mM AMP, 10 mM cGMP, 100 mM cGMP, 10 mM cAMP, and then 100 mM cAMP (*see* **Note 16**).
2. For cGMP-agarose baits the sequential elution strategy is as follows: 100 mM ADP, 100 mM GMP, 10 mM cAMP, 100 mM cAMP, 10 mM cGMP, and then 100 mM cGMP.

3. For the EtOH-NH-agarose negative control perform the sequential elution according to the strategy for either the cAMP- or the cGMP-agarose baits.
4. Prepare the elution buffer stock solutions in assay buffer: 0.5 M ADP; 0.5 M AMP; 0.5 M cAMP; 0.5 M GDP; 0.5 M GMP; and 0.2 M cGMP. A 150 μL volume of 0.5 M ADP, 0.5 M AMP, 0.5 M GDP, and 0.5 M GMP is required to perform sequential elutions on four CN-baits and the negative control, while 300 μL of 0.5 M cAMP and 750 μL of 0.2 M cGMP is required.
5. Prepare the elution buffers by diluting the stock solutions with assay buffer: 100 mM GDP; 100 mM ADP; 100 mM AMP; 100 mM GMP; 10 mM cGMP; 10 mM cAMP; 100 mM cGMP; and 100 mM cAMP. A 650 μL volume of 100 mM GDP, ADP, AMP, and GMP elution buffers and 1,300 μL of 10 mM and 100 mM cAMP and cGMP elution buffers are sufficient for five pull-down assays on four CN-baits and the negative control.
6. Perform the first elution by adding 200 μL of either 100 mM GDP for cAMP-agarose baits or 100 mM ADP for cGMP-agarose baits and incubating on a rotator at 40 rpm at 4 °C for 5 min (*see* **Note 17**). Centrifuge the mixture in a microfuge at 100 × *g* for 30 s and collect the supernatant as the first elution faction.
7. Perform an intermittent wash step by adding 1 mL of wash buffer to the beads and incubating on a rotator at 40 rpm at 4 °C for 5 min. Centrifuge the mixture in a microfuge at 100 × *g* for 30 s and remove the supernatant as wash fraction 7 (*see* **Note 18**).
8. Perform the second elution by repeating the first elution procedure, using either 100 mM AMP for cAMP-agarose baits or 100 mM GMP for cGMP-agarose baits.
9. Repeat the intermittent wash step and collect the supernatant (wash 8).
10. Perform the third elution by repeating the first elution procedure, using either 10 mM cGMP for cAMP-agarose baits or 10 mM cAMP for cGMP-agarose baits.
11. Repeat the intermittent wash step and collect the supernatant (wash 9).
12. Perform the fourth elution by repeating the first elution procedure, using either 100 mM cGMP for cAMP-agarose baits or 100 mM cAMP for cGMP-agarose baits.
13. Repeat the intermittent wash step and collect the supernatant (wash 10).

14. Perform the fifth elution by repeating the first elution procedure, using either 10 mM cAMP for cAMP-agarose baits or 10 mM cGMP for cGMP-agarose baits.
15. Repeat the intermittent wash step and collect the supernatant (wash 11).
16. Perform the sixth and final elution by repeating the first elution procedure, using either 100 mM cAMP for cAMP-agarose baits or 100 mM cGMP for cGMP-agarose baits.
17. Repeat the intermittent wash step and collect the supernatant (wash 12).

3.5 Protein Precipitation and Sample Preparation for SDS-PAGE

The elution fractions will contain very few proteins; therefore proteins in these fractions must be precipitated in order to concentrate them so that they can be visualized on the gel (*see* **Note 19**).

1. Add 800 μL acetone to 200 μL of each elution fraction and mix by inverting the tube.
2. Incubate overnight at −20 °C.
3. The following day, pellet the precipitated proteins by centrifuging in a microfuge at 14,000 × *g* for 10 min at room temperature and discard the supernatant. Repeat the centrifugation for a further 5 min to completely remove any residual acetone.
4. Air-dry the pellet for 5–10 min at room temperature.
5. Set a heating block to 95 °C.
6. Resuspend the precipitated proteins and the beads in 25 μL 1× SAB.
7. Boil the samples in SAB for 5 min at 95 °C to denature proteins.
8. Spin down the samples and proceed immediately to loading the gel (Subheading 3.6, **step 7**).

3.6 SDS-PAGE

Proteins are separated using one-dimensional SDS-PAGE according to the method of Laemmli [30].

1. Clean gel plates and combs with 70 % (v/v) ethanol and allow to dry.
2. Assemble the protein mini-gel apparatus.
3. Prepare the separating gel: 12 % (w/v) acrylamide/bis, 375 mM Tris pH 8.8, 0.1 % (w/v) SDS, 0.05 % (v/v) TEMED, 0.05 % (w/v) APS. To prepare five separating gels (one for each of the four baits and the negative control), add 8.5 mL Milli-Q water, 10 mL 30 % (w/v) acrylamide/bis 29:1, 6.25 mL 1.5 M Tris pH 8.8, 125 μL 20 % (w/v) SDS, 12.5 μL TEMED, and 125 μL 10 % (w/v) APS to a beaker, gently mix,

and pour immediately into the five gel rigs, leaving room for the stacking gel (*see* **Note 20**). Carefully overlay the surface with Milli-Q water to ensure a smooth interface between the separating and stacking gels. Allow to set (20–30 min) and remember to remove the water layer before adding the stacking gel.

4. Prepare the stacking gel: 4 % (w/v) acrylamide/bis, 125 mM Tris pH 6.8, 0.1 % (w/v) SDS, 0.1 % (v/v) TEMED, 0.05 % (w/v) APS. To prepare five stacking gels add 7.688 mL Milli-Q water, 1.675 mL 30 % (w/v) acrylamide/bis 29:1, 3.125 mL 0.5 M Tris pH 6.8, 62.5 μL 20 % (w/v) SDS, 12.5 μL TEMED, and 62.5 μL 10 % (w/v) APS to a beaker, gently mix, then pour immediately on top of the separating gel, and insert the combs (15 well). Allow to set (20–30 min).
5. Remove the comb and insert the gel into the tank. Fill the tank with 1× running buffer so that all the wells are covered.
6. Load 5 μL of protein ladder.
7. Load the entire 25 μL volume for each of the elution fraction and bead samples, leaving a lane gap between each of the samples (*see* **Note 21**). In the gap lanes load 25 μL 1× SAB.
8. Top up the gel apparatus with 1× running buffer to the required level.
9. Electrophorese at 100 V until the dye front has migrated approximately 2 cm into the separating gel (about 30–40 min) (*see* **Note 22**).
10. Switch off the power pack and dismantle the gel apparatus. Place the gel into staining solution and incubate for at least 1 h to overnight on a shaker.
11. Next, pour off the stain solution, replace with destain, and return to the shaker. Change the destain solution every 30 min–1 h until background staining has been reduced and protein bands can be clearly visualized (*see* **Note 23**).

3.7 Excision of Proteins from SDS-PAGE

1. Rehydrate the gel in HPLC-grade water for 30 min. Replace with fresh water and repeat this step.
2. Remove the gel from the water and place onto a clean glass slide.
3. Using a clean scalpel blade, excise the entire lane for each of the elution fraction and bead samples, excluding as much excess gel as possible. Cut the lane into four equally sized gel slices (*see* **Note 24**).
4. Further divide each gel slice into 1 × 1 mm squares, place these into an Eppendorf tube, and cover with 200 μL HPLC water. Store at 4 °C until ready for IGTD.

3.8 In-Gel Tryptic Digest

Gloves must be worn at all times and hair kept tied back (*see* **Note 25**).

1. Prepare 10 mL of 10 mM DTT in 100 mM NH_4HCO_3.
2. Prepare 10 mL of 50 mM IOA in 100 mM NH_4HCO_3 and protect the solution with foil.
3. Set heating blocks to 56 and 37 °C.
4. Remove the HPLC water from the gel pieces by centrifuging the tubes at maximum speed in a microfuge for 15 s and aspirating off the water.
5. Wash the gel pieces by adding 100 μL of 50 mM NH_4HCO_3 to each tube. Collect the washed gel pieces to the bottom of the tube by centrifuging in a microfuge for 15 s at maximum speed and discard the wash solution. Repeat this step to perform two washes.
6. Destain the gel pieces by adding 100 μL of 25 mM NH_4HCO_3, 50 % (v/v) ACN to the tubes and incubating for 30 min with intermittent vortexing. Centrifuge the tubes at maximum speed for 15 s and discard the destain solution.
7. Dehydrate the gel pieces by incubating them in 200 μL ACN for 10 min. Centrifuge the tubes at maximum speed for 15 s and discard the ACN solution. Repeat this step to fully dehydrate the gel pieces.
8. Next, dry the gel pieces in a speedvac for 20 min.
9. Reduce the cysteine residues by rehydrating the gel pieces in 50 μL 10 mM DTT in 100 mM NH_4HCO_3 (or sufficient volume to cover them) and incubating them at 56 °C for 45 min. Cool the tubes and centrifuge to remove the DTT solution.
10. Alkylate the cysteine residues by adding 50 μL (or an equal volume to that of DTT) of 50 mM IOA in 100 mM NH_4HCO_3 and incubating in the dark for 30 min. Discard the IOA solution after centrifugation.
11. Next, wash the gel pieces in 200 μL of 100 mM NH_4HCO_3 and discard the solution.
12. Add 200 μL ACN to the tubes, vortex, and incubate for 10 min. Centrifuge and remove the ACN.
13. Add 100 μL of 100 mM NH_4HCO_3, incubate for 10 min, and then discard the NH_4HCO_3 solution.
14. Repeat the ACN incubation (**step 12**).
15. Dry the gel pieces in a speedvac for 20 min.
16. Rehydrate the gel pieces by covering them with 20–50 μL of digestion buffer. It may be necessary to add a little extra 50 mM NH_4HCO_3 to ensure this. Incubate the tryptic digest at 37 °C overnight.

17. The following day, centrifuge to collect the digestion solution to the bottom of the tube. Add 20 μL of 20 mM NH_4HCO_3, vortex, and incubate for 10 min. Centrifuge and collect the extracted peptides into an Eppendorf tube.
18. Perform a second extraction step by incubating the gel pieces with 20 μL 5 % (v/v) FA, 50 % (v/v) ACN for 10 min, then centrifuge, and remove the supernatant into a second Eppendorf tube. Repeat the FA extraction, collecting this into the same tube to combine both FA extracts.
19. Finally, dry both the extracted peptides from the digestion solution and the FA extracts in a speedvac until approximately 1 μL of liquid remains (*see* **Note 26**). Store the peptides at −20 °C until MS analysis.

3.9 Protein Identification

1. Perform LC-MS/MS analysis according to the standard operating procedures of the proteomics facility to which the samples have been submitted in order to identify the peptides.
2. Search the MS/MS data against Arabidopsis proteins using Mascot and compile the results using Scaffold. Filter the data to remove common contaminant proteins and perform statistical analysis.
3. Consider matches positive if a protein is represented by at least 2 peptides with 95 % probability.
4. For each pull-down, combine the results for the four gel slices and then compare the elution and bead fractions to assess whether proteins are present in one or multiple fractions (*see* **Note 27**).
5. From each experiment, subtract the proteins identified in the negative control.
6. Compare the results from each of the baits containing the same CN (cAMP or cGMP) in different orientations to determine the overlap between the results (*see* **Note 28**).
7. Compare results of the cAMP and cGMP baits to identify proteins that bind specifically to either CN or those that have dual affinity for both (*see* **Note 29**).
8. Assess the resultant proteins for binding sites for other nucleotide-like compounds, particularly ATP/ADP, GTP/GDP, NAD^+, NADH, DNA, and RNA.

4 Notes

1. This technique can only be performed on sequenced plant species since its success depends on the identification of isolated proteins. Proteins are identified through matching the sequence

of experimentally derived peptides to theoretical peptides produced by in silico digestion of all proteins in the searched database. Without accurate sequence information it is nearly impossible to make such matches [31]. The importance of this is demonstrated in the three previous plant CN affinity studies, in which positive identification of proteins was only successful in experiments using tobacco and Arabidopsis whose genomes are sequenced but failed in *Avena sativa*, because its genome is not yet sequenced [26–28]. Leaf tissue is preferred over callus tissue even though the leaf proteome is dominated by highly abundant chloroplast proteins which could obscure the identification of low-abundance proteins involved in signal transduction, because the callus proteome is rich in proteins involved in RNA metabolism and these pose a greater problem due to the known affinity of RNA-binding proteins for CN-baits [6, 32].

2. The reasons for inclusion of each component in the extraction buffer are discussed. Tris buffers the solution at a neutral pH where most proteins should be soluble. Sucrose osmotically cushions proteins and stabilizes membranes. Magnesium is a cofactor required for many enzymes and stabilizes protein structures while potassium is important for maintaining enzyme function and ionic homeostasis. The dual function of EDTA is to inhibit metalloproteases and reduce oxidative damage. Ascorbic acid prevents oxidation of polyphenols which can then form aggregates with proteins [33].
3. PMSF is a serine protease inhibitor.
4. PVPP complexes with polyphenols, removing them from solution. This has been shown to dramatically improve protein identification by MS [34].
5. The CN-baits are synthetic CNs that are immobilized on agarose beads. The products differ in linker lengths and positions at which the bead is attached to the CN moiety. BioLog offer a number of options for the type of support, linker length, and attachment position. The choice of support will depend largely on the equipment available for pull-down. Agarose beads are the easiest to use since they can be separated from solution using simple centrifugal techniques. Other commonly used supports include Biacore chips for surface plasmon resonance and biotin-linked CNs [4, 35], a discussion of which falls beyond the scope of this study as they require specialized equipment. The use of two or more baits attached to the same support at different positions with variable length linkers allows for flexibility in the orientation in which the CN is presented to the prey proteins and a more complete probing of the interaction space [5]. Results from animal studies have shown that while there is overlap between proteins pulled down by

CN-baits that differ in linker length and attachment position, there are also distinct proteins identified [6, 7]. In animals, baits with longer hexyl linkers attached to position two of the nucleotide have been found to pull down more proteins—most likely a result of the CN binding to the CNBD in its target protein in an orientation that does not require interaction between amino acids in the protein and position two of the CN. Thus attachment of the CN to the bead at this position imposes the least steric hindrance on the majority of interactions between CN and CNBP [5, 35]. Since plant CNBPs may contain unique CNBDs, it is possible that they bind CNs in a different orientation. Therefore, it is recommended that a number of different baits are used and the results compared. Free beads must be specially requested; otherwise the agarose baits will be provided as pre-packed columns.

6. The negative control is used to detect proteins that bind nonspecifically to the bait [36]. These proteins are later subtracted from the experimental results to give a more accurate account of true CNBPs [4].
7. Add β-ME in a fume cupboard. Aliquots of 2× SAB can be stored at −20 °C, in which case, add after thawing.
8. APS works best when prepared fresh but can be stored at −20 °C for up to 1 month.
9. Performing the extraction at 4 °C inhibits proteases. Changes in pH and temperature could result in precipitation and nonspecific loss of proteins [31].
10. Use a single preparation of assay buffer for pre-equilibration of the beads, extraction of proteins, and wash and elution steps to ensure that conditions (especially pH and salt concentration) remain constant. Similarly, perform one large-scale protein extraction to eliminate variability arising from the use of different pools of tissue.
11. The assay buffer is designed to maximize the number of prey proteins, maintaining them in their native conformation while eliminating proteases and reducing the presence of interfering compounds that are abundant in plant tissue [31]. Typical of extraction buffers used in other pull-down assays (including animal and plant CN affinity pull-downs), it is based on a Tris buffer at neutral pH [6, 7, 26–28, 37]. Importantly, commonly used reducing agents such as β-ME and DTT are avoided since they denature proteins and thus inhibit their interaction with the bait. Add the protease inhibitor cocktail and PMSF to the buffer just prior to use to ensure maximal activity.
12. If the absorbance reading of the sample does not fall within the range of the standard curve, a new set of standards and a different dilution of the sample must be prepared and the

measurements repeated. A protein concentration of approximately 1.5 mg/mL is expected from 2.5 g leaf tissue in 6 mL extraction buffer.

13. The beads are supplied at a ligand concentration of 6 μmol/mL settled resin for 2-AHA-cAMP-agarose and 8-AEA-cAMP-agarose and 4–5 and 6.5 μmol/mL settled resin for 2-AH-cGMP-agarose and 8-AET-cGMP-agarose, respectively. Thus 200 μL resin equates to 0.8–1.3 μmol CNs per pull-down experiment. This is well in excess of endogenous levels of CNs both in plants and animals [1]. One can adjust the volume of 2-AH-cGMP-agarose and 8-AET-cGMP-agarose to challenge prey proteins with equivalent amounts of these CNs—that is, 267 μL of 2-AH-cGMP-agarose and 185 μL of 8-AET-cGMP-agarose.
14. Only remove the pre-equilibration buffer when ready to use the bait for the pull-down. Removing the buffer beforehand runs the risk of the resin drying out which could compromise the integrity of the bait.
15. Ideally, one should incubate an equal amount of prey protein with an equal amount of ligand in an equal volume for every experiment. Thus, if repeating the experiment the protein concentration will need to be adjusted to account for differences between protein extractions.
16. In animal CN affinity assays it has been shown that many proteins bind the CN-baits with low affinity including, highly abundant ATP/ADP-, GTP/GDP-, NAD^+-, NADH-, DNA-, and RNA-binding proteins [6, 7]. A sequential elution technique has been developed to improve the proportion of true CNBPs collected in the final elution fractions and retained on the beads. The principle of this is to competitively displace proteins that bind the bait with low affinity through their higher affinity for similar nucleotides such as noncyclic nucleotide triphosphates and nucleotide diphosphates [6]. Therefore the protein–CN-bait complex is challenged with a series of elution buffers containing excesses of such nucleotides. In animal CN affinity assays, regardless of whether the bait is cAMP and cGMP, the sequential elution strategy follows the same basic formula: 10 mM ADP, 10 mM GDP, 5–20 mM cGMP, and 10–200 mM cAMP. Here we have adapted the strategy to increase specificity for cAMP and cGMP baits by adding the elution buffers in the inverse order of their affinity to the bait so that high-affinity proteins are eluted in the final fractions or retained on the bead. Stepwise concentration increases in the CN elution buffers serve a similar purpose—an approach routinely used in other affinity purification procedures, for example affinity chromatography.

17. Adding 200 μL of 100 mM nucleotide elution buffer equates to an amount of 20 μmol of that nucleotide—a level 20 times greater than that of the CN ligand attached to the bait which is suspended in the same volume during the elution. Likewise, 200 μL of 10 mM nucleotide equates to 2 μmol—twice the amount of the ligand in the same volume.
18. Performing intermittent wash steps reduces carryover and cross-contamination between elution fractions and is necessary since it is not possible to completely separate the buffer from the bead after each elution.
19. The protein extraction solution and the flow through and first two wash fractions will be of sufficient protein concentration to be visualized on the gel; however subsequent wash and elution fractions will contain very little proteins and thus proteins in these fractions must be precipitated to concentrate them so that they can be visualized on a gel. There is no need to precipitate the bead-bound proteins as these are already concentrated in a small volume. It is only necessary to run the eluted and bead-bound proteins on the gel; therefore precipitation is only necessary for the eluted proteins. Evaluation of the protein extraction, flow through, and washes may be desired, in which case precipitation should also be performed on wash fractions 3–12. Acetone precipitation concentrates proteins and provides a cleanup step by removing components of the extraction buffer. However it may result in protein loss as some proteins can be denatured and become difficult to resuspend.
20. Handle TEMED in a fume cupboard. TEMED and APS should be added last as these initiate polymerization.
21. The gap lanes prevent cross-contamination between samples. It is important to load SAB in the empty lanes; otherwise the lanes containing proteins will expand into the spaces left by the empty lanes.
22. Fractionation of proteins by one-dimensional SDS-PAGE serves the dual purpose of removing any residual components from the extraction buffer and precipitation and separates proteins according to size.
23. Bands may not be visible in all elution fractions. Staining and destaining fix the proteins. It is then possible to store the gel at 4 °C. Add water to prevent the gel from dehydrating which could lead to tearing.
24. Dividing the lane into four gel slices ensures that high-abundance proteins will only have a masking affect in the gel slice that they are present rather than obscuring all proteins in the lane. The masking of low-abundance proteins by high-abundance proteins occurs because high-abundance peptides can be preferentially ionized during MS.

25. Wearing gloves and keeping hair tied back minimize contamination by keratin—a protein found in skin and hair that is a common contaminant in MS analysis [38].
26. It is best not to completely dry the peptide samples as this may hamper reconstitution for MS analysis.
27. Proteins that are displaced in the final elutions or remain bound to the beads after the final elution have high affinity for the bait while those that are completely eluted early in the sequential elution procedure bind the bait with low affinity. Proteins that are identified in three or more elutions are likely to be nonspecific contaminants [4].
28. Greater confidence can be had in proteins that bind to multiple variations of a given bait as these are likely to bind the CN specifically. Proteins that only bind to one variant of the bait may bind the CN in a steric-specific manner. Alternatively, these proteins could have affinity to the linker and therefore should be considered with caution.
29. For proteins that bind both cAMP and cGMP baits, the elution fraction in which the proteins are displaced can give an indication of whether the protein has greater specificity for one or the other CN.

References

1. Newton RP, Roef LUC, Witters E, Van Onckelen H (1999) Tansley review no. 106. New Phytol 143:427–455
2. Martinez-Atienza J, Van Ingelgem C, Roef L, Maathuis FJ (2007) Plant cyclic nucleotide signalling: facts and fiction. Plant Signal Behav 2:540–543
3. Newton RP, Smith CJ (2004) Cyclic nucleotides. Phytochemistry 65:2423–2437
4. Visser NF, Scholten A, van den Heuvel RH, Heck AJ (2007) Surface-plasmon-resonance-based chemical proteomics: efficient specific extraction and semiquantitative identification of cyclic nucleotide-binding proteins from cellular lysates by using a combination of surface plasmon resonance, sequential elution and liquid chromatography-tandem mass spectrometry. Chembiochem 8:298–305
5. Kim E, Park JM (2003) Identification of novel target proteins of cyclic GMP signaling pathways using chemical proteomics. J Biochem Mol Biol 36:299–304
6. Scholten A, Poh MK, van Veen TA, van Breukelen B, Vos MA, Heck AJ (2006) Analysis of the cGMP/cAMP interactome using a chemical proteomics approach in mammalian heart tissue validates sphingosine kinase type 1-interacting protein as a genuine and highly abundant AKAP. J Prot Res 5:1435–1447
7. Scholten A, van Veen TA, Vos MA, Heck AJ (2007) Diversity of cAMP-dependent protein kinase isoforms and their anchoring proteins in mouse ventricular tissue. J Proteome Res 6:1705–1717
8. Kolb A, Busby S, Buc H, Garges S, Adhya S (1993) Transcriptional regulation by cAMP and its receptor protein. Annu Rev Biochem 62:749–795
9. Francis SH, Corbin JD (1999) Cyclic nucleotide-dependent protein kinases: intracellular receptors for cAMP and cGMP action. Crit Rev Clin Lab Sci 36:275–328
10. Francis SH, Corbin JD (1994) Structure and function of cyclic nucleotide-dependent protein kinases. Annu Rev Physiol 56:237–272
11. Conti M, Beavo J (2007) Biochemistry and physiology of cyclic nucleotide phosphodiesterases: essential components in cyclic nucleotide signaling. Annu Rev Biochem 76:481–511
12. Cook NJ, Hanke W, Kaupp UB (1987) Identification, purification, and functional reconstitution of the cyclic GMP-dependent channel from rod photoreceptors. Proc Natl Acad Sci USA 84:585–589

13. Nakamura T, Gold GH (1987) A cyclic nucleotide-gated conductance in olfactory receptor cilia. Nature 325:442–444
14. Craven KB, Zagotta WN (2006) CNG and HCN channels: two peas, one pod. Annu Rev Physiol 68:375–401
15. Bos JL (2003) Epac: a new cAMP target and new avenues in cAMP research. Nat Rev 4:733–738
16. Goldberg JM, Bosgraaf L, Van Haastert PJ, Smith JL (2002) Identification of four candidate cGMP targets in Dictyostelium. Proc Natl Acad Sci USA 99:6749–6754
17. Shabb JB, Corbin JD (1992) Cyclic nucleotide-binding domains in proteins having diverse functions. J Biol Chem 267:5723–5726
18. Kannan N, Wu J, Anand GS, Yooseph S, Neuwald AF et al (2007) Evolution of allostery in the cyclic nucleotide binding module. Genome Biol 8:R264
19. Zoraghi R, Corbin JD, Francis SH (2004) Properties and functions of GAF domains in cyclic nucleotide phosphodiesterases and other proteins. Mol Pharmacol 65:267–278
20. Leng Q, Mercier RW, Yao W, Berkowitz GA (1999) Cloning and first functional characterization of a plant cyclic nucleotide-gated cation channel. Plant Physiol 121:753–761
21. Hoshi T (1995) Regulation of voltage dependence of the KAT1 channel by intracellular factors. J Gen Physiol 105:309–328
22. Bridges D, Fraser ME, Moorhead GB (2005) Cyclic nucleotide binding proteins in the *Arabidopsis thaliana* and *Oryza sativa* genomes. BMC Bioinf 6:6
23. Johnson JL, Leroux MR (2010) cAMP and cGMP signaling: sensory systems with prokaryotic roots adopted by eukaryotic cilia. Trends Cell Biol 20:435–444
24. Talke IN, Blaudez D, Maathuis FJ, Sanders D (2003) CNGCs: prime targets of plant cyclic nucleotide signalling? Trends Plant Sci 8:286–293
25. Isner JC, Nuhse T, Maathuis FJ (2012) The cyclic nucleotide cGMP is involved in plant hormone signalling and alters phosphorylation of *Arabidopsis thaliana* root proteins. J Exp Bot 63:3199–3205
26. Dubovskaya LV, Volotovsky ID (2004) Affinity chromatography isolation and characterization of soluble cGMP binding proteins from *Avena sativa* L. seedlings. Bulg J Plant Physiol 30:14–24
27. Dubovskaya LV, Bakakina YS, Kolesneva EV, Sodel DL, McAinsh MR et al (2011) cGMP-dependent ABA-induced stomatal closure in the ABA-insensitive Arabidopsis mutant abi1-1. New Phytol 191:57–69
28. Laukens K, Roef L, Witters E, Slegers H, Van Onckelen H (2001) Cyclic AMP affinity purification and ESI-QTOF MS-MS identification of cytosolic glyceraldehyde 3-phosphate dehydrogenase and two nucleoside diphosphate kinase isoforms from tobacco BY-2 cells. FEBS Lett 508:75–79
29. Bradford MM (1976) A rapid and sensitive method for the quantitation of microgram quantities of protein utilizing the principle of protein-dye binding. Anal Biochem 72:248–254
30. Laemmli UK (1970) Cleavage of structural proteins during the assembly of the head of bacteriophage T4. Nature 227:680–685
31. Rose JK, Bashir S, Giovannoni JJ, Jahn MM, Saravanan RS (2004) Tackling the plant proteome: practical approaches, hurdles and experimental tools. Plant J 39:715–733
32. Baerenfaller K, Grossmann J, Grobei MA, Hull R, Hirsch-Hoffmann M et al (2008) Genome-scale proteomics reveals Arabidopsis thaliana gene models and proteome dynamics. Science 320:938–941
33. Peckham GD, Bugos RC, Su WW (2006) Purification of GFP fusion proteins from transgenic plant cell cultures. Protein Expr Purif 49:183–189
34. Charmont S, Jamet E, Pont-Lezica R, Canut H (2005) Proteomic analysis of secreted proteins from Arabidopsis thaliana seedlings: improved recovery following removal of phenolic compounds. Phytochemistry 66:453–461
35. Luo Y, Blex C, Baessler O, Glinski M, Dreger M et al (2009) The cAMP capture compound mass spectrometry as a novel tool for targeting cAMP-binding proteins: from protein kinase A to potassium/sodium hyperpolarization-activated cyclic nucleotide-gated channels. Mol Cell Prot 8:2843–2856
36. Aye TT, Mohammed S, van den Toorn HW, van Veen TA, van der Heyden MA et al (2009) Selectivity in enrichment of cAMP-dependent protein kinase regulatory subunits type I and type II and their interactors using modified cAMP affinity resins. Mol Cell Prot 8: 1016–1028
37. Dubovskaya LV, Molchan OV, Volotovsky ID (2002) Cyclic GMP-binding activity in Avena sativa seedlings. Russ J Plant Physiol 49: 216–220
38. Shevchenko A, Tomas H, Havlis J, Olsen JV, Mann M (2006) In-gel digestion for mass spectrometric characterization of proteins and proteomes. Nat Protoc 1:2856–2860

Chapter 12

Structural and Functional Characterization of Receptor Kinases with Nucleotide Cyclase Activity

Victor Muleya, Janet I. Wheeler, and Helen R. Irving

Abstract

There has been an increase in the identification and characterization of plant receptor kinases possessing nucleotide cyclase activity. This has necessitated the development of robust methodologies for the structural and functional characterization of this biologically important family of proteins. Here we outline some of the techniques that can be effectively used in the characterization of this bifunctional family of proteins.

Key words Nucleotide cyclase activity, Guanylate cyclase activity, Kinase activity, Structure, Function, Cloning, Phosphorylation assays

1 Introduction

Receptor kinases possessing nucleotide cyclase activity constitute a family of catalytically active membrane-bound proteins that play crucial roles in signal transduction across biological membranes. In humans, this family of proteins has been well characterized [1–5] and current understanding of their biological function and its regulation is at a relatively advanced stage. Contrary to this, the identification and characterization of plant receptor kinases with nucleotide cyclase activity are currently at a relatively rudimentary stage, with only four plant proteins belonging to this family being characterized thus far [6–9]. Just like their human orthologs, the domain architecture in plant receptor kinases with nucleotide cyclase activity consists of an extracellular ligand-binding domain, a single transmembrane-spanning domain, and an intracellular cytoplasmic domain consisting of a kinase domain and a nucleotide cyclase catalytic domain. Despite their striking resemblance in domain organization to their mammalian counterparts, the domain conferring nucleotide cyclase activity in plant receptor kinases is encapsulated within the kinase domain (reviewed in [10]).

Chris Gehring (ed.), *Cyclic Nucleotide Signaling in Plants: Methods and Protocols*, Methods in Molecular Biology, vol. 1016, DOI 10.1007/978-1-62703-441-8_12,

This is not the case in mammalian receptor kinases, as the kinase domain is separated from the nucleotide cyclase catalytic domain via a linker region and it is often an inactive kinase [11].

More often than not, when studying receptor kinases with nucleotide cyclase activity, it is important to carry out a structure–function analysis so as to gain molecular insight as to how these proteins function. The importance of relating structure–function cannot be overemphasized as the two are inextricably intertwined in understanding the crucial aspects of biological function of proteins in general. First and foremost, the preliminary identification of candidate receptor kinases with putative nucleotide cyclase activity using computational bioinformatics is usually a good starting point (extensively discussed in chapter 17). Following this preliminary identification, it is necessary to obtain some structural insight as to how the domains conferring these two catalytic functions are organized within the protein. In cases where structural data from X-ray crystallography or nuclear magnetic resonance is unavailable for structure determination, protein structure prediction by homology modelling is the best route to take in understanding the domain organization of the protein being characterized. This approach has been successfully applied in the characterization of many proteins, for instance, the structural characterization of a receptor-like kinase in Arabidopsis called STRUBBELIG [12] furnishes a good example. The structural analysis may help to identify any structural peculiarities like specific residues and protein motifs which may guide the inference of biological function of the protein being characterized. Furthermore, some of the observed structural peculiarities may be used in rationalizing the functional mechanism of action of the protein.

Once the amino acid sequence information of the protein being studied seems to suggest that the protein in question is a candidate kinase with a putative nucleotide cyclase catalytic domain, the next thing to do is to experimentally validate the presence of these two catalytic functions. Since these receptor kinases are membrane proteins they are often difficult to express as full-length proteins in vitro. However, for most of these proteins, the kinase and nucleotide cyclase domain are located on the part of the protein occurring in the cytoplasm. As a result of this convenient domain organization, the recombinant in vitro expression of the cytoplasmic domain is often not as challenging as expressing the full-length protein; therefore, it is often expedient to just express the cytoplasmic domain. This enables the in vitro characterization of the nucleotide cyclase catalytic center and the kinase domain of the candidate protein being studied. An investigation of the oligomeric state of the cytoplasmic domain can give clues to the mechanism of action of the protein being investigated. Chemical cross-linking is a good technique for undertaking this task, and this chapter gives a detailed outline of this procedure. Furthermore, functional studies can also be

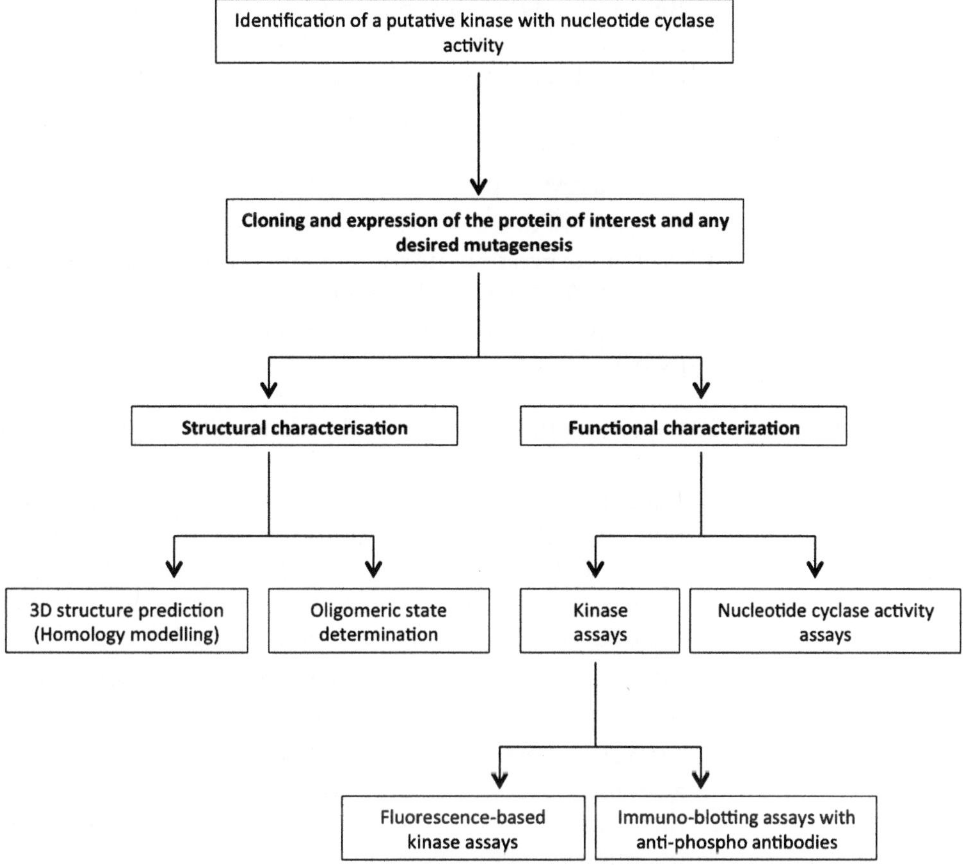

Fig. 1 An outline showing the experimental design of the structural and function characterization of receptor kinase possessing nucleotide cyclase activity

done in planta, in vivo, or in vitro depending on the convenience of the chosen experimental design as shown in Fig. 1. This chapter gives a detailed outline of some of the methods that can be used in the structural and functional characterization of receptor kinases with nucleotide cyclase activity.

2 Materials

All reagents used should be of analytical (or molecular biology) grade and must be prepared in ultrapure water unless otherwise stated. Prepare and store all reagents at room temperature (20–26 °C) unless otherwise stated. Dispose of all waste materials following the regulations and recommendations of your institution.

2.1 Cloning

1. Basic molecular biology reagents and equipment for agarose gel electrophoresis, a thermocycler, BP and LR clonases (Life Technologies), *E. coli* strains, various growth media, and antibiotics.

2. Receptor cDNA in plasmid such as pDONR207 (e.g., *Arabidopsis thaliana* phytosulfokine receptor1 (AtPSKR1) [7]) or *Arabidopsis thaliana* genomic DNA or cDNA.
3. Two forward primers incorporating a start codon (ATG) for the full-length and cytoplasmic domain of your gene of interest as well as a reverse primer incorporating a STOP codon. All primers incorporating relevant gateway recombination sites.
4. Protein expression destination vector such as pDEST17 or pDEST15 (Life Technologies) for bacteria and a vector designed for high-copy protein expression in plant cells such as p2GW7,0 [13].

2.2 Recombinant Protein Expression and Purification

1. Super Optimal Broth with Catabolite repression (SOC) medium (Life Technologies).
2. Luria Broth (LB) medium: Dissolve 20 g of LB powder in 800 ml of distilled water and then pH the solution to a pH of 7.5. Fill up the solution to 1 L with distilled water before autoclaving at 120 °C for 20 min. Supplement this medium with 10 mM $MgCl_2$ and 0.4 % glucose before use.
3. 20 % L-arabinose: Dissolve 10 g of L-arabinose in 50 ml of distilled water and sterilize by filtration.
4. LB plates: Weigh out 10 g of LB powder and 15.5 g of bacterial agar and dissolve these in 500 ml of distilled water. Autoclave this solution, cool to ~55 °C, and add the required amount of antibiotic before pouring out in Petri dishes.
5. Protein expression vector containing the gene of interest such as pDEST17PSKRcds, *E. coli*-competent cells for protein expression such as BL21-A1.
6. Water bath, 37 °C incubator and shaking incubator, spectrophotometer, cuvettes, vortex, 1.5–50 ml tubes, flasks with cotton wool stopper up to 2,500 ml.
7. High-speed centrifuge with rotors capable of holding volumes from 30 to 250 ml such as the Beckman Avanti™ centrifuge J-25 model (Beckman Coulter, Inc.) equipped with JA14 and JA25.5 rotors.
8. 250 ml centrifuge bottles for use with the JA14 rotor and 40 ml tubes that are compatible with the JA25.5 rotor or equivalent tubes and rotors.
9. Cell lysis/wash buffer: 100 mM NaH_2PO_4, 300 mM NaCl, 45 mM imidazole, pH 8.0 in the presence of 1 tablet of EDTA-free cocktail of protease inhibitors (Roche).
10. 10 mg/ml lysozyme stock solution: Dissolve 0.1 g of lyophilized lysozyme in 10 ml of distilled water. Aliquot this solution into 2 ml tubes and keep at −20 °C.
11. Probe Sonicator.

12. Ni-NTA agarose beads for purification of His-tagged proteins.
13. Rotary shaker.
14. Single-beam spectrophotometer.
15. Centrifugal concentrators (we have found that the Vivaspin® 20 (Sartorius Stedim Biotech) concentrators are good for this purpose).
16. Protein elution buffer: 100 mM NaH_2PO_4, 300 mM NaCl, 250 mM imidazole, pH 8.0.
17. 100 mM PMSF stock solution: Dissolve 0.174 g PMSF in 10 ml of isopropanol, aliquot this into 2 ml tubes, and store at −20 °C.
18. Protein storage buffer: 20 mM Tris pH 7.5, 1 mM PMSF. Protein storage conditions differ for different proteins.

2.3 SDS-PAGE

1. SDS gels such as 12 % or 4–10 % Mini-PROTEAN® TGX™ precast gels (Bio-Rad Laboratories) or prepare your own gels.
2. 1× SDS-PAGE running buffer: 25 mM Tris, 192 mM glycine, 0.1 % SDS. Weigh out 14.4 g of glycine, 3.02 g of Tris, and 1 g SDS into a clean beaker and add 1 L of water. Mix by stirring on a magnetic stirrer until there are no visible particles in suspension (a concentrated 10× stock can be prepared and diluted on the day of use).
3. Protein molecular weight marker (can be commercially obtained from the supplier of choice).
4. 2× SDS-PAGE sample loading buffer: 4 % SDS, 20 % glycerol, 0.12 M Tris pH 6.8, and 10 % β-mercaptoethanol. For 5 ml, mix 1.8 ml of 10 % SDS, 0.9 ml of 100 % glycerol, 0.54 ml 1 M Tris pH 6.8, 1.26 ml water, 0.1 g bromophenol blue, 0.5 ml of β-mercaptoethanol.
5. Mini-PROTEAN® tetra cell system (Bio-Rad Laboratories), and a heating block capable of heating at 95 °C.
6. InstantBlue™ (Expedeon Inc., Harston, UK).

2.4 Cross-Linking

1. Water-soluble homobifunctional cross-linker: 100 mM BS^3 bis[sulfosuccinimidyl] suberate (Thermo Scientific). Weigh out 2 mg of BS^3 into a clean microfuge tube and add 35 μl of water. Mix until all the cross-linker is totally dissolved.
2. Protein sample of known concentration.
3. Reaction buffer (amine-free buffer) at pH 7–9 such as phosphate-buffered saline (PBS): 0.1 M sodium phosphate, 0.15 M NaCl (*see* **Note 1**).
4. Quenching buffer: 1 M Tris–HCl pH 7.5.
5. Snake Skin™ dialysis tubing (Thermo Scientific Fisher).

2.5 Kinase Assay

1. Protein of known concentration.
2. Fluorescent Omnia® Ser/Thr Peptide 1 kit: Kinase assay kit with the SOX peptide (Life Technologies), including 10 mM ATP stock and 2 mM dithiothreitol (DTT) stock.
3. FluoroNunc™ Maxisorp™ white 96 well microtiter plate (Thermo Scientific); alternatively black plates may be used.
4. A fluorescence microplate reader capable of measuring fluorescence at excitation and emission wavelengths of 360 and 485 nm, respectively. (The EnVision™ 2101 plate reader from PerkinElmer® is suitable for this purpose.)

2.6 Cyclic GMP Assay

All reagents should be prepared in plasticware.

1. Protein of known concentration or known amount of Arabidopsis leaf mesophyll protoplasts (*see* **Note 2**).
2. 50 mM Tris pH 7.5: For a 50 ml solution, dissolve 0.3 g Tris in 30 ml of water and mix the solution on a magnetic stirrer. Adjust the pH of the solution to 7.5 and then make up with water to a final volume of 50 ml.
3. 10 mM GTP stock solution: Weigh out 5.23 mg of GTP into a 1.5 ml microcentrifuge tube and add 1 ml of water. Mix by vortexing till all the suspended particles have dissolved.
4. 0.5 M stock solution of $MgCl_2$: Dissolve 2.54 g $MgCl_2$ $6H_2O$ (if you are using the hexahydrate form) into 25 ml of water in a 50 ml Falcon™ tube and mix by vortexing.
5. 0.5 M stock solution of $MnCl_2$: Dissolve 4.95 g $MnCl_2$ $6H_2O$ (if you are using the hexahydrate form) into 25 ml of water in a 50 ml Falcon™ tube and mix by vortexing.
6. cGMP enzymeimmunoassay biotrak (EIA) kit (GE Healthcare).
7. Optional: 50 mM stock solution of isobutylmethylxanthine (IBMX): Weigh out 22.2 mg of IBMX into a 2 ml Eppendorf tube. Add 2 ml of dimethyl sulfoxide (DMSO) and mix the solution by vortexing. IBMX is a potent nonspecific inhibitor of cAMP and cGMP phosphodiesterases and only really necessary for in vivo studies.
8. Optional: 5× lysis reagent: cell culture lysis reagent (CCLR) (Promega) for lysis of protoplasts.

2.7 Phosphorylated Protein Assay

1. Pro-Q® Diamond phosphoprotein gel stain (Life Technologies).
2. Fixing solution for Pro-Q® diamond staining: 50 % methanol, 10 % acetic acid.
3. Destaining solution for Pro-Q® diamond staining: 50 mM sodium acetate pH 4.0, 20 % acetonitrile.
4. Ultrapure water for washing Pro-Q®-stained gels.
5. Orbital shaker and rocking platform.

6. PeppermintStick™ phosphoprotein molecular weight marker (Life Technologies).
7. Phospho-imager for viewing Pro-Q®-stained gels (the Typhoon Trio from GE Healthcare is suitable for this purpose).
8. Nitrocellulose membrane (e.g., Amersham Hybond ECL nitrocellulose membrane, GE Healthcare Life Sciences), electrophoretic transfer system (e.g., Mini-Trans Blot® Electrophoretic Transfer Cell (BioRad) for wet transfer), filter papers, and fiber pads. Transfer buffer: 25 mM Tris, 192 mM glycine, 20 % methanol pH 8.3 (weigh out 3.03 g Tris, 14.4 g glycine, add 200 ml methanol, dilute to 1 L with distilled water, and dissolve before storing at 4 °C as it needs to be cold) or the buffers recommended by the manufacturer of your transfer system.
9. Antisera specific for phospho-amino acids such as rabbit anti-pThreonine and rabbit anti-pSerine (Life Technologies) or mouse anti-pTyrosine (Merck) and phosphothreonine, phosphoserine, and phosphotyrosine amino acids to confirm specificity of antibodies in your system.
10. Secondary antisera and a detection system. The Odyssey® Infrared imaging system (Li-Cor, Lincoln, Nebraska, USA) offers a good platform to obtain electronic files of western data with reduced background. Goat-anti-mouse IRDye680 or 800CW and anti-rabbit IRDye680 or 800CW are suitable secondary antibodies (available from Li-Cor). Odyssey® blocking buffer (Li-Cor) and PBS (1× PBS: 137 mM NaCl, 2.7 mM KCl, 4.3 mM Na_2HPO_4, and 1.4 mM KH_2PO_4) and PBSTween (1× PBST: 1× PBS to which 0.1 % Tween 20 has been added) are required for the western protocol (*see* **Note 3**).

2.8 In Planta Experiments

1. Col-0 and mutants of your gene (e.g., *pskr1/pskr2/psyr*) mutant seed, Murashige and Skoog (MS) agar, plates and reagents required for Arabidopsis protoplast induction and polyethylene glycol (PEG) transfection [14]; *see* **Note 2**.
2. Plasmid containing your favorite gene for high protein expression in plant cells (e.g., p2GW7-PSKR1 plasmid [7]).

3 Methods

3.1 Cloning of GC Kinase for Protein and Plant Expression

The researcher should be familiar with standard molecular biology techniques [15].

1. Design primers to amplify the full-length and cytoplasmic domain or your chosen GC/kinase such as PSKR1. The full-length coding region is used for high protein expression in plant cells. As expression of membrane-spanning proteins in bacteria is problematic only the cytoplasmic domain will be expressed for

a

Primary primer pair to amplify AtPSKR region from genomic DNA

AtPSKR1-fwd 5′ ccc gtt tcg tat act ctc ag 3′

AtPSKR1-rev 5′ tga aaa tca ctc cct aca aaa 3′

b

Primers to incorporate gateway recombination sequences

AtPSKR1-f-fwd-gw

5′ gggg aca agt ttg tac aaa aaa gca ggc ttc gaa gga gat aga ATG CGT GTT CAT CGT TTT TGT GTG A 3′

AtPSKR1-cd-fwd-gw

5′ gggg aca agt ttg tac aaa aaa gca ggc ttc ATG cgt gct cgt aga cgg tca gga gaa gtt g 3′

AtPSKR1-rev-gw-stop

5′ gggg ac cac ttt gta caa gaa agc tgg gtc cta gtc gtt ggc ctc tgt ttc ggg ttt t 3′

Fig. 2 Examples of primers used to amplify the AtPSKR1. (**a**) Primers used to amplify the full-length receptor from genomic DNA (this gene has no introns, so genomic DNA could be used as a source; if introns were present DNA would be required). (**b**) Primers used to incorporate the gateway recombination sequences into the full-length and cytoplasmic domain of PSKR1. Key: *Green*=guanine residues, *Brown*=att recombination sequence, *Red*=Shine–Dalgarno sequence, *Black*=gene-specific sequence, *Blue*=fill in bp for correct framing and the start and stop bp are underlined

recombinant protein expression in bacteria. Gateway recombination sites should also be incorporated into the primer ensuring that the GC/kinase coding region is in frame (i.e., the resulting amino acid for each nucleotide triplicate is as expected) as well as start (ATG) and stop codons. We have found that if amplifying from genomic DNA or cDNA it is best if primary primers (spanning ~30–100 bp either side of the required sequence) are used to amplify the region of interest. An example is shown in Fig. 2. This product is then used as the template for the incorporation of the gateway recombination sites and any tags you may wish to add (*see* **Note 4**).

2. If you are not using plasmid template use the primary primer pair with genomic DNA or cDNA to amplify your gene of interest. No more than 25 amplification cycles. Then use PCR to amplify full-length and cytoplasmic domain PSKR1 fragments incorporating the gateway recombination sites, purify, and recombine in a BP reaction into pDONR207 using standard molecular biology techniques.
3. Resulting plasmids should be sequenced to ensure fidelity before LR recombination reaction of the cytoplasmic domain entry vector into bacterial protein destination vector such as pDEST17 or pDEST15 to make for example pDEST17PSKRcds and the full-length entry vector into high protein expression in plant cell destination vector such as p2GW7.0 to make p2GW7-PSKRfls.
4. Resulting plasmids can be sequenced across the 5′ recombination site to ensure that the GC/kinase sequence is in frame.

3.2 Recombinant Protein Expression

3.2.1 Pilot Expression

1. Transform your bacterial expression plasmid (e.g., pDEST17P-SKRcds) into BL21-AI cells and plate out onto LB with 200 μg/ml carbenicillin selection at 37 °C overnight.
2. Culture 9 isolated colonies for each plasmid in 5 ml using augmented LB medium (*see* **Note 5**) with 0.4 % glucose, carbenicillin 200 μg/ml, and 5 mM $MgCl_2$ (or 5 mM $MnCl_2$ depending on guanylate cyclase ion selectivity) at 37 °C overnight in an orbital shaking incubator at 200 rpm. All incubations of liquid cultures for the pilot expression are done at this speed unless otherwise stated.
3. For each overnight culture use a 1 in 20 dilution to inoculate ~20 ml augmented LB medium with 0.4 % glucose, carbenicillin 200 μg/ml, and 5 mM $MgCl_2$ (or 5 mM $MnCl_2$ depending on guanylate cyclase ion selectivity). Record OD_{600}. Place at 37 °C in the shaker until OD_{600} is between 0.4 and 0.5.
4. When OD_{600} is between 0.4 and 0.5 make five seed stock tubes (750 μl culture into sterile screw top tubes containing 250 μl glycerol, vortex, and snap freeze in liquid nitrogen immediately) and take a 1 ml T0 (time zero) sample. Then add filter-sterilized L-arabinose to 0.2 % and put culture in shaker at 27 °C.
5. Take further 1 ml samples at T3, T4, and T5 (time 3, 4, and 5 h). Culture samples can be stored on ice and spun down together. The pellets can be stored at −20 °C.

3.2.2 Checking Protein Expression of Clones

1. Culture pellets are resuspended in 100 μl lysis buffer, vortexed for 15 s, and centrifuged in a benchtop centrifuge at maximum speed for 1 min. Unused portions can be stored at −20 °C.
2. Take 8 μL of supernatant and 8 μL of 2× SDS loading buffer and put into a fresh tube and incubate at 95 °C for 5 min and then on ice for 1 min.
3. Load samples (T0, T3, T4, T5, etc.) with appropriate ladder onto SDS-PAGE gel and run gel at 200 V for 40 min (*see* Subheading 3.3).
4. Stain with ~20 ml InstantBlue™ for 45–60 min on rocker, then rinse with distilled water, and take picture of gel.
5. Identify clones that show a high level of recombinant protein expression after 3, 4, or 5 h. PSKR1cd is approximately 40 kDa and is shown in Fig. 3.

3.2.3 Large-Scale Expression

1. Inoculate 10–50 ml augmented LB medium with 0.4 % glucose, carbenicillin 200 μg/ml, and 5 mM $MgCl_2$ (or 5 mM $MnCl_2$ depending on guanylate cyclase ion selectivity; *see* **Note 5**) with a single seed stock of a high-expressing clone and incubate overnight at 37 °C in shaker at 200 rpm. All incubations of liquid cultures for large-scale expression are done at this speed unless otherwise stated.

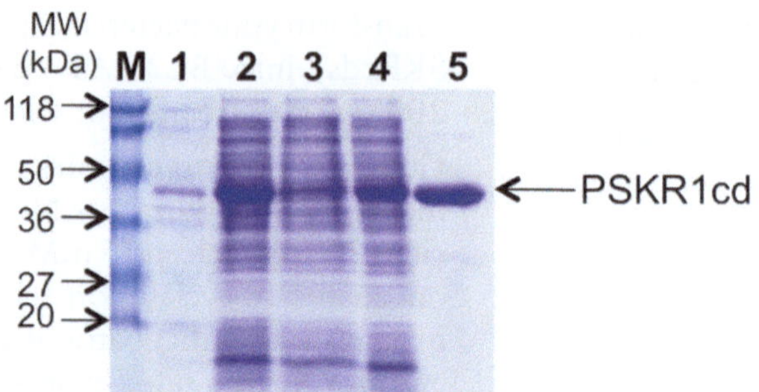

Fig. 3 SDS-PAGE analysis of protein fractions during affinity purification. *Lane 1* shows the pellet fraction, *lane 2* corresponds to the crude lysate fraction, *lane 3* shows the flow-through fraction, *lane 4* shows the column wash fraction, and *lane 5* corresponds to the fraction eluted from the Ni-NTA agarose column. *Lane M* shows the protein molecular weight marker

2. Use a 1 in 20 dilution to inoculate 200–500 ml augmented LB medium with 0.4 % glucose, carbenicillin 200 μg/ml, and 5 mM $MgCl_2$ (or 5 mM $MnCl_2$ depending on guanylate cyclase ion selectivity). Record the optical density at 600 nm (OD_{600}) and then incubate at 37 °C in a shaking incubator until OD_{600} of 0.4–0.5 is reached.
3. When OD_{600} is between 0.4 and 0.5 add L-arabinose to 0.2 % and incubate the culture in shaker at 27 °C for optimal number of induction hours.

3.2.4 Recombinant Protein Extraction

1. Harvest cells from the large-scale expression by centrifugation at 8,000×*g* for 10 min using a high-speed centrifuge (we usually use the Beckman Avanti™ centrifuge J-25 model).
2. Discard the supernatant and resuspend the cells by gently vortexing in cell lysis buffer (use 10 ml of lysis buffer per 500 ml culture). Add 1 ml of the 10 mg/ml stock of lysozyme to this mixture and incubate on ice for 30 min.
3. Sonicate on ice using 10 sets of 10-s bursts at 300 W with 10-s cooling periods in between each burst.
4. Centrifuge the lysate for 30 min at 4 °C at 10,000×*g* in order to pellet cell debris (we usually use the JA-25.5 rotors in a Beckman Avanti™ centrifuge J-25 model). You can fill up the sonicated mixture with distilled water in order to meet centrifuge volume requirements.
5. Save the supernatant (crude lysate) and keep 5 μl of this and a grain-sized part of the pellet for SDS-PAGE analysis.

3.2.5 Purification of Recombinant Protein

1. Add 2 ml of 50 % of equilibrated Ni-NTA slurry to 4 ml or more of cleared crude lysate (*see* **Note 6**).
2. Mix gently by shaking on a rotary shaker at 4 °C for at least 1 h (overnight shaking can also be done to maximize binding).

3. Load the lysate and Ni-NTA mixture into a plastic column with a bottom cap.
4. Remove the bottom outlet and collect the flow-through fraction. Save 5 μl of the flow-through fraction for SDS-PAGE analysis.
5. Wash twice with 5 ml wash buffer and collect the wash fractions. Keep 20 μl of these fractions for SDS-PAGE analysis.
6. Elute protein with 4 ml of elution buffer and keep 20 μl for SDS-PAGE analysis.
7. Analyze all the purification factions by SDS-PAGE including the pellet and crude lysate fractions from the extraction procedure so as to assess the expression profile of the recombinant protein (*see* Fig. 3).

3.3 SDS-Polyacrylamide Gel Electrophoresis

1. Place a freshly opened Mini-PROTEAN® TGX™ gel into the gel tank of an appropriately assembled Mini-PROTEAN® system.
2. Fill the gel tank with 1 L of 1× SDS-PAGE running buffer.
3. Prepare the samples in 2× SDS-PAGE sample loading buffer. Boil the samples in a heating block at 95 °C for 5 min.
4. Load the samples including a protein molecular weight marker into appropriate wells of the TGX™ gel.
5. Run the samples on SDS-PAGE at a constant voltage of 100 V for 60 min.
6. When the run is done, remove the gel from the TGX™ plates and place into a clean empty container. Stain the gel by adding InstantBlue™ and leave at room temperature for at least 45 min (*see* **Note 7**).

3.4 Chemical Cross-Linking

1. Prepare protein in reaction buffer using dialysis in a SnakeSkin™ dialysis tubing (*see* **Note 1**).
2. Add freshly prepared cross-linker solution to the protein sample to a final concentration of 0.25–5 mM.
3. Incubate the reaction mixture at room temperature for 30 min.
4. After 30 min, stop the reaction by adding quenching buffer to a final concentration of 50 mM and leave the reaction for 15 min at room temperature.
5. Set up a negative control with just the protein in reaction buffer without the cross-linker.
6. In order to analyze the cross-linking reaction perform an SDS-PAGE of the cross-linked samples and the appropriate controls (*see* **Note 8**).

3.5 Fluorescence-Based Kinase Assay

1. Prepare stock solutions for the kinase reaction using reagents provided in the Omnia® Ser/Thr Peptide 1 kit. Preparation of stock solutions can be done as outlined in Table 1.

Table 1
Preparation of the components of the kinase assay

Component	Volume required (μl)	Volume of water added	Concentration of stock solution
SOX peptide	7	63 μl	100 μM
ATP	7	63 μl	10 mM
DTT	2	998 μl	2 mM
Kinase reaction buffer	500	4.5 ml	1×

Table 2
Preparation of master mixes used in phosphorylation assay

Component	Control reaction (μl)	Experimental reaction (μl)
Kinase reaction buffer	30	30
ATP	30	30
DTT	30	30
SOX	30	30

2. Prepare 2 sets of 4× master mix solutions: one set for 3 control reactions and the other set for 3 experimental reactions. This master mix is sufficient for triplicate sets of control and experimental reactions and should also account for the inevitable pipetting errors that may result. The final volume of each reaction can be made up to 75 μl to ensure that the bottom of each well in the microtiter plate or microplate is completely covered by reaction components. This is necessary in order to obtain consistent fluorescence readings and also minimizes variation between technical replicates (*see* **Note 9**). The 4× master mix can be prepared as outlined in Table 2.
3. Since it is a 4× master mix, divide the total volume of the master mix by 4 and the amount you obtain will be the amount of master mix you need to add in each of the three wells.
4. Determine the amount of protein that should be used in each well. Each reaction should contain 1 μg of protein (i.e., 1 μg of protein per 75 μl reaction; *see* **Note 10**). Remember it is to be done in triplicate.
5. Once the amount of protein that goes into each well has been determined, and the volume of the master mix to be added into each well is known, you can fill up the reaction mixture in each well to 75 μl. There is no need to add the protein in the

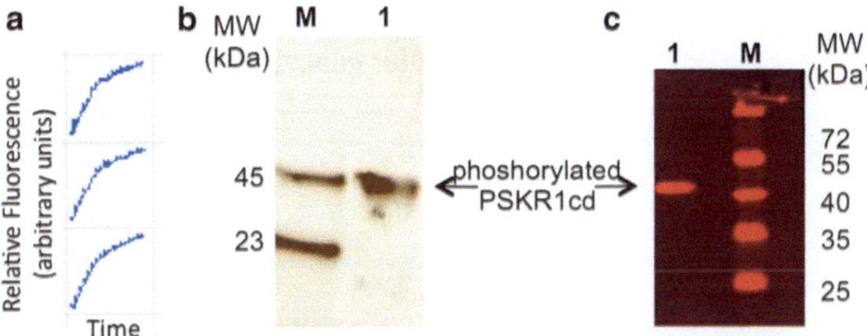

Fig. 4 Kinase activity and detection of (auto)phosphorylated protein. (**a**) Time course of phosphorylation of the SOX substrate by AtPSKR1cd showing 3 separate (technical) reactions of one biological replicate. (**b**) Pro-Q® diamond phosphostaining of ATP-treated PSKR1cd. *Lane 1* represents ATP-treated PSKR1cd (phospho-PSKR1cd) and *lane M* shows the molecular weight marker (PeppermintStick™ ladder). Only phospho-proteins are visible upon staining with ProQ®-diamond phosphostain. The two phosphorylated protein standards in the PeppermintStick™ ladder (45 and 23 kDa) are the only bands that become visible upon phosphostaining with Pro-Q® diamond. The other three bands corresponding to the 18, 66, and 116 kDa protein standards of the PeppermintStick™ ladder are not visible upon phosphostaining with Pro-Q® diamond because they are not phosphorylated; however they become visible upon staining with the non-phospho-specific stain like SYPRO® Ruby. (**c**) Phosphorylated AtPSKR1cd detected by anti-phosphothreonine antisera (Life Technologies). The membrane was probed with rabbit anti-phosphothreonine antisera as described in the text and the secondary antisera was goat anti-rabbit IRDye680 which was detected on the 700 nm channel that also detects the blue color of the prestained ladder (PageRuler™ Fermentas) corresponding to 25, 35, 40, and 55 kDa but is less efficacious at detecting the red marker at ~70 kDa

control reaction (although it is sometimes worthwhile to see the baseline values in the presence of protein).

6. The protein has to be added immediately before you start measuring fluorescence on the microplate reader. Make sure that the microplate reader is ready to measure fluorescence before the protein is added to the reaction mixture. The excitation and emission wavelengths on the microplate reader need to be set at 360 and 485 nm, respectively.
7. If doing an end point kinase assay you only need the initial and the final values of the measured fluorescence units, and the difference between these two values is known as relative fluorescence unit (RFU). The kinase reaction is usually allowed to carry on for 5–15 min, but longer reaction times of up to 60 min can also be done. You can also generate Michaelis–Menten kinetics if fluorescence units are recorded throughout the progress of the reaction (*see* Fig. 4a).

3.6 CyclicGMP Assay

Guanylate cyclase (GC) activity in vitro can be measured using recombinant protein such as AtPSKRcds whereas guanylate cyclase

Table 3
Master mix for guanylate cyclase assay

Component	Volume	Final concentration (mM)
1 M Tris, pH 7.5		50
0.5 M $MnCl_2$		5
10 mM GTP		1
50 mM IBMX		2
5 mM $CaCl_2$		10
Water		–
Total volume	$(n+1) \times 50$ µl	–

n is equal to the total number of reaction you need to carry out, including controls

activity in planta can be measured using plant protoplasts expressing high levels of your full-length recombinant protein such as AtPSKRfls. For preparation and transfection of protoplasts see ref. 14 (*see* **Note 2**).

3.6.1 Recombinant Protein Guanylate Cyclase Activity

1. Determine the total number of reactions you need to carry out in the assay (*see* **Note 11**).
2. Prepare a master mix for all your reactions as outlined in Table 3. Add 50 µl of the master mix into all the reaction tubes (except the GTP control, *see* **Note 11**), ideally 1.5 ml microcentrifuge tubes.
3. Determine the amount of protein that is supposed to be added into each reaction. This can be done as outlined in **step 4** of Subheading 3.5. However, your final volume in this case becomes 100 µl and the final protein concentration can be 2–5 µg per 100 µl reaction. Set up the reaction for each tube so that it contains 50 µl of master mix, and protein at 2–5 µg (determine by calculation), and is made up to 100 µl with water. The protein should be added last as a means of starting the reaction.
4. Allow the reaction to carry on for 5–20 min at room temperature.
5. Stop the reaction with 10 mM EDTA final concentration. This can be done by adding 10 µl of a 100 mM EDTA stock solution.
6. The terminated reactions are then clarified through centrifugation at 19,500 × *g* for 10 min.
7. Collect the respective supernatants and measure the amount of cGMP in each tube (*see* **Note 12**).

8. The amount of cGMP can be determined using the cGMP enzymeimmunoassay biotrak (EIA) kit (GE Healthcare) following protocols 2 or 4.

3.6.2 Plant Protoplasts Expressing High Levels of Your Full-Length Recombinant Protein Guanylate Cyclase Activity

1. After transfection of your high-expressing full-length GC recombinant protein (*see* Subheading 3.1) following the protocols described in [14] allow your protoplasts to rest for 2–18 h in the dark at 23 °C for time to express the protein.
2. Before you start protoplast collection allow 5× lysis reagent CCLR to defrost at room temperature, and prepare 1× lysis reagent by adding 4 volumes of water to 1 volume of 5× lysis reagent (CCLR).
3. For each protoplast sample gently swirl protoplasts to resuspend and using a fresh transfer pipette move to a round-bottom tube (e.g., use a 50 ml Falcon tube but not a 15 ml tube) being careful not to crush the protoplasts.
4. Spin down protoplasts at 100 g for 5 min.
5. Take off and discard supernatant.
6. Add 100 μl of 1× CCLR, vortex for 15 s, and spin down at 14,000 ×*g*.
7. Transfer supernatant to a fresh tube.
8. Then assess the protein concentration, and proceed to the cGMP assay (as above) or store at −80 °C until ready to measure.

3.7 Detection of Phosphorylated Proteins

3.7.1 Phosphoprotein Analysis Using ProQ® Diamond Gel Stain

1. Run phosphorylated protein on an SDS-PAGE including the PeppermintStick™ phosphoprotein molecular weight marker.
2. Transfer the SDS-PAGE gel into a clean container and add 100 ml of Pro-Q® fixing solution and place this in an orbital shaker at 35 rpm for 30 min. All the shaking is to done at this speed. Discard the fixing solution and repeat this fixing step to a total of two times with the addition of fresh fixing solution each time.
3. Discard the fixing solution and wash the gel twice with 100 ml of deionized water.
4. Stain the gel in the dark with 65 ml of 3× dilute solution of Pro-Q® diamond stain for 2 h. All the following steps are to be done in the dark and this can be achieved by covering the container with aluminum foil to protect the gel from light.
5. Pour out the staining solution and add 100 ml of destaining solution and place on a shaking orbital shaker for 30 min. Repeat the destaining step for a total of four times with addition of fresh destaining solution each time.
6. Wash the gel twice with 100 ml of deionized water for 5 min each time.
7. View the phosphostained gel using a phosphoimager; an example of a Pro-Q® diamond-stained gel is shown in Fig. 4b.

3.7.2 Phosphoprotein Analysis Using Specific Antiphospho-Amino Acid Antiserum

1. Follow the manufacturer's instructions for your electrophoretic transfer system, obtain your nitrocellulose membrane with your proteins of interest, and air-dry the blot for at least 2 h (*see* **Note 13**).
2. Prepare blocking buffer (Odyssey® blocking buffer: 1× PBS (1:1, v/v) which can be reused). Saturate membranes in blocking buffer for 1 h on a rocking platform at room temperature.
3. Dilute primary antibody to appropriate concentration in Odyssey® Blocking buffer: 1×PBST (1:1) (e.g., 1:1,000 = 10 μl of antibody in 10 ml blocking buffer; this can also be reused). Probe with primary antibody for 2 h at room temperature (or overnight at 4 °C, *see* **Note 14**) on the rocking platform.
4. Wash membrane three times for 5 min with 1× PBST on the rocking platform.
5. Prepare secondary antibody (e.g., goat anti-rabbit conjugated to IRDye680 or 800CW) in Odyssey® Blocking buffer: 1× PBST (1:1) and cover in foil to protect from light (see below). Probe with secondary antibody for 1 h (protect from light by covering with aluminum foil for example as the conjugated secondary antibodies are light sensitive; incubation can be 30–60 min but no more) on the rocking platform.
6. Wash three times for 5 min with 1× PBST on the rocking platform.
7. Wash once for 5 min in 1× PBS (no Tween) on the rocking platform.
8. Detect using the Odyssey® Infrared imaging system and an example is shown in Fig. 4c. The membrane can be imaged wet or dry but it is easier to manipulate if wet, so store in 1× PBS protected from light at 4 °C.
9. Analyze image intensity using Odyssey® analysis programs or export as TIF files and analyze using ImageJ which can be downloaded from http://rsbweb.nih.gov/ij/ (*see* **Note 15**).

3.8 In Planta Experiments

In vitro assays using recombinant protein allow determination of isolated processes such as the kinase activity or GC activity of our favorite GC/kinase enabling us to discover if the recombinant protein is capable of catalyzing these reactions under specific conditions. However, these conditions may or may not exist within the plant cell. It is therefore important to test the activity of our favorite GC/kinase in a plant cell to assess the activity of the protein in the living cell.

There are two themes that can be explored with in planta experiments; the first is mutant complementation. GC and kinase activity from protoplasts derived from the wild-type and your favorite GC/kinase mutant plants (in the case of AtPSKR1 the single (*pskr1*) or the triple mutant (*pskr1/pskr2/psy1*) [16]) can be compared where the loss of GC or kinase activity in the mutant is assumed to be due to the lack of your favorite GC/kinase. Transient transfection of the mutant protoplasts with the high protein expression destination vector (such as p2GW7-PSKRfls, *see* Subheading 3.1 above) could be used to reevaluate GC and kinase activity compared to wild-type levels. After complementation has been shown, site-directed mutagenesis can be used to assess the contribution of particular amino acids to either the GC or the kinase activity of your favorite GC/kinase.

4 Notes

1. Since the BS^3 cross-linker reacts with primary amines in the lysine residues and N-terminus of proteins, the reaction buffer has to be a non-amine-containing buffer; otherwise it will compete with the protein for the cross-linking reagent. It is important to change the buffer of the protein to a non-amine-containing buffer. The protein sample has to be free of amine-containing components like Tris-based buffers and imidazole. Dialysis in a SnakeSkin™ dialysis tubing or diafiltration in centrifugal filters can be used for efficient buffer exchange. At least three 2 L buffer exchanges may be necessary to remove any traces of amine-containing components in the protein sample.
2. Preparation and transient transfection of protoplasts are beyond the scope of this chapter and the reader is referred to Yoo et al. [14] for an excellent description of the procedures involved. However, we do wish to highlight that in our hands as well, improved preparations of protoplasts are obtained from plants grown under short days.
3. In the western analysis using antisera to phosphorylated amino acids, it is important to ensure that blocking and washing buffers do not contain phosphorylated proteins; therefore do not use milk powder products in the blocking process. Stock buffers of PBS and PBST can also be made at 10× concentration and diluted on the day of use to 1× buffers.
4. To identify the full cytoplasmic domain of your receptor GC kinase for prime design, the transmembrane domain of the GC kinase needs to be determined using protein data available through the TAIR Web site (www.TAIR.com) or SWISS PROT (www.SWISSPROT.com) and this can be checked using any of the free software available. You could also incorporate various tags by using alternative vectors or the tags could be integrated into the primer design. For incorporation of a C-terminal tag be sure that no stop codon is included in the reverse primer before the tag and that the sequence encoding the tag is in frame with your gene of interest. Possible tags include C-MYC, V5, FLAG, AU1, HA, or HIS. For further details see ref. 15.

5. Augmentation of LB medium should be done on the day of use and as follows. Firstly, adjust pH of culture medium to pH 7.5, and then add glucose to a final percentage of 0.4 % after autoclaving (do not add glucose to medium before autoclaving) and before you inoculate your cultures. Medium may be further supplemented with the known cofactor as this may increase protein solubility. In the case of AtPSKR either magnesium or manganese chloride salts within a range 1–10 mM final concentration. For other receptor GC kinases, the preferred metal for the GC activity should be determined by experimentation (*see* Subheading 3.6). Insert cotton wool on the neck of the flask rather than foil (this is to increase aeration which may contribute towards protein solubility).
6. It is important to ensure that the Ni-NTA solution is equilibrated with buffer as it is often stored in a 20 % ethanol solution. Wash beads with deionized sterile water using 5 column volumes (i.e., wash 2 ml resin with 10 ml water). Pellet beads (brief spin), remove water, and wash again with filter-sterilized water. Equilibrate the beads with lysis buffer using 5 column volumes and pellet beads before adding fresh lysis buffer and finalizing the equilibration process.
7. Protein bands are usually visible after 15 min of incubation but it is advisable to allow the staining process to go on for at least 45 min. Longer periods of staining can also be done but they have no significant benefit to the staining process.
8. On the SDS-PAGE gel, if the protein is dimeric in its native conformation the cross-linked sample will assume a molecular weight that is approximately double that of the control. It is imperative that DTT be present in your protein samples so as to make sure that any oligomerization that is observed is not due to cysteine dimer formation. That is why it is necessary to use a non-thiol-cleavable cross-linker like BS^3.
9. Biological replicates (at least 3 independent replicates is advisable) for the kinase assays can be performed with different clones of the protein albeit using the same assay parameters.
10. Calculation of the amount of protein required can be done using the formula $C_1V_1 = C_2V_2$, where C_1 = the concentration of protein sample (determined using a protein quantification method of choice); C_2 = the final concentration needed in the reaction (i.e., 1 μg of protein per 75 μl reaction) = 13.3 μg/ml (1 μg of protein per 75 μl reaction when converted to μg/ml using cross multiplication should be 13.3 μg/ml which is your C_2 value); and V_1 = the amount of protein you are required to add per well of your kinase reaction. This is the value we are looking for. V_2 is the final volume of the reaction per well which is 75 μl.
11. It is advisable to carry out reactions in triplicate for technical replicates and using different clones for biological replicates in

the assay. In the GC activity assays, control reactions may include a set without the protein and another set without GTP (GTP control—GTP is a substrate of the guanylate cyclase reaction). The GTP control can have its own master mix prepared separately from the master mix of the other reaction tubes. It needs to be determined in preliminary experiments whether Mg or Mn chloride salts are best metal cofactor first (basically you need to test both cofactors using separate master mixes). IBMX is a nonspecific phosphodiesterase inhibitor and theoretically not necessary if protein preparations are pure but is often important when using biological material such as protoplasts or plant extracts.

12. If reactions are to be stored and used at a later stage, the clarified reactions can be snap-frozen in liquid nitrogen before being stored at −80 °C. At the time of use, these can then be thawed on ice and briefly pulse-spun before the supernatants are collected for cGMP quantification.
13. It is best to let the membranes air-dry overnight as it gives the proteins a chance to slightly refold before blocking and probing [17].
14. Specificity and the best dilutions of the primary antibodies should be determined in preliminary experiments. The best dilution of primary antibody needs to be determined empirically using several dilutions in the range of 1:250 to 1:2,000 and selecting for the best sensitivity and reduced background in your detection system. Specificity can be demonstrated by blocking binding of the primary antibody in the presence of the phospho-amino acid (concentrations of 10–20 mM are typically used). It is also important to demonstrate that a different phospho-amino acid does not block binding (e.g., phospho-threonine should block binding of anti-phosphothreonine antisera but phospho-serine should not). The incubation with the primary antibody can also be determined empirically but we usually use 2 h at room temperature although others use overnight incubations at 4 °C.
15. Data can be analyzed on the Odyssey Infrared imaging system or exported as TIF files for analysis using ImageJ (Web site: http://rsbweb.nih.gov/ij/) to determine relative intensity of images. It is important to export and analyze as TIF files as export of JPG files can result in pixel data being compressed or lost.

Acknowledgments

This work was supported by the Australian Research Council's Discovery project funding scheme (DP0878194 and DP110104164). V.M. is supported by a scholarship from the Monash Institute of Pharmaceutical Sciences, Monash University.

References

1. Duda T, Yadav P, Sharma RK (2011) Allosteric modification, the primary ATP activation mechanism of atrial natriuretic factor receptor guanylate cyclase. Biochem 50:1213–1225
2. Misono KS, Philo JS, Arakawa T, Ogata CM, Qiu Y, Ogawa H et al (2011) Structure, signalling mechanism and regulation of the natriuretic receptor guanylate cyclase. FEBS J 278:1818–1829
3. Pattanaik P, Fromondi L, Ng KP, He J, van den Akker F (2009) Expression, purification, and characterization of the intra-cellular domain of the ANP receptor. Biochimie 91:888–893
4. Potter LR (2011) Regulation and therapeutic targeting of peptide-activated receptor guanylyl cyclases. Pharmacol Therap 130:71–82
5. Sharma RK (2010) Membrane guanylate cyclase is a beautiful signal transduction machine: overview. Mol Cell Biochem 334:3–36
6. Kwezi L, Meier S, Mungur L, Ruzvidzo O, Irving H, Gehring C (2007) The *Arabidopsis thaliana* brassinosteroid receptor (AtBRI1) contains a domain that functions as a guanylyl cyclase in vitro. PLoS One 2:e449
7. Kwezi L, Ruzvidzo O, Wheeler JI, Govender K, Iacuone S, Thompson PE et al (2011) The phytosulfokine (PSK) receptor is capable of guanylate cyclase activity and enabling cyclic GMP-dependant signaling in plants. J Biol Chem 286:22580–22588
8. Meier S, Ruzvidzo O, Morse M, Donaldson L, Kwezi L, Gehring C (2010) The Arabidopsis wall associated kinase-like 10 gene encodes a functional guanylyl cyclase and is co-expressed with pathogen defense related genes. PLoS one 5:e8904
9. Qi Z, Verma R, Gehring C, Yamaguchi Y, Zhao Y, Ryan CA et al (2010) Ca^{2+} signaling by plant *Arabidopsis thaliana* Pep peptides depends on AtPepR1, a receptor with guanylyl cyclase activity, and cGMP-activated Ca^{2+} channels. Proc Natl Acad Sci 107: 21193–21198
10. Irving HR, Kwezi L, Wheeler JI, Gehring C (2012) Moonlighting kinases with guanylate cyclase activity can tune regulatory signal networks. Plant Sig Behav 7:201–204
11. Biswas KH, Shenoy AR, Dutta A, Visweswariah SS (2009) The evolution of guanylyl cyclases as multidomain proteins: conserved features of kinase-cyclase domain fusions. J Mol Evol 68:587–602
12. Vaddepalli P, Fulton L, Batoux M, Yadav RK, Schneitz K (2011) Structure-function analysis of STRUBBELIG, an Arabidopsis atypical receptor-like kinase involved in tissue morphogenesis. PLoS One 6:e19730
13. Karimi M, Inze D, Depicker A (2002) Gateway™ vectors for *Agrobacterium*-mediated plant transformation. Trends Plant Sci 7:193–195
14. Yoo S-D, Cho Y-H, Sheen J (2007) Arabidopsis mesophyll protoplasts: a versatile cell system for transient gene expression analysis. Nat Prot 2:1565–1572
15. Ausubel FM, Brent R, Kingston RE, Moore DD, Seidman JG, Smith JA et al (eds) (2002) Short protocols in molecular biology. Wiley, New York
16. Amano Y, Tsubouchi H, Shinohara H, Ogawa M, Matsubayashi Y (2007) Tyrosine-sulfated glycopeptide involved in cellular proliferation and expansion in *Arabidopsis*. Proc Natl Acad Sci 104:18333–18338
17. Van Dam A (1994) Transfer and blocking conditions in immunoblotting. In: Dunbar BS (ed) Protein blotting: a practical approach. IRL, Oxford, pp 73–85

Chapter 13

Computational Identification of Candidate Nucleotide Cyclases in Higher Plants

Aloysius Wong and Chris Gehring

Abstract

In higher plants guanylyl cyclases (GCs) and adenylyl cyclases (ACs) cannot be identified using BLAST homology searches based on annotated cyclic nucleotide cyclases (CNCs) of prokaryotes, lower eukaryotes, or animals. The reason is that CNCs are often part of complex multifunctional proteins with different domain organizations and biological functions that are not conserved in higher plants. For this reason, we have developed CNC search strategies based on functionally conserved amino acids in the catalytic center of annotated and/or experimentally confirmed CNCs. Here we detail this method which has led to the identification of >25 novel candidate CNCs in *Arabidopsis thaliana*, several of which have been experimentally confirmed in vitro and in vivo. We foresee that the application of this method can be used to identify many more members of the growing family of CNCs in higher plants.

Key words Cyclic nucleotide cyclase, Adenylyl cyclase, cAMP, Guanylyl cyclase, cGMP, Catalytic center, Motif search, Homology modeling, Basic local alignment search tool, *Arabidopsis thaliana*

1 Introduction

Adenylyl cyclases (ACs) and guanylyl cyclases (GCs) are cyclic nucleotide cyclases (CNCs) that catalyze the reaction from ATP and GTP, respectively, to the messengers cAMP or cGMP. Particularly the role of cGMP in many plant responses is well documented and includes responses to light [1], hormones and signaling peptides [2–4], salt and drought stress [5, 6], and ozone and pathogens [7, 8]. Cyclic nucleotides can also directly affect cellular ion homeostasis by gating an entire class of ion channels, the cyclic nucleotide-gated channels (CNGCs) [9]. Given the importance of the role of cyclic nucleotides it is not surprising that there is considerable interest in the enzymes that generate these molecules.

However, plant molecules with CNC activity are outside the detection limit of BLAST searches and other biochemical tools such as specific antibodies against CNCs from, e.g., bacteria or animals

Chris Gehring (ed.), *Cyclic Nucleotide Signaling in Plants: Methods and Protocols*, Methods in Molecular Biology, vol. 1016, DOI 10.1007/978-1-62703-441-8_13, © Springer Science+Business Media New York 2013

because of the high level of divergence and complexity of CNCs which often combine two or more different domains [10, 11]. Examples of complex GCs with several domains are the plant leucine-rich receptor kinases, many of which contain extracellular ligand-binding domains and intracellular kinase and GC domains [4, 12]. We have therefore proposed and tested a search strategy that uses search motifs based on conserved amino acids in the catalytic center of experimentally tested CNCs [10, 12, 13]. The residues include in position 1 the amino acid that does the hydrogen bonding with ATP or GTP, in position 3 the amino acid that confers substrate specificity, and in position 14 the amino acid that stabilizes the transition state (ATP to cAMP or GTP to cGMP) and the C-terminal Mg^{2+}/Mn^{2+}-binding site [14] (Fig. 1). Additional search conditions like the presence of a glycine-rich N-terminal domain or the presence

a

```
                      Glycine-rich    Catalytic
                         domain         center      ⇩
T. rubripes       N'--GSVLAGVVGVKM-PRYCLFGNNVTLANKFESCSQ--C'
D. melanogaster   --GEVVTGVIGNRV-PRYCLFGNTVNLTSRTETTGV--
C. elegans        --GPCVAGVVGKTM-PRYTLFGDTVNTASRMESNGE--
D. discoideum     --GPVVGGIIGKKKL-SWHLFGDTINTSSRMASHSS--
C. reinhardtii    --GPATSGVVGQKM-PRFCLFGDTVNTASRMESTGR--
Synechocystis     --GEVVVGNIGSEKRTKYGVVGAQVNLTYRIESYTT--
                                   1-----------14

Original GC
search motif:[RSK][YFW][CTGH][VIL][FV]X[DNA]X[VIL]X{4}[KR]
              1    2    3     4    5  6  7   8  9        14

AtGC1             --G-ILDGN-GDSTFPRYCLFDDPLVSDGKYRDAGL–
```

b

```
Modified AC
search motif:[RSK][YFW][DE][VIL][FV]X{8}[KR]X{1,3}[DE]
              1    2    3    4    5       14

Relaxed AC
search motif:[RSK]X[DE]X{9,11}[KR]X{1,3}[DE]
              1   2 3          14
```

Fig. 1 Example of an alignment of CNC catalytic centers and the building of search motifs for candidate CNCs. (**a**) Edited ClustalX alignment of catalytic centers from annotated GCs from different species. In the deduced 14 amino acid motif the substitutions are in square brackets ([ ]), "X" stands for any amino acid and the gap size is marked in curly brackets ({ }). The underlined amino acids have been added to the motif because of their chemical similarity to the amino acid at this position. The amino acids in positions 1, 3 and 14 are functionally annotated [10]. The *open arrow* (⇩) signifies the glutamic acid (E) implicated in Mg^{2+} respectively Mn^{2+} binding (not included in the original search motif). (**b**) Modifications of the motif for the discovery of candidate ACs. The residues in position 3 confer substrate specificity and have been changed to D or E to recognize ATP rather than GTP. The Mg^{2+} respectively Mn^{2+} binding residue is on the C-terminal side downstream of the core motif

of additional motifs such as an H-NOX motif [15] diagnostic for gas binding will add stringency to the candidate protein selection and allow for the identification of specific classes of functionally defined candidate CNCs. Confidence in identified candidate CNCs increases if firstly, structure modeling indicates that the catalytic center can assume a fold that is compatible with the functional requirements and secondly, if reciprocal BLAST with closely related species confirms that orthologous sequences also contain the conserved motifs and the catalytic centers.

A predictive 3D structure of the candidate CNCs can be constructed using homology modeling of candidate CNCs with known structures deposited in the Protein Data Bank (PDB) (http://www.rcsb.org/pdb/home/home.do). The key steps involved in protein homology modeling are template structure selection, the construction of protein models based on selected template, and the verification of these models [16]. A structural model can provide important clues to the protein function, which in this case is the CNC activity. Here we propose the use of the "Modeller" software [17] to construct 3D models for candidate CNCs. Orthologous sequences found in reciprocal BLAST provide additional information on the evolutionary conservation of these catalytic centers considered essential for CNC function and together with the predicted 3D models, they can further support the search for novel candidate CNCs in plants.

The method outlined here is particularly straightforward when working with *Arabidopsis thaliana* mainly because of the many online tools that are freely available in the public domain. However, in principle it can be applied successfully to many other species as long as a significant amount of sequence data of the species is available.

2 Materials

2.1 Homology Modeling and Search for Orthologs

1. Download and install the "Modeller 9.10" software from the Web site http://salilab.org/modeller/download_installation.html.
2. The candidate example proteins modeled in Subheadings 3.2 and 3.3 are the brassinosteroid receptor AtBRI-GC (At4g3900), an annotated monooxigenase AtNOGC1 (At1g62580), and a diacylglycerol kinase AtDGK4 (At5g57690).
3. Download the respective protein crystal structures from the PDB Web site at http://www.rcsb.org/pdb/home/home.do. The reference crystal structures in Subheadings 3.2 and 3.3 are the catalytic domain of a eukaryotic guanylate cyclase (PDB entry: 3ET6), the bacterial nitric oxide sensor (PDB entry: 1XBN), and the *Escherichia coli* lipid kinase (*YegS*) (PDB entry: 2BON).
4. Download the "UCSF Chimera" software at http://www.cgl.ucsf.edu/chimera/download.html.

3 Methods

The method detailed here is designed to identify candidate CNCs in *Arabidopsis thaliana*; however ACs and GCs in other species can be inferred if the orthologs of these candidate CNCs also contain the search motif or relaxed search motifs.

3.1 Search for Candidate CNCs in Arabidopsis

1. Download annotated CNC sequences from protein data repositories (e.g., NCBI (http://www.ncbi.nlm.nih.gov) or UniProtKB/Swiss-Prot (http://web.expasy.org/docs/swiss-prot_guideline.html)) and select entries of the functional class of CNC of interest (*see* **Note 1**). Consider both entries from closely related and distantly related species (Fig. 1a).
2. Identify the catalytic center (or other domains of interest) of the annotated CNC, align them with an alignment program (e.g., "ClustalX" available at http://www.clustal.org/clustal2/), and curate the alignment by hand so that a motif can be built (Fig. 1a).
3. The search motif/search pattern is built by including all amino acids in the vertical alignment. Gaps of various lengths can be included and undefined amino acids are marked as "X" (*see* **Note 2**).
4. Once a search motif (pattern) has been built, open the TAIR Web site (www.arabidopsis.org), pull down the "Tools" menu, and then go to the "Patmatch" function. In the "Patmatch" function choose peptide sequence of pattern and enter the search pattern. In the first instance the default options should be applied, searching "TAIR10" proteins with "no mismatch" allowed. The searches should begin with the most stringent motif and in subsequent searches the stringency can be relaxed, e.g., by increasing the gaps or omitting residues (Fig. 1b) that may not be essential for catalysis (*see* **Note 3**).
5. Particularly given that the number of hits for a relaxed motif can be large, it is worth considering additional secondary search criteria to narrow the search. These may include, e.g., a glycine-rich domain immediately N-terminal of the catalytic center and/or a pyrophosphate (PPi)-binding motif that consists of an arginine (R) flanked by aliphatic amino acids (20-30 AA) N-terminal of the catalytic center.
6. A search with two or more motifs, e.g., for the detection of candidate gas sensing CNC, can be performed (*see* **Note 4**).

3.2 Search for Ortholog Sequences

1. Open the NCBI Web site (http://www.ncbi.nlm.nih.gov), insert the name of the desired protein in the search column, and search against the protein database.
2. On the results page, click on the correct protein match and then select "run BLAST" function at the right column.

3. On the BLAST page, select the "non-redundant protein sequences" option under the search set database and the "BLASTP" (protein–protein BLAST) option under the program selection. Then hit the BLAST button (*see* **Note 5**).
4. To increase the stringency of the search to identify a particular domain(s) within these search results (for example, the CNC and/or the H-NOX domains), repeat the BLAST search selecting the "phi-BLAST" (Pattern Hit Initiated BLAST) option in the program selection.
5. Insert the predefined CNC and/or H-NOX motifs, and hit BLAST (*see* **Note 6**). This will identify ortholog sequences that also harbor the predicted CNC and/or H-NOX domains (*see* **Note 7**).
6. Analyze the ortholog list and make rational inferences by, for example, considering the species of the orthologs and the evolutionary conservation of the predicted domain in the orthologs. For example, a phi-BLAST search with the H-NOX motif within the ortholog list (approximately 100 orthologs) of a full-length AtDGK4 (TAIR entry: At5g57690) BLAST search returns only eight candidates, seven of which are plant dicots. This suggests a highly conserved kinase catalytic center across species, but a gas-binding domain that is unique only to plants and specifically the dicots. If this is true, then the plant DGKs (at least in the dicots) have evolved to be gas-sensing molecules.

3.3 Homology Modeling

1. Open the NCBI Web site (http://www.ncbi.nlm.nih.gov), insert the name of the desired protein in the search column, and perform a search against the protein database.
2. In the results page, choose the correct protein match and select "FASTA" function beneath the protein title. Save the FASTA file of the desired protein by clicking the "Send to" option on the right of the protein title (*see* **Note 8**).
3. Also in the previous BLAST result page, choose the correct protein match, and then analyze the protein sequence by selecting the "run BLAST" function at the right column of the Web site.
4. Perform a BLAST search of the desired protein sequence against the "Protein Data Bank proteins" database and with the "BLASTP" (protein–protein BLAST) program selection. Hit the BLAST button and the Web site will return a list of known related crystal structures arranged by default in descending order of the "Expect value (E-value)." One can also sort the list in the order of "max score," "max identity," or "query coverage" by clicking on the respective title headers. In the case of CNC candidates, modeling is done against templates with the following criteria in descending order of priority: (1) sequence identity, (2) sequence coverage, and (3) max score (*see* **Note 9**).

5. Select the desired template structures from the previous BLAST search, then go to the "Protein Data Bank" Web site (http://www.rcsb.org/pdb/home/home.do), and download the respective template PDB text files. To do this, insert the PDB IDs of the templates on the search bar and in the resulting page, pull down the "download files" function on the right column and select PDB file (text) to save the template protein structures (*see* **Note 10**).
6. Download the software "Modeler 9.10" and register for a license at http://salilab.org/modeller/download_installation.html (*see* **Note 11**).
7. Open the "Modeller" application file, and run Scripts 1–5 for a complete modeling of the candidate protein. The "Modeller" scripts are written in the python programming language. Examples of scripts can be downloaded from http://salilab.org/modeller/tutorial/basic.html (*see* **Note 12**).
8. Follow the instructions detailed on the "Modeller" tutorial page at http://salilab.org/modeller/tutorial/basic.html to run the "Modeller" scripts. Script 1 instructs "Modeller" to search for template structures related to the protein of interest; script 2 allows the user to select one or more suitable templates; script 3 aligns the protein of interest to the selected template; script 4 builds the models, while script 5 validates the quality of the constructed models.
9. View and assess the "best" model in "UCSF Chimera" or with other protein structural visualization software. The "best" model can be determined by assessing the molpdf, DOPE, and GA341 scores from the log file of script 5. In the case of the candidate CNCs, additional model evaluation using the "Ramachandran" plot can be performed by uploading the PDB file of the "best" model at http://mordred.bioc.cam.ac.uk/~rapper/rampage.php (*see* **Note 13**).
10. These modeling procedures are for model building when only the amino acid sequence of the protein of interest is known. In reality, template structures and/or alignments may have already been performed in another program. It is at the user's discretion to skip one or more steps accordingly.
11. When assessing the 3D model, highlight the predicted functional domains (e.g., catalytic center of CNCs, H-NOX domain, or ATP-binding site) and analyze the structural properties such as the shape and conformation of the function domain, spatial and hydrophobic interactions, and domain organizations within the molecule to determine functional compatibility. Also, assess these properties against known structures for further verifications. For example, based on the structural model of the candidate GC AtBRI-GC (TAIR entry: At4g3900), the functionally assigned

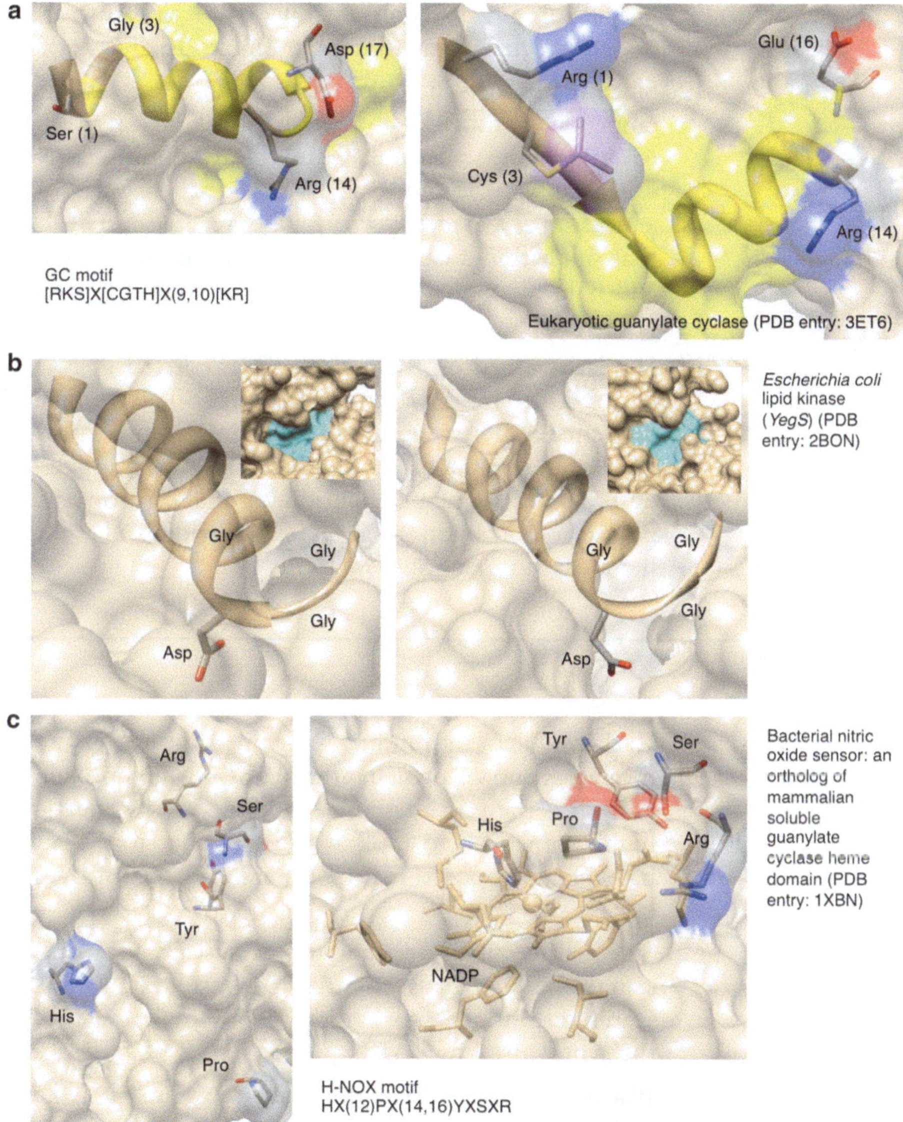

Fig. 2 Examples of protein models constructed using the "Modeller" software and how they compare to their respective reference crystal structures. (**a**) The AtBRI-GC catalytic center (*left*) was modeled against the *Pseudomonas syringae* AvrPtoB (PDB entry: 3TL8) and compared to the reference structure (*right*). The GC domain is highlighted in *yellow* and the functionally assigned residues of the GC catalytic center are detailed. The amino acid residue in position one (R/K/S) of the GC motif does the hydrogen bonding with GTP; the residue in position three (C/G/T/H) confers substrate specificity, while the position 14 amino acid stabilizes the transition state of GTP to cGMP. The 16th/17th residue (D/E) at the C-terminal of the 14-amino acid GC motif suggests binding to Mg^{2+}/Mn^{2+}. (**b**) The AtDGK4 kinase domain (*left*) was modeled against the crystal structure of *Salmonella typhimurium YegS* (PDB entry: 2P1R). The ATP-binding site highlighted in cyan and forms part of a helical coil that is buried in a cavity (insets) which is in agreement to that of the reference structure (*right*). (**c**) The AtNOGC1 gas-binding domain (*left*) was modeled against the crystal structure of a bacterial flavin-containing monooxygenase (PDB entry: 2VQ7). Functionally important residues are represented and their interactions with NADP are shown in the reference structure (*right*). All images are created using the "UCSF Chimera" software. Note that the AtBRI-GC has GC activity in vitro [12]; the AtDGK4 is annotated as diacylglycerol kinase (TAIR) and has an ATP-binding site that reflects the "GXGG" nucleotide-binding consensus sequence and the AtNOGC1 has GC and gas-binding activity in vitro

residues are organized in a manner that is similar to the reference crystal structure of a eukaryotic soluble GC (Fig. 2a).

12. Since many identified candidate CNCs have catalytic domains embedded within a kinase, the kinase region can also be modeled to evaluate the structural features of the multifunctional protein. For example, the model for a candidate GC AtDGK4 (TAIR-annotated as a diacylglycerol kinase) reveals an ATP-binding site that sits within a cavity presumably ideal for catalytic function, and is in agreement with other known lipid kinase structures (Fig. 2b).
13. Similarly, based on the model, the predicted H-NOX domain of AtNOGC1 (TAIR entry: At1g62580) is compared to the reference NO-sensing crystal structure from bacteria (Fig. 2c). The arrangement of the functionally important residues suggests a plausible heme environment that can conceivably incorporate a porphyrin ring, hence providing the rationale for heme–gas interactions (*see* **Note 14**).

4 Notes

1. Cyclic nucleotide-binding proteins and CNC are a highly diverse group of proteins that has been functionally classified [18]. It is therefore essential to study the classification and it is highly beneficial to look at the sequence logos (motifs) that have been proposed for the different classes of the nucleotide cyclase superfamily.
2. Since different search engines and search tools vary in the syntax they use, it is essential to consult the sample motif or search term for each program. The examples given (Fig. 1) use the syntax used by TAIR.
3. If a motif search does not return any hits, it is advisable to include chemically related amino acids into the search pattern. An example of such inclusions/substitutions of a chemically related amino acid is isoleucine (I in position 4) and leucine (L in position 9) (Fig. 1a). The latter has led to the discovery of AtGC1 (TAIR entry: At5g05930.1).
4. The Protein Information Resource (PIR) (http://pir.georgetown.edu/pirwww/search/pattern.shtml) also allows online pattern searches in UniProt and is particularly useful for a combined search with more than one motif/pattern. Note that the pattern should be entered in both orientation (A–B and B–A) and that the gap between the two motifs can be chosen. Note that the syntax is different from the one used in TAIR; hence consult the instructions in "user-defined pattern."

5. A BLAST with the full-length CNC will return sequences across species, hence providing information about the evolutionary diversity of the protein of interest. To locate specific domains within these results, the respective motifs must be included in the phi-BLAST function.
6. The phi-BLAST function is sensitive to only a specific set of pattern syntax rules that can be obtained at http://www.ncbi.nlm.nih.gov/blast/html/PHIsyntax.html.
7. The presence of CNC or H-NOX domains in the ortholog sequences provides a degree of confidence in the ability of the candidate proteins to perform their predicted functions. The type of species of the orthologs will provide clues about the evolutionary conservation or diversification of these predicted domains.
8. The amino acid sequence of the candidate protein is required for subsequent modeling procedures. The downloaded FASTA file of the protein sequence must be converted to PIR database format (http://salilab.org/modeller/9v8/manual/node454.html), which is recognized by the "Modeller" software. For an extensive tutorial on how to use the "Modeller" program, see http://salilab.org/modeller/tutorial/basic.html.
9. For a detailed description about the BLAST scores, please refer to the Fall/Winter 2006/07 (Vol. 15, Issue 2) NCBI newsletter available online at http://www.ncbi.nlm.nih.gov/Web/Newsltr/V15N2/BLView.html. Templates may have high percentage identity but low sequence coverage to the protein of interest and vice versa. In principle, templates with identity percentage of >50 % can generate good-quality models. However, an identity percentage of 16–30 % (depending on individual genomes) may be sufficient to construct reasonable models [16]. If a particular domain of the protein is of high importance to the overall function of the protein, then it is advisable to select templates with higher identity percentage to the region of interest of the protein. However, the selection of suitable templates is at the user's discretion.
10. The crystal structure of templates (in text file) is required for subsequent modeling procedures. Alternatively, template alignment and selection can also be performed by running "Script 1" on "Modeller." For a complete tutorial on how to use the "Modeller" program, please see http://salilab.org/modeller/tutorial/basic.html.
11. The "Modeller" software is free. After installation, a registration using an institutional e-mail address is required. For further download, installation, and registration instructions please see http://salilab.org/modeller/download_installation.html.
12. For noncomputer scientists/bio-informaticians, the downloaded scripts can be modified and applied for most modeling

applications. To do this, first change the script file extension from ".py" to ".txt." Secondly, open the text script files and replace the names of the default protein and templates to user-specified protein file names. Then, save the edited script files in python format (.py), readable by "Modeller."

13. In the log file for script 5, the "best" model is determined by the lowest Modpdf and DOPE scores or the highest GA431 score. The GA341 score ranges from 0.0 (worst) to 1.0 (native-like). The Modpdf and DOPE scores are not absolute measures as they only indicate relative model quality, that is, they only rank models calculated from the same alignment. For assessment using the "Ramachandran" plot, a percentage of >90 % of residues falling in the allowed region usually indicates good-quality models.
14. The ability of AtBRI-GC to function as a GC has been proven experimentally [12] while in vitro evidence also confirmed AtNOGC1 to be both a GC and a gas-sensing molecule biased towards NO [15]. The AtDGK4 has kinase catalytic domain that is highly conserved across species and has ATP-binding site that reflects the consensus sequence of "GXGG."

References

1. Neuhaus G, Bowler C, Hiratsuka K, Yamagata H, Chua NH (1997) Phytochrome-regulated repression of gene expression requires calcium and cGMP. EMBO J 16:2554–2564
2. Pharmawati M, Billington T, Gehring CA (1998) Stomatal guard cell responses to kinetin and natriuretic peptides are cGMP dependent. Cell Mol Life Sci 54:272–276
3. Gehring CA, Irving HR (2003) Natriuretic peptides—a class of heterologous molecules in plants. Int J Biochem Cell Biol 35:1318–1322
4. Kwezi L, Ruzvidzo O, Wheeler JI, Govender K, Iacuone S, Thompson PE, Gehring C, Irving HR (2011) The phytosulfokine (PSK) receptor is capable of guanylate cyclase activity and enabling cyclic GMP-dependent signaling in plants. J Biol Chem 286:22580–22588
5. Maathuis FJ, Sanders D (2001) Sodium uptake in Arabidopsis roots is regulated by cyclic nucleotides. Plant Physiol 127:1617–1625
6. Donaldson L, Ludidi N, Knight MR, Gehring C, Denby K (2004) Salt and osmotic stress cause rapid increases in Arabidopsis thaliana cGMP levels. FEBS Lett 569:317–320
7. Pasqualini S, Meier S, Gehring C, Madeo L, Fornaciari M, Romano B, Ederli L (2009) Ozone and nitric oxide induce cGMP-dependent and -independent transcription of defence genes in tobacco. New Phytol 181:860–870
8. Qi Z, Verma R, Gehring C, Yamaguchi Y, Zhao Y, Ryan CA, Berkowitz GA (2010) Ca^{2+} signaling by plant *Arabidopsis thaliana* Pep peptides depends on AtPepR1, a receptor with guanylyl cyclase activity, and cGMP-activated Ca^{2+} channels. Proc Natl Acad Sci USA 107:21193–21198
9. Leng Q, Mercier RW, Yao W, Berkowitz GA (1999) Cloning and first functional characterization of a plant cyclic nucleotide-gated cation channel. Plant Physiol 121:753–761
10. Ludidi N, Gehring C (2003) Identification of a novel protein with guanylyl cyclase activity in Arabidopsis thaliana. J Biol Chem 278:6490–6494
11. Meier S, Seoighe C, Kwezi L, Irving H, Gehring C (2007) Plant nucleotide cyclases: an increasingly complex and growing family. Plant Signal Behav 2:536–539
12. Kwezi L, Meier S, Mungur L, Ruzvidzo O, Irving H, Gehring C (2007) The Arabidopsis thaliana brassinosteroid receptor (AtBRI1) contains a domain that functions as a guanylyl cyclase in vitro. PloS One 2:e449
13. Gehring C (2010) Adenyl cyclases and cAMP in plant signaling—past and present. Cell Commun Signal 8:15
14. Liu Y, Ruoho A, Rao V, Hurley J (1997) Catalytic mechanisms of the adenyl and guanylyl cyclases: modelling and mutational analysis. Proc Natl Acad Sci USA 94:13414–13419
15. Mulaudzi T, Ludidi N, Ruzvidzo O, Morse M, Hendricks N, Iwuoha E, Gehring C (2011)

Identification of a novel Arabidopsis thaliana nitric oxide-binding molecule with guanylate cyclase activity in vitro. FEBS Lett 585:2693–2697

16. Krieger E, Nabuurs SB, Vriend G (2005) Homology modeling. Structural bioinformatics. Wiley, New York, pp 509–523
17. Eswar N, Webb B, Marti-Renom MA, Madhusudhan MS, Eramian D, Shen MY, Pieper U, Sali A (2001) Comparative protein structure modeling using MODELLER. Current protocols in protein science. Wiley, New York, Chapter 2, Unit 2.9.
18. McCue L, McDonough K, Lawrence C (2000) Functional classification of cNMP-binding proteins and nucleotide cyclases with implications for novel regulatory pathways in Mycobacterium tuberculosis. Genome Res 10:204–219

Chapter 14

Identification of Cyclic Nucleotide Gated Channels Using Regular Expressions

Alice K. Zelman, Adam Dawe, and Gerald A. Berkowitz

Abstract

Cyclic nucleotide-gated channels (CNGCs) are nonselective cation channels found in plants, animals, and some bacteria. They have a six-transmembrane/one-pore structure, a cytosolic cyclic nucleotide-binding domain, and a cytosolic calmodulin-binding domain. Despite their functional similarities, the plant CNGC family members appear to have different conserved amino acid motifs within corresponding functional domains than animal and bacterial CNGCs do. Here we describe the development and application of methods employing plant CNGC-specific sequence motifs as diagnostic tools to identify novel candidate channels in different plants. These methods are used to evaluate the validity of annotations of putative orthologs of CNGCs from plant genomes. The methods detail how to employ regular expressions of conserved amino acids in functional domains of annotated CNGCs and together with Web tools such as PHI-BLAST and ScanProsite to identify novel candidate CNGCs in species including *Physcomitrella patens*.

Key words Cyclic nucleotide-gated channel, Cyclic nucleotide-binding domain, Hinge domain, Protein motif, Regular expressions, ScanProsite, PHI-BLAST, Ortholog identification, *Physcomitrella patens*

1 Introduction

Eukaryotic genomes encode diverse gene channel families, both those unique to certain kingdoms and those shared by all organisms. Cyclic nucleotide-gated channels (CNGCs) are cation-conducting channels found in animals, plants, and some bacteria [1–3] that conduct calcium, potassium, and sodium ions [4]. Plant CNGCs were first identified by their structural similarity to animal CNGCs and because they have a cyclic nucleotide-binding domain (CNBD) [5]. They are members of the P-loop superfamily, which is found in all organisms [6, 7]. The role of the pore as a selectivity filter of CNGCs has been studied extensively. This selectivity filter is not found in any other family of channels [8]. Animal and plant CNGCs contain six transmembrane (TM) domains, a pore, a

Chris Gehring (ed.), *Cyclic Nucleotide Signaling in Plants: Methods and Protocols*, Methods in Molecular Biology, vol. 1016, DOI 10.1007/978-1-62703-441-8_14,

CNBD, and a calmodulin-binding domain (CaMBD). These domains have been used to investigate the evolutionary relationships between *Arabidopsis thaliana* CNGCs (AtCNGCs). On the basis of phylogenetic analyses the CNGCs have been divided into 5 groups: I, II, III, IVa, and IVb [9]. AtCNGC2 and AtCNGC4 form a clade based on alignments of the full amino acid sequences, the pore sequence, and the pore plus the sixth TM domain [9, 10].

When cyclic nucleotides (cyclic GMP and cyclic AMP) bind the CNBD, they allosterically cause the channel to open [11, 12]. In plant CNGCs, the CaMBD and CNBD partially overlap at the C-terminal cytosolic region. In the presence of calcium, calmodulin binds to the CaMBD and prevents the binding of cyclic nucleotides to the CNBD. Calmodulin is proposed to occlude the binding site of the cyclic nucleotide, and thereby prevent the activation of CNGCs by cyclic nucleotides [11, 13]. CNGC gene products are subunits of tetrameric functional channels. In animals, all CNGCs are heterotetramers formed by translation products of at least two different CNGC genes. They are presumed to form heterotetrameric channels in plants as well [14, 15]. Within the CNBD, C-terminal to the highly conserved phosphate-binding site (PBS), is a hinge motif that is the most highly conserved region of the CNGC sequences. The hinge motif is proposed to influence ligand selectivity and binding affinity [16].

Most experimental evidence for plant CNGCs comes from studies in Arabidopsis. In *A. thaliana*, of 56 open reading frames predicted to encode subunits of cation channels, twenty are CNGCs [6]. CNGCs have also been characterized in moss, barley, and tobacco [15, 17, 18]. Most annotations of plant CNGCs are based on sequence homology and particularly BLAST searches [5, 6, 9, 15, 18–20] (*see* **Note 1**). Pairwise alignment tools such as BLAST yield a wealth of information about protein similarities. However, identifications of homologous genes using these tools are not always correct, and pairwise alignments may miss evolutionarily distant family members [21]. Members of protein families often share conserved motifs that are unique and which can be identified and used to search for as yet unidentified family members in databases. Some bioinformatic tools for querying protein databases predict orthologous sequences based on the presence of specific domains and motifs. ScanProsite [22], hosted by ExPASy, and PHI-BLAST (Pattern Hit Initiated BLAST), hosted by the National Center for Biological Information (NCBI), perform searches using regular expressions in protein sequences. Regular expressions can be used to describe and find motifs with specific conserved amino acids, or sets thereof, at particular positions while also allowing for the presence of any amino acid at non-conserved positions. They therefore facilitate searches where amino acids at different positions can be given a different weighting. Because regular expressions are constructed based only on known sequences, the constructed pattern may be found too stringent when new family members are characterized. In this case revisions to the motif can be made, based on rational criteria such as the inclusion of chemically similar amino acids into the motif, to make it more permissive [21]. The regions matching the motifs can be aligned and analyzed to investigate evolutionary relationships between family members.

To ensure that novel CNGC family members are found, it is advantageous to use several search methods to identify orthologs. This chapter outlines the use of the *phmmer* [23] and BLASTp programs publicly accessible on Web servers to find putative CNGCs. Two other programs, ScanProsite and PHI-BLAST, can be used to query databases for the presence of CNGC-specific motifs. In this chapter we also construct regular expressions as queries to ScanProsite and PHI-BLAST. We create and analyze phylogenetic trees from alignments of regions identified as putative homologs by the regular expressions to evaluate the relationships between CNGCs. We use the moss *Physcomitrella patens* as an example, but these methods are also suitable for analyzing other plant species if there is a substantial amount of transcriptome or proteome data available.

2 Materials

The searches are done on protein sequences rather than DNA sequences. Amino acids carry the conserved biological function, and the difference in codon usage would make DNA pattern searches unnecessarily complex.

1. The reference sequences in Subheadings 3.2 and 3.3 below are the 20 *A. thaliana* CNGC proteins. From The Arabidopsis Information Resource (TAIR) at www.arabidopsis.org [24] download the twenty Arabidopsis CNGC sequences by searching for "CNGC" as a search term with the "genes" option selected. Select "get all sequences," which directs to the Sequence Bulk Download and Analysis page. The 20 CNGC "Locus/Gene Model Identifiers or Sequences" field will list the AGI locus identifiers: AT5G53130 (CNGC1), AT5G15410 (CNGC2), AT2G46430 (CNGC3), AT5G54250 (CNGC4), AT5G57940 (CNGC5), AT2G23980 (CNGC6), AT1G15990 (CNGC7), AT1G19780 (CNGC8), AT4G30560 (CNGC9), AT1G01340 (CNGC10), AT2G46440 (CNGC11), AT2G46450 (CNGC12), AT4G01010 (CNGC13), AT2G24610 (CNGC14), AT2G28260 (CNGC15), AT3G48010 (CNGC16), AT4G30360 (CNGC17), AT5G14870 (CNGC18), AT3G17690 (CNGC19), and AT3G17700 (CNGC20).

In the Dataset field, choose "AGI protein sequences" in the drop-down menu, and in Output Options choose Fasta as the output format. Choose the option for "Get one sequence per locus (representative gene model/splice form only)." Additionally, download the sequence for AtAKT1 (AT2G26650.1) to use as an outgroup for the phylogenetic analysis in Subheading 3.2. The AKT channels are six TM channels with a pore in the same position as the CNGC pore, and they are members of the P-loop superfamily [9].

2. Download the following moss sequences from the sequence retrieval resource (under the "Genome" menu option) on the *P. patens* bioinformatics resource cosmoss.org (http://cosmoss.org/bm/retrieval?type=1). The following are the names assigned [15] and the accession IDs (in parentheses—use these for retrieval) from cosmoss.org [25]: CNGCa (Pp1s68_102V6.1), CNGCb (Pp1s90_245F4.1), CNGCc (Pp1s37_149V6.1), CNGCd (Pp1s99_40E1.1), CNGCe (Pp1s183_83E1.1), CNGCf (Pp1s189_85V6.1 and Pp1s189_85V2.1), CNGCg (Pp1s204_120E1.1), and CNGCh (Pp1s211_70E2.1). For this sequence list, ensure that the "P.patens.all_models_proteins" option is selected for the database field. This database was frozen on 13.09.2010 as of this writing. Bioinformatic approaches have been used recently [15] to annotate these sequences as CNGCs with functional studies in some cases confirming the assignment.

3. Download MEGA5 from http://www.megasoftware.net/. MEGA5 [26] is the phylogenetic analysis program used in Subheading 3.2. Other software may be used if desired.

3 Methods

3.1 Identification of Candidate CNGCs

1. In order to facilitate the identification of orthologous CNGCs in other species, it is useful to find the common patterns present in all members of the family and to search for these patterns in protein databases to find additional family members in unannotated sequences. ScanProsite and PHI-BLAST both search for hits that match a specified regular expression. PHI-BLAST also requires the input of a representative sequence and will only return hits that share similar domains to this sequence. Choose a representative CNGC to use as a query for PHI-BLAST. AtCNGC2 and AtCNGC4 have a pore which is the most different from other types of channels in plants and animals, so AtCNG2 was chosen as a representative CNGC for this example. On the PHI-BLAST search page (http://blast.ncbi.nlm.nih.gov/Blast.cgi?PAGE=Proteins), also found under "protein blast" on the NCBI BLAST tools

page, select "PHI-BLAST" as the "Algorithm" to use and enter the representative AtCNGC sequence into the protein sequence template entry box. Check the boxes to search against the nr database, which will search both confirmed and putative polypeptide products. In the pattern entry box, enter the regular expression [LI]-X(2)-[GS]-X-[FYIVS]-X-G-X(0,1)-[DE]-LL-X(8,25)-S-X(10)-E-X-F-X-[IL], which was recently described as a regular expression that spanned the hinge and PBC in the CNBD of rice, Arabidopsis, and Selaginella. This motif identifies CNGCs and no other proteins [27]. Listed in the square brackets "[]" are the amino acids allowed at this position of the motif. Round brackets "()" indicate the number of amino acids in between conserved residues. "X" represents any amino acid. Two numbers appearing in the round brackets indicate a range. For example, "X(2,9)" indicates that a number of amino acids between two and nine may be considered a match. Limit the taxonomy to "Physcomitrella" by entering this in the "Organism" field. Perform the search. This search yields, at present, four sequences. Save the sequences in text or Fasta format. (Under the "Alignments" section, tick off the "Select All" box. Then choose "Get selected sequences." Click on "Send to" and choose "FASTA" format and click on "Create File.")

2. Choose a representative CNGC to use as a query for *phmmer* and upload its amino acid sequence to the query box on the HMMER Webpage (http://hmmer.janelia.org/search/phmmer) search tab (*see* **Note 2**). Submit this sequence to *phmmer* using the default options (Click "Start" button). The output will be a list of sequences. Restrict the results to *P. patens* in the taxonomy tab. The search will yield results similar to Fig. 1a. Select the results and follow the links of the top hits to their UniProt pages. Clicking on the ">" sign in the leftmost column expands the results to provide more detail for sequences of interest. Download these sequences. Some hits will be annotated. From this example two of the hits are annotated as AKT channels. AKT channels could be expected to appear in the results, because they are closely related to CNGCs and are members of the P-loop superfamily [9]. The search described here yielded an additional hit (a short putative protein) with a lower *E*-value than the AKT channels (NCBI Reference Sequence: XP_001772381.1). In Subheading 3.2 these hits will be aligned to confirm that they are not CNGCs.
3. Align the cosmoss.org sequences downloaded in Subheading 3.1 and the *phmmer* result protein sequences in MEGA5, using MUSCLE [28]. Note whether they are identical.
4. Use BLASTp (http://blast.ncbi.nlm.nih.gov/Blast.cgi?PAGE=Proteins) to repeat the search on the representative AtCNGC

a

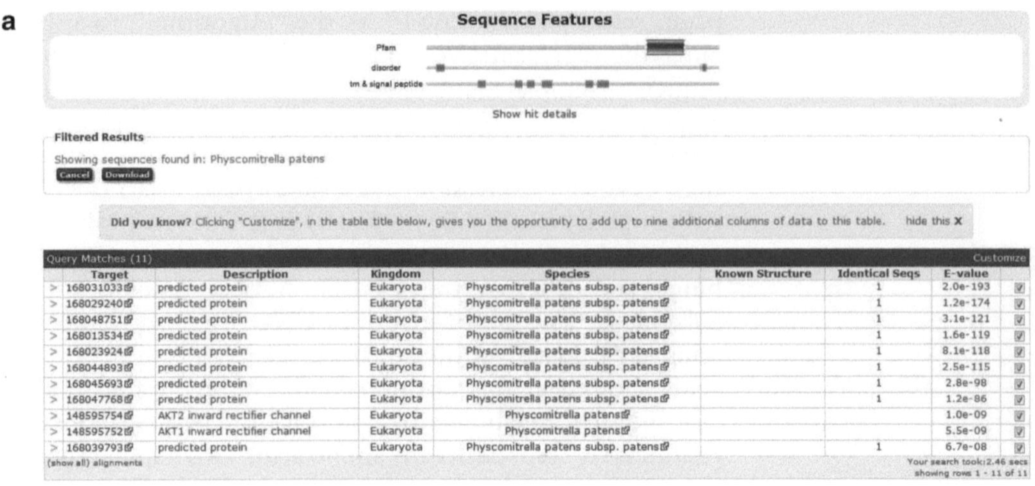

Target	Description	Kingdom	Species	Known Structure	Identical Seqs	E-value
168031033	predicted protein	Eukaryota	Physcomitrella patens subsp. patens		1	2.0e-193
168029240	predicted protein	Eukaryota	Physcomitrella patens subsp. patens		1	1.2e-174
168048751	predicted protein	Eukaryota	Physcomitrella patens subsp. patens		1	3.1e-121
168013534	predicted protein	Eukaryota	Physcomitrella patens subsp. patens		1	1.6e-119
168023924	predicted protein	Eukaryota	Physcomitrella patens subsp. patens		1	8.1e-118
168044893	predicted protein	Eukaryota	Physcomitrella patens subsp. patens		1	2.5e-115
168045693	predicted protein	Eukaryota	Physcomitrella patens subsp. patens		1	2.8e-98
168047768	predicted protein	Eukaryota	Physcomitrella patens subsp. patens		1	1.2e-86
148595754	AKT2 inward rectifier channel	Eukaryota	Physcomitrella patens			1.0e-09
148595752	AKT1 inward rectifier channel	Eukaryota	Physcomitrella patens			5.5e-09
168039793	predicted protein	Eukaryota	Physcomitrella patens subsp. patens		1	6.7e-08

b [LI]-X(2)-[GS]-X-[VFIYS]-X-G-X(0,1)-[DE]-L-[LI]-X-[WN]-X(6,32)-[SA]-X(9)-[VTI]-[EN]-[AG]-F-X-[LI]

Fig. 1 (**a**) An example of results from a *phmmer* query using an HMM generated from the results of a search on AtCNGC10. The sequences resulting from the query are analyzed in Subheadings 3.2 and 3.3. (**b**) The region of the eight moss sequences and twenty Arabidopsis sequences that matched the regular expression above was aligned in MEGA5 using ClustalW and a Gonnet matrix, and colored with BOXSHADE. This region spans the PBC and hinge domain and is the most highly conserved region of the putative CNGCs

on the nr database (select "blastp" as the "Algorithm" in the "Program Selection" section). This search will yield a greater number of alignments than *phmmer*. Select a cutoff expectation (*E*) value (under "Algorithm Parameters," "Expect threshold") to discount all alignments falling above it. *E*-values are the number of sequences expected to align with the query sequence with a given score by chance, so lower numbers represent an increased likelihood of homology. While an *E*-value below 0.001 can be considered significant, these hits are not likely to all be family members. The top hits of the BLASTp search should be identical to the results of the *phmmer* search. If additional hits are found that produced alignments with a high degree of coverage and similarity to the query template, it is recommended to download these hits as well to test in Subheading 3.2. On the BLASTp page, regions of high similarity between the query and target are represented by red segments in the graphical display shown at the top of the results page.

5. Construct an alignment of the cosmoss.org sequences downloaded in Subheading 2.2 and the BLASTp result protein sequences. Select the sequences with no missing regions to use for further analysis. At the time of writing the cosmoss.org-derived sequences were the longer of the two sets, demonstrating that it is advantageous to check several database sources before attempting analysis (especially species-specific genomic/proteomic databases where available). Here the cosmoss-derived full-length sequences were used for the identification of putative moss CNGCs.
6. Discard results that are fragments or duplicates.

3.2 Using Phylogenetic Analysis to Investigate Relationships of CNGCs

1. Align the full-length sequences of the selected moss and Arabidopsis proteins from Subheading 3.1. Construct a cladogram using MEGA5. From this cladogram discount any sequences that group with AtAKT1 (the outgroup) or that are basal to the outgroup. The PpAKT and XP_001772381.1 sequences are in this category. They can be discarded from further analysis.
2. Phylogenies constructed from functional domain sequences, as opposed to full-length sequences, can be informative. Construct an alignment of the full-length Arabidopsis CNGCs and the moss sequences downloaded in Subheading 2.2 (*see* **Note 3**). From this alignment, note the most conserved regions of the CNGC sequences which correspond to the pore, CNBD, and CaMBD. Select the region spanning the regular expression in Subheading 3.1 in all sequences (Fig. 1b). The motif is not present in four of the moss sequences, but the region can be selected in the multiple alignment in MEGA5. Realign and generate a phylogenetic tree from the alignment spanning just

this segment. Several phylogenetic analysis methods should be used to determine the reliability of the tree [21]. Here we have generated trees using Maximum Likelihood (Fig. 2) and Neighbor-Joining methods (Fig. 3). In both of these trees, the topology indicates that there are two groups of moss CNGCs: four sequences align with Group IV, and the other four form a separate group that is distributed within a clade with Arabidopsis groups. In the trees constructed by the majority of methods, AtCNGC4, AtCNGC2, PpCNGCb, PpCNGCd, and PpCNGCf consistently form a clade, as do PpCNGCg, AtCNGC19, and AtCNGC20. PpCNGCa, PpCNGCc, PpCNGCe, and PpCNGCh form a clade that is interspersed with Arabidopsis CNGCs. Trees were also constructed using the Maximum Parsimony method. These trees did not accord with the trees constructed by the other methods (not shown). The trees depicted in the figures are representative of the topology seen in the majority of trees. The illustrated trees seem to be more reliable, as they place AtCNGC2 and AtCNGC4 in a clade containing no other AtCNGCs, as previously published [9].

3. Generate cladograms from alignments of the pore region (Fig. 4) and from the CNBD (Fig. 5), and compare them to the cladograms generated from just the motif sequence segment. The trees are identical and the topologies of the motif-only, pore, and CNBD trees accord with the full-length sequence tree. This may suggest that the CNGC-specific motif and the CNBD of which it is a part have evolved similarly to the full-length sequences.
4. The cladograms based on just the motif region, as well as those based on the pore region and CNBD, are almost congruent to the tree published previously when complete sequences of the Arabidopsis and moss candidate CNGCs were aligned [15]. A previously published cladogram grouped four moss CNGCs (PpCNGCa, PpCNGCc, PpCNGCe, and PpCNGCh) as an outgroup to the rest of the Arabidopsis and moss sequences. However, in the trees in Figs. 2, 3, 4, and 5, PpCNGCa, PpCNGCc, PpCNGCe, and PpCNGCh form a clade that is interspersed with Arabidopsis sequences and separate from the clade containing PpCNGCb, PpCNGCg, PpCNGCd, PpCNGCf, AtCNGC2, and AtCNGC4. PpCNGCi was discarded because it lacks the functional domains critical to CNGC function. It was suggested that CNGCs predate the divergence of land plants [15]. The phylogenetic trees in Figs. 2, 3, 4, and 5 agree with this hypothesis. This offers evidence that the predicted moss sequences are indeed CNGCs. PpCNGCa, PpCNGCc, PpCNGCe, and PpCNGCh may comprise a new subfamily of plant CNGCs aside from Groups I, II, II, IV-A, and IV-B proposed earlier [9].

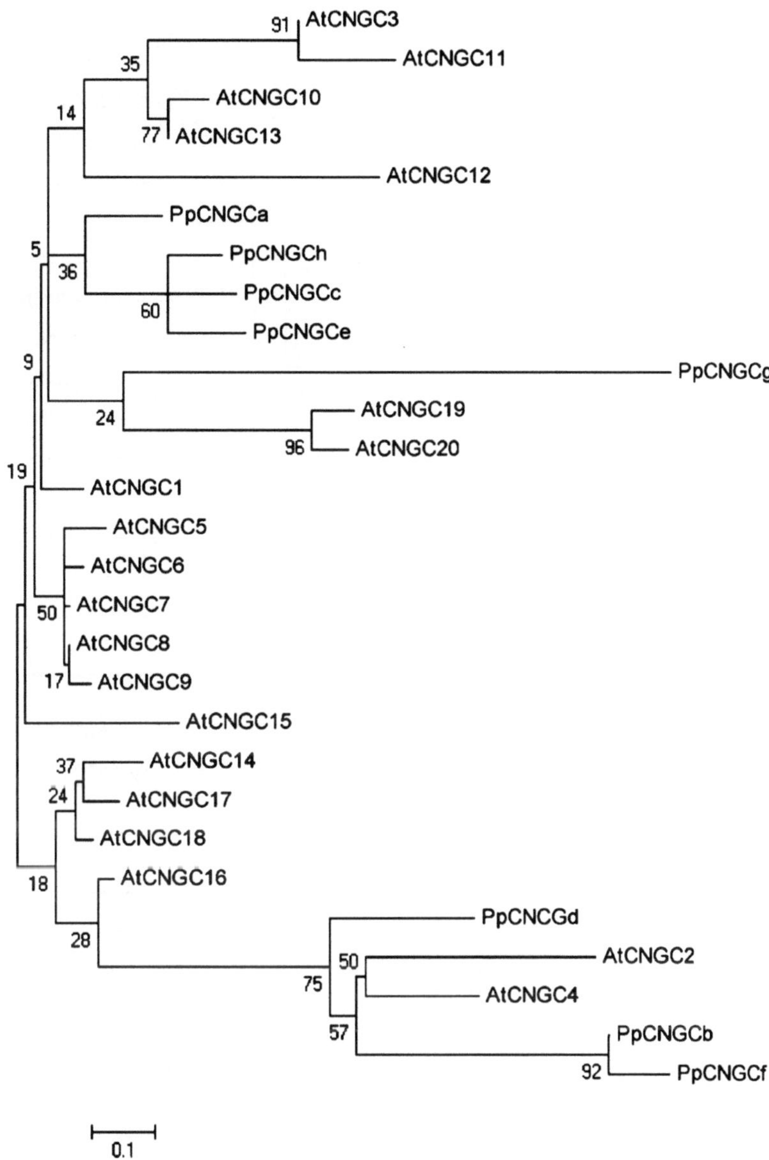

Fig. 2 Molecular phylogenetic evolution of the region spanning a CNGC-specific motif inferred by the Maximum Likelihood method based on the JTT matrix-based model [29]. Sequences were aligned using MUSCLE [28]. The bootstrap consensus tree inferred from 500 replicates is taken to represent the evolutionary history of the taxa analyzed [30]. Branches corresponding to partitions reproduced in less than 50 % bootstrap replicates are collapsed. The percentage of replicate trees in which the associated taxa clustered together in the bootstrap test (500 replicates) is shown next to the branches. Initial tree(s) for the heuristic search was obtained automatically as follows. When the number of common sites was <100 or less than one-fourth of the total number of sites, the Maximum Parsimony method was used; otherwise BIONJ method with MCL distance matrix was used. The tree is drawn to scale, with branch lengths measured in the number of substitutions per site. The analysis involved 28 amino acid sequences. All positions containing gaps and missing data were eliminated. There were a total of 34 positions in the final dataset

Fig. 3 Molecular phylogenetic evolution of the region spanning a CNGC-specific motif inferred by the Neighbor-Joining method [31]. Sequences were aligned using MUSCLE [28]. The optimal tree with the sum of branch length = 6.04757075 is shown. The percentage of replicate trees in which the associated taxa clustered together in the bootstrap test (500 replicates) is shown next to the branches [30]. The tree is drawn to scale, with branch lengths in the same units as those of the evolutionary distances used to infer the phylogenetic tree. The evolutionary distances were computed using the JTT matrix-based method [29] and are in the units of the number of amino acid substitutions per site. The analysis involved 28 amino acid sequences. All positions containing gaps and missing data were eliminated. There were a total of 125 positions in the final dataset

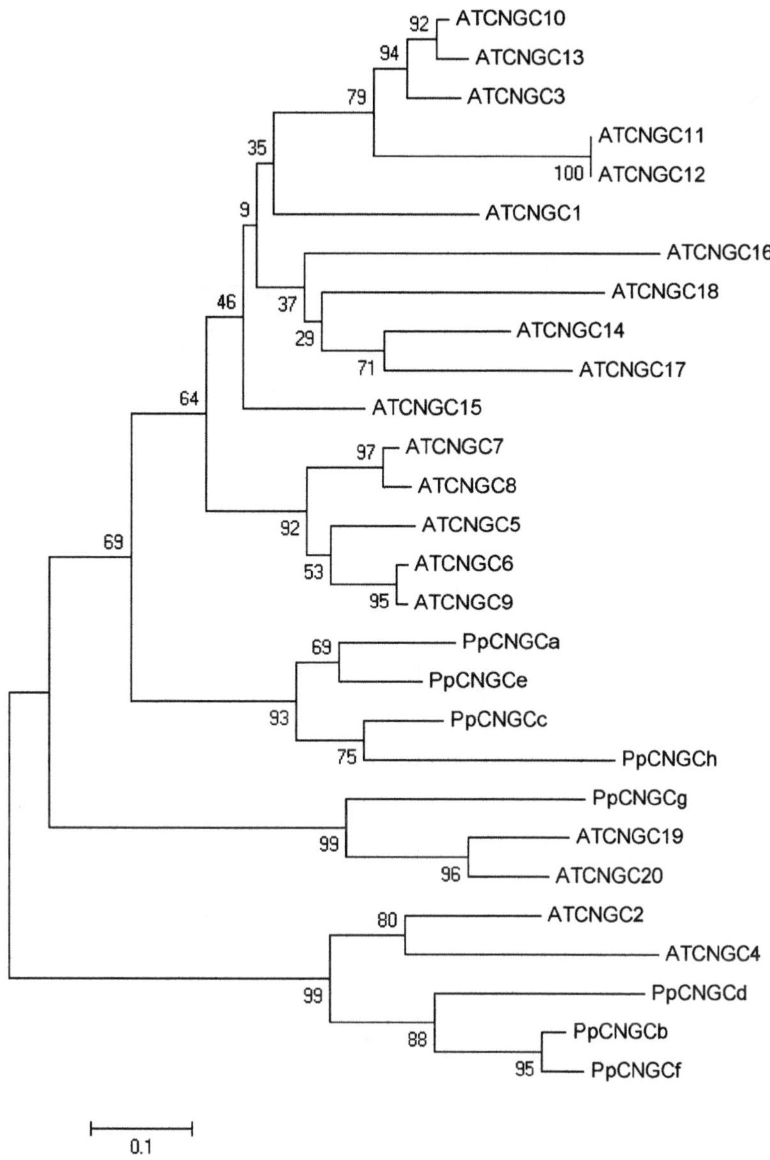

Fig. 4 Molecular phylogenetic evolution of the region spanning the pore and S6 region of plant CNGCs. The pore region was delineated according to Leng et al. analysis [8]. Sequences were aligned using MUSCLE [28]. The evolutionary history was inferred using the Neighbor-Joining method [31]. The optimal tree with the sum of branch length = 5.40139676 is shown. The percentage of replicate trees in which the associated taxa clustered together in the bootstrap test (500 replicates) is shown next to the branches [30]. The tree is drawn to scale, with branch lengths in the same units as those of the evolutionary distances used to infer the phylogenetic tree. The evolutionary distances were computed using the JTT matrix-based method [29] and are in the units of the number of amino acid substitutions per site. The analysis involved 28 amino acid sequences. All positions containing gaps and missing data were eliminated. There were a total of 53 positions in the final dataset

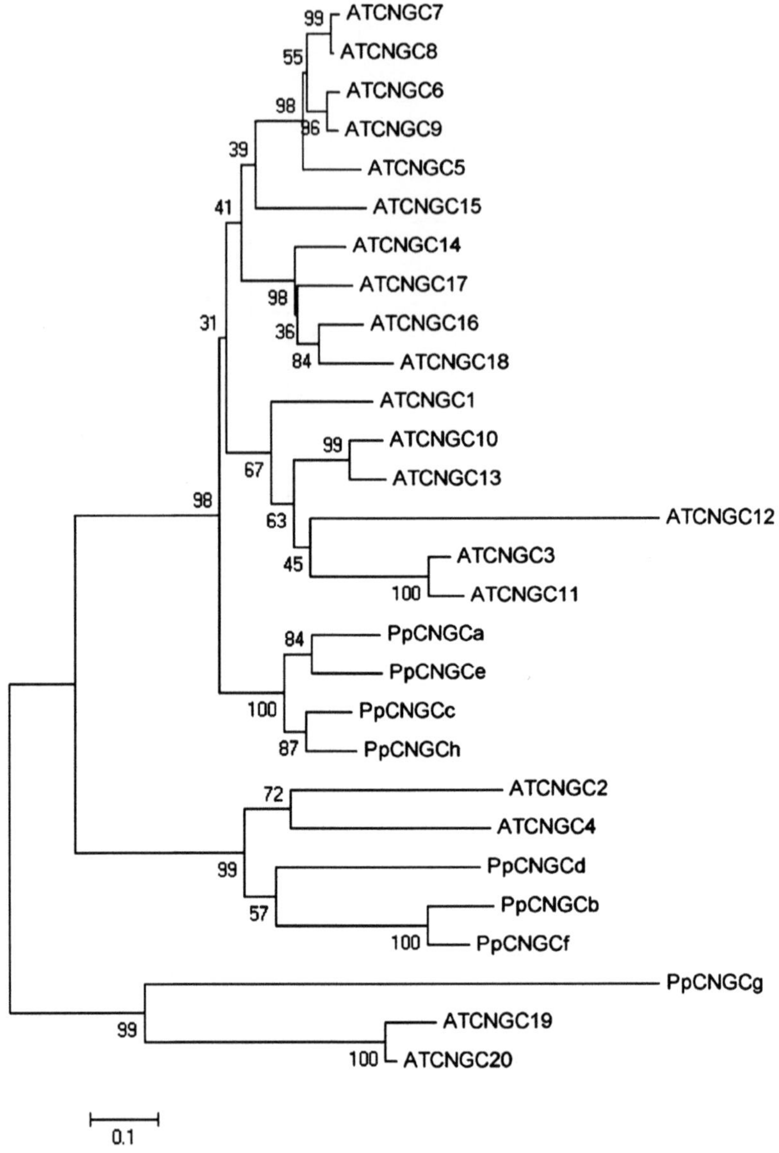

Fig. 5 Evolutionary relationships of CNBD regions of moss and Arabidopsis CNGCs. The CNBD region was taken from the sequence of PDB entry 1wgp [32], the CNBD from AtCNGC6. Sequences were aligned using MUSCLE [28]. The tree was inferred using the Neighbor-Joining method [31]. The optimal tree with the sum of branch length = 6.04757075 is shown. The percentage of replicate trees in which the associated taxa clustered together in the bootstrap test (500 replicates) is shown next to the branches [30]. The tree is drawn to scale, with branch lengths in the same units as those of the evolutionary distances used to infer the phylogenetic tree. The evolutionary distances were computed using the JTT matrix-based method [29] and are in the units of the number of amino acid substitutions per site. The analysis involved 28 amino acid sequences. All positions containing gaps and missing data were eliminated. There were a total of 125 positions in the final dataset

3.3 Relaxing a Stringent Motif

1. The corroboration of the putative moss CNGC sequences, based on the motif alignment and phylogram, necessitates the expansion of the CNGC-specific motif to a regular expression that recognizes the moss sequences in addition to the Arabidopsis sequences. There are many possible motifs of varying lengths that can be constructed, and longer regular expressions will provide greater selectivity. To expand the motif, choose residues that are highly conserved. The new motif we selected here is [LI]-X(2)-[GS]-X-[VFIYS]-X-G-X(0,1)-[DE]-L-[LI]-X-[WN]-X(6,32)-[SA]-X(9)-[VTI]-[EN]-[AG]-F-X-[LI]. Where longer gaps occur in the multiple sequence alignment of the motif region, consider expanding the range permitted to capture CNGC target sequences with longer insertions, as has been done for the regular expression designated here.
2. PHI-BLAST only returns hits that share similar domain architecture with a query protein sequence, while ScanProsite will return every sequence with an occurrence of the pattern. The aim of this step is to determine whether the motif is CNGC specific. Hence it is necessary to find every instance of the pattern. Test if the motif is stringent but sensitive enough to match all twenty AtCNGCs and all eight moss CNGCs, but no other proteins in these genomes. If necessary modify the motif. Search the UniProt and UniProt/TrEMBL databases for moss sequences containing the motif. At the top of the results page, the expected number of random sequences matching the motif is given. This number should be less than one. The ScanProsite server (http://prosite.expasy.org/scanprosite/) allows searches that will return any sequence containing the motif, and can be restricted to specific taxa (*see* **Note 4**). The cosmoss.org database lists a different sequence for PpCNGCg, which contains the new CNGC-specific motif. Currently the UniProtKB/TrEMBL and nr databases used by ScanProsite and PHI-BLAST contain a version of the PpCNGCg sequence that does not contain the motif and which is probably incorrect (*see* **Note 5**). Thus, seven results will be returned by ScanProsite. Follow the links for each result and check if these sequences are PpCNGCa, PpCNGCb, PpCNGCc, PpCNGCd, PpCNGCe, PpCNGCf, and PpCNGCh. This can be performed quickly by aligning the sequences from cosmoss with the sequences from the ScanProsite results and generating a cladogram.
3. If desired, test the motif against other plant genomes of interest. On the PHI-BLAST page, enter the regular expression describing the motif and a representative CNGC. The regular expression above identifies four CNGCs from pincushion spikemoss, *Selaginella moellendorffii*, and 44 from rice (*Oryza sativa*). These results, while suggestive, do not prove that the

CNGCs identified are acting as cyclic nucleotide-gated cation channels in plants, but these candidates can be studied experimentally following identification (*see* **Note 6**).

4 Notes

1. Bridges et al. [19] used a combination of BLAST and domain identification using the Simple Modular Architecture Research Tool (SMART) [33]. Ward et al. [6] used position-specific scoring matrices (PSSMs) to identify the number of CNGCs in Arabidopsis, rice, poplar, and the alga Chlamydomonas. One of the variants of BLAST, PSI-BLAST [34, 35], available as a Web interface or as a stand-alone in the BLAST2.27.2+ package from NCBI, constructs a PSSM from an input sequence and uses it to query a protein sequence database. The PSI-BLAST stand-alone is more versatile than the Web version. For example, it accepts multiple sequences or a multiple alignment as input to construct the PSSM for the first iteration. This improves the accuracy and sensitivity of the search.
2. The HMMER suite is available on the Web (http://hmmer.janelia.org) and includes programs that construct PSSMs from multiple sequence alignments and use profile Hidden Markov Models (HMMs). HMMs are statistically modelled machine-learning algorithms that are often used for sequence alignment. The HMMER suite includes tools for protein, HMM, multiple sequence alignment, and iterative search versus protein database searches, as well as protein sequence versus profile HMM searches.
3. A program, algorithm, and matrix of choice can be substituted, for example ClustalW using the BLOSUM or Gonnet algorithms. In the examples tested, using MUSCLE or ClustalW did not change the resulting phylogenetic analysis.
4. Submitting CNGC sequences to ScanProsite and searching for motifs in the library do not yield the CNBD, CaMBD, hinge, or PBS in the results. Other programs may be used to search for the CNBD, CaMBD, and TM domains. The latter can be predicted by TMpred, available from http://www.ch.embnet.org/software/TMPRED_form.html [36], and TMHMM from http://www.cbs.dtu.dk/services/TMHMM [37]. SMART is among the most reliable domain prediction programs for regulatory domains. The current versions of BLAST display conserved domain database search results (CD-Search) as an intermediate result before the search is completed and displayed [38]. When CNGC sequences are submitted to PSI-BLAST or BLASTp, 5 TM domains, a cNMP-binding domain, and a flexible hinge are predicted as conserved domains by

CD-Search. The current version of CD-Search fails to recognize the remaining TM domain and the CaMBD. ScanProsite is offered by ExPASy, which hosts the PROSITE database. It allows searches against the PROSITE database, which contains both regular expressions and profiles, but the database is incomplete [39]. The ScanProsite program is a resource available as either a stand-alone tool or on the Web. This program is used in several ways. It can either find sequences with a particular motif, query sequences against the existing PROSITE database of family-specific motifs, or find a motif in a set of user-supplied sequences. ScanProsite will output a motif with PROSITE syntax. A description of PROSITE syntax is available at http://prosite.expasy.org/scanprosite/scanprosite-doc.html. On the same server, the program PRATT is offered to construct regular expressions that describe a consensus sequence pattern that captures the most conserved amino acid residues of the input. PROSITE excels at finding short motifs that often correspond to functional residues, for example, phosphorylation sites, active sites, and docking sites [40], but the motifs themselves carry no information about function, only sequence conservation. Unless prior knowledge of the function of the residues in question exists, these motifs must be studied experimentally to deduce any biological relevance. For the purpose of studying CNGCs, the existing information about the CNBD and CaMBD regions complements motif generation and pattern searching. When a match to a CNGC-specific motif is found in a new sequence, the presence of a hinge within a PBS in the CNBD matching the regular expression would corroborate this identification.

5. UniProtKB/TrEMBL is a database that contains automatically added, computer-generated, annotated translations of predicted protein-coding regions. nr comprises nonidentical protein sequences from several protein databases.

6. Typically, CNGCs are functionally characterized by expression in heterologous systems, or by analysis of mutant plants which lack specific CNGC genes and which have cation-related phenotypes. Native plant membranes have been the subject of electrophysiological studies showing that, for example, AtCNGC2 is capable of conducting K^+, Li^+, Cs^+, and Rb^+, and that Ca^{2+} blocks K^+ conductance [8]. Studies of plasma membranes of native wild-type Arabidopsis and *dnd1* mutants, which carry a premature stop codon in AtCNGC2, showed that channel complexes containing CNGC2 were Ca^{2+} channels activated by cyclic AMP [41]. The membranes of yeast, bacteria, *Xenopus laevis* oocytes, or HEK293 cells heterologously expressing plant CNGCs have also been used for voltage clamping studies [42–45]. These studies have elucidated the

gating characteristics, voltage dependence, and selectivity of CNGCs. Measurements of various cation concentrations in the tissues of mutants versus wild-type plants reveal roles of CNGCs in cation accumulation. For example, measuring the contents of *Atcngc1*- mutants showed that they accumulate less Ca^{2+} and K^+ and that they are less sensitive to salt stress [46–49]. Similarly, *Atcngc1*- mutant shoots were less sensitive to salt and K^+. Hampton et al. [46] showed that in shoots, AtCNGC2, AtCNGC3, AtCNGC16, AtCNGC19, and AtCNGC20 mutants had reduced Cs^+ content whereas AtCNGC1, AtCNGC9, and AtCNGC12 mutants had increased Cs^+. The array of tools available for studying CNGCs provides the means to strengthen the tentative annotations assigned by bioinformatic analysis, and to expand our understanding of CNGCs' roles in plants.

Acknowledgements

The authors would like to thank Chris Gehring of the Division of Chemistry at King Abdullah University of Science and Technology for his valuable comments and editing. The authors would also like to thank Pascal LaPierre of the University of Connecticut Bioinformatics Facility for his advice and assistance with phylogenetic analyses. The authors also thank Ari Lawson for proofreading the manuscript.

References

1. Craven KB, Zagotta WN (2006) CNG and HCN channels: two peas, one pod. Annu Rev Physiol 68:375–401
2. Kaplan B, Sherman T, Fromm H (2007) Cyclic nucleotide-gated channels in plants. FEBS Lett 581:2237–2246
3. Cukkemane A, Seifert R, Kaupp UB (2011) Cooperative and uncooperative cyclic nucleotide gated ion channels. Trends Biochem Sci 36:55–64
4. Ma W, Yoshioka K, Gehring CA, Berkowitz GA (2010) The function of cyclic nucleotide gated channels in biotic stress. Ion channels and plant stress responses. Springer, Berlin, pp 159–174
5. Talke IN, Blaudez D, Maathuis FJ, Sanders D (2003) CNGCs: prime targets of plant cyclic nucleotide signalling? Trends Plant Sci 8:286–293
6. Ward JM, Mäser P, Schroeder JI (2009) Plant ion channels: gene families, physiology, and functional genomics analyses. Ann Rev Physiol 71:59–82
7. Zhorov BS, Tikhonov DB (2004) Potassium, sodium, calcium and glutamate-gated channels: pore architecture and ligand action. J Neurochem 88:782–799
8. Leng Q, Mercier RW, Hua B-GG, Fromm H, Berkowitz GA (2002) Electrophysiological analysis of cloned cyclic nucleotide-gated ion channels. Plant Physiol 128:400–410
9. Mäser P, Thomine S, Schroeder JI, Ward JM, Hirschi K et al (2001) Phylogenetic relationships within cation transporter families of Arabidopsis. Plant Physiol 126:1646–1667
10. Hua B-GG, Mercier RW, Leng Q, Berkowitz GA (2003) Plants do it differently. A new basis for potassium/sodium selectivity in the pore of an ion channel. Plant Physiol 132:1353–1361
11. Hua B-G, Mercier RW, Zielinski RE, Berkowitz GA (2003) Functional interaction of calmodulin with a plant cyclic nucleotide gated cation channel. Plant Physiol Biochem 41:945–954
12. Lemtiri-Chlieh F, Berkowitz GA (2004) Cyclic adenosine monophosphate regulates calcium

channels in the plasma membrane of Arabidopsis leaf guard and mesophyll cells. J Biol Chem 279:35306–35312

13. Yuen CYL, Christopher DA (2010) The role of cyclic nucleotide-gated channels in cation nutrition and abiotic stress. Ion channels and plant stress responses. Springer, Berlin, pp 137–158
14. Ma W, Berkowitz GA (2011) Ca^{2+} conduction by plant cyclic nucleotide gated channels and associated signaling components in pathogen defense signal transduction cascades. New Phytol 190:566–572
15. Finka A, Quendet AFH, Maathuis FJM, Saidi Y, Goloubinoff P (2012) Plasma membrane cyclic nucleotide gated calcium channels control land plant thermal sensing and acquired thermotolerance. Plant Cell 24:3333–3348
16. Young EC, Krougliak N (2004) Distinct structural determinants of efficacy and sensitivity in the ligand-binding domain of cyclic nucleotide-gated channels. J Biol Chem 279:3553–3562
17. Schuurink RC, Shartzer SF, Fath A, Jones RL (1998) Characterization of a calmodulin-binding transporter from the plasma membrane of barley aleurone. Proc Natl Acad Sci U S A 95:1944–1949
18. Arazi T, Sunkar R, Kaplan B, Fromm H (1999) A tobacco plasma membrane calmodulin-binding transporter confers Ni^{2+} tolerance and Pb^{2+} hypersensitivity in transgenic plants. Plant J 20:171–182
19. Bridges D, Fraser ME, Moorhead GB (2005) Cyclic nucleotide binding proteins in the *Arabidopsis thaliana* and *Oryza sativa* genomes. BMC Bioinf 6:6
20. Verret F, Wheeler G, Taylor AR, Farnham G, Brownlee C (2010) Calcium channels in photosynthetic eukaryotes: implications for evolution of calcium-based signalling. New Phytol 187:23–43
21. Higgs PG, Attwood TK (2005) Bioinformatics and molecular evolution. Blackwell Science Ltd., Malden, MA
22. De Castro E, Sigrist CJA, Gattiker A, Bulliard V, Langendijk-Genevaux PS et al (2006) ScanProsite: detection of PROSITE signature matches and ProRule-associated functional and structural residues in proteins. Nucleic Acids Res 34:W362–W365
23. Finn RD, Clements J, Eddy SR (2011) HMMER web server: interactive sequence similarity searching. Nucleic Acids Res 39:W29–W37
24. Lamesch P, Berardini TZ, Li D, Swarbreck D, Wilks C, Sasidharan R et al (2011) The Arabidopsis information resource (TAIR): improved gene annotation and new tools. Nucleic Acids Res 40:D1202–D1210
25. Lang D, Eisinger J, Reski R, Rensing SA (2005) Representation and high-quality annotation of the *Physcomitrella patens* transcriptome demonstrates a high proportion of proteins involved in metabolism in mosses. Plant Biol 7:238–250
26. Tamura K, Peterson D, Peterson N, Stecher G, Nei M et al (2011) MEGA5: molecular evolutionary genetics analysis using maximum likelihood, evolutionary distance, and maximum parsimony methods. Mol Biol Evol 28: 2731–2739
27. Zelman AK, Dawe A, Gehring C, Berkowitz GA (2012) Evolutionary and structural perspectives of plant cyclic nucleotide-gated cation channels. Front Plant Sci 3:95
28. Edgar RC (2004) MUSCLE: a multiple sequence alignment method with reduced time and space complexity. BMC Bioinf 5:113
29. Jones DT, Taylor WR, Thornton JM (1992) The rapid generation of mutation data matrices from protein sequences. Comput Appl Biosci 8:275–282
30. Felsenstein J (1985) Confidence limits on phylogenies: an approach using the bootstrap. Evolution 39:783–791
31. Saitou N, Nei M (1987) The neighbor-joining method: a new method for reconstructing phylogenetic trees. Mol Biol Evol 4:406–425
32. Chikayama E, Nameki N, Kigawa T, Koshiba S, Inoue M, Tomizawa T, Kobayashi N, Yokoyama S (2004) Solution structure of the cNMP-binding domain from Arabidopsis thaliana cyclic nucleotide-regulated ion channel. http://www.metalife.com/PDB/26282
33. Letunic I, Doerks T, Bork P (2012) SMART 7: recent updates to the protein domain annotation resource. Nucleic Acids Res 40: D302–D305
34. Altschul SF, Madden TL, Schäffer AA, Zhang J, Zhang Z et al (1997) Gapped BLAST and PSI-BLAST: a new generation of protein database search programs. Nucleic Acids Res 25:3389–3402
35. Schäffer AA, Aravind L, Madden TL, Shavirin S, Spouge JL et al (2001) Improving the accuracy of PSI-BLAST protein database searches with composition-based statistics and other refinements. Nucleic Acids Res 29:2994–3005
36. Hofmann K, Stoffel W (1993) TMbase—a database of membrane spanning proteins segments. Biol Chem Hoppe-Seyler 374:166
37. Krogh A, Larsson B, von Heijne G, Sonnhammer ELL (2001) Predicting transmembrane protein topology with a hidden Markov model:

application to complete genomes. J Mol Biol 305:567–580
38. Derbyshire MK, Lanczycki CJ, Bryant SH, Marchler-Bauer A (2012) Annotation of functional sites with the conserved domain. Database, ID bar058 (doi:10.1093/database/bar058)
39. Sigrist CJA, Cerutti L, de Castro E, Langendijk-Genevaux PS, Bulliard V et al (2010) PROSITE, a protein domain database for functional characterization and annotation. Nucleic Acids Res 38:161–166
40. Ofran Y, Rost B (2005) Predictive methods using protein sequences. Bioinformatics: a practical guide to the analysis of genes and proteins. Wiley, Hoboken, NJ, Part 3, Chapter 2
41. Ali R, Ma W, Lemtiri-Chlieh F, Tsaltas D, Leng Q, von Bodman S, Berkowitz GA (2007) Death don't have no mercy and neither does calcium: Arabidopsis cyclic nucleotide gated channel2 and innate immunity. Plant Cell 19: 1081–1095
42. Ali R, Zielinski RE, Berkowitz GA (2006) Expression of plant cyclic nucleotide-gated cation channels in yeast. J Exp Bot 57:125–138
43. Yuen CYL, Christopher DA (2010) The role of cyclic nucleotide-gated channels in cation nutrition and abiotic stress. Springer, Berlin, pp 137–157
44. Urquhart W, Gunawardena AH, Moeder W, Ali R, Berkowitz GA, Yoshioka K (2007) The chimeric cyclic nucleotide-gated ion channel ATCNGC11/12 constitutively induces programmed cell death in a Ca2+ dependent manner. Plant Mol Biol 65:747–761
45. Gobert A, Park G, Amtmann A, Sanders D, Maathuis FJM (2006) *Arabidopsis thaliana* cyclic nucleotide gated channel 3 forms a non-selective ion transporter involved in germination and cation transport. J Exp Bot 57:791–800
46. Hampton CR, Bowen HC, Broadley MR, Hammond JP, Mead A et al (2004) Cesium toxicity in Arabidopsis. Plant Physiol 136:3824–3837
47. Ma W, Ali R, Berkowitz GA (2006) Characterization of plant phenotypes associated with loss-of function of AtCNGC1, a plant cyclic nucleotide gated cation channel. Plant Physiol Biochem 44:494–505
48. Frietsch S, Wang YF, Sladek C, Poulsen LR, Romanowsky SM et al (2007) A cyclic nucleotide-gated channel is essential for polarized tip growth of pollen. Proc Natl Acad Sci USA 104:14531–14536
49. Maathuis FJ (2006) The role of monovalent cation transporters in plant responses to salinity. J Exp Bot 57:1137–1147

Chapter 15

Inferring Biological Functions of Guanylyl Cyclases with Computational Methods

May Alqurashi and Stuart Meier

Abstract

A number of studies have shown that functionally related genes are often co-expressed and that computational based co-expression analysis can be used to accurately identify functional relationships between genes and by inference, their encoded proteins. Here we describe how a computational based co-expression analysis can be used to link the function of a specific gene of interest to a defined cellular response. Using a worked example we demonstrate how this methodology is used to link the function of the Arabidopsis *Wall-Associated Kinase-Like 10* gene, which encodes a functional guanylyl cyclase, to host responses to pathogens.

Key words *Arabidopsis thaliana*, Guanylyl cyclase, cGMP, Computational analysis, Co-expression, Functional annotation, Functional/biological process

1 Introduction

In plants, the cyclic nucleotide, guanosine 3′,5′-cyclic monophosphate (cGMP) has been shown to be an essential signaling molecule in many diverse physiological processes [1] including development [2, 3] and biotic [4] and abiotic stress responses [5]. The identification of guanylyl cyclases (GCs), enzymes that catalyze the synthesis of cGMP from guanosine triphosphate (GTP), was advanced in higher plants following the design of a specific search motif that is based on functionally assigned amino acids present in the catalytic center of annotated GCs in lower and higher eukaryotes [6]. Querying the Arabidopsis proteome with GC motifs has resulted in the identification of a number of candidate GCs in *Arabidopsis thaliana* and several of these, including Wall-Associated Kinase-Like 10 (AtWAKL10; At1g79680), have been cloned and their purified recombinant proteins have been shown to synthesize cGMP in vitro [6–10] and in vivo [11]. However, the specific functional processes in which many of the identified GCs function

Chris Gehring (ed.), *Cyclic Nucleotide Signaling in Plants: Methods and Protocols*, Methods in Molecular Biology, vol. 1016,
DOI 10.1007/978-1-62703-441-8_15,

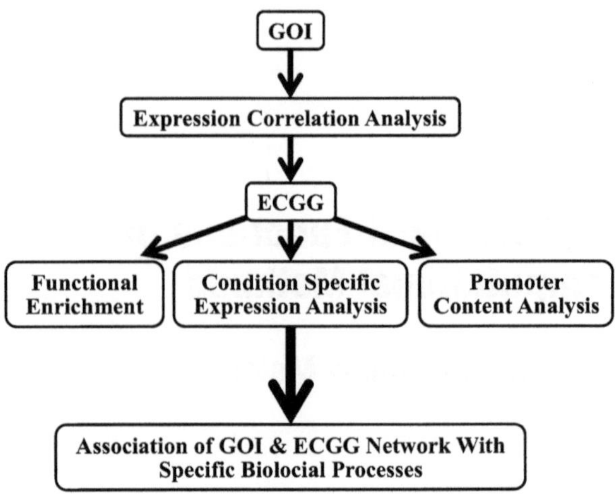

Fig. 1 Overview illustration of the work flow of the analysis. An expression correlation analysis is performed to identify genes that most highly co-expressed with a specific gene of interest (GOI). The determined expression correlated gene group (ECGG) is subsequently subjected to a functional enrichment analysis, stimulus-specific expression analysis, and promoter content analysis

remain unknown. Here we detail how a computational based method that is centered around a co-expression analysis and the sequential use of a number of complementary online tools can be used to associate the function of a specific gene or protein with a specific biological response and we do so using *AtWAKL10* as an example.

It is well established in eukaryotes that cellular responses are not mediated by a single gene, but rather require the coordinated participation of networks of functionally related genes. The development of near full genome transcriptome analysis technologies such as microarray over a decade ago has led to the accumulation of masses of data in public repositories. Many freely available analysis tools have been developed that can mine these data and identify biologically relevant patterns in gene expression. A number of studies have shown that functionally related genes are often co-expressed and consequently co-expression analysis has proved useful in identifying genes that function in common cellular responses [12–14]. When a gene of unknown function is determined to be co-expressed with genes with a well-defined role in a specific biological process, the function of the unknown gene can be related to that process.

The outline of the analysis is illustrated in Fig. 1 and begins with an expression correlation analysis that is performed over a large number of diverse, near full genome, microarray experiments to identify genes that are most highly co-expressed with a specific gene of interest (GOI), in this case *AtWAKL10.* A functional enrichment analysis is then performed on the top co-expressed genes (≥50), hereby referred to as the expression correlated gene group (ECGG), in order to determine if there is any enrichments in functional annotation terms associated with the group. The transcriptional responses of the genes are next examined across a range of microarray experimental conditions. This serves to both confirm that the genes are indeed co-expressed and identify conditions that induce expression of the genes. Finally, a promoter content analysis is performed to determine if the promoters of the ECGG contain enrichments in any known plant transcription factor-binding sites (TFBS) that may be causative for their co-expression.

In the example presented here, *AtWAKL10* is determined to be highly co-expressed with a group of genes that contain a functional enrichment in a number of annotation terms associated with pathogen defense. It was subsequently confirmed that the expression of *AtWAKL10* and its ECGG are strongly and coordinately induced in response to a range of biotic stress-related conditions. Finally, the promoters of the ECGG were found to contain an enrichment in putative binding sites for WRKY TFs that are known to regulate the expression of pathogen defense-related genes [15]. In summary, these results provide complementary evidence which supports that AtWAKL10 functions in biotic stress responses.

2 Materials

All the tools used in this methodology are available online in the public domain unless otherwise stated. The analysis and selection of data can be performed from any computer supporting Office programs with a particular requirement for Excel or equivalent programs. The following method is limited to sequenced and well-studied organisms since it requires quality sequence data and functional annotation information as well as substantial amounts of global transcriptomics data in the public domain. *Arabidopsis thaliana* is used in this example.

3 Methods

3.1 Expression Correlation

1. A number of co-expression tools are freely available in the public domain that can be used to identify genes that are most highly co-expressed with a specific GOI (*see* **Note 1**). The Arabidopsis Co-expression tool (ACT; http://www.arabidopsis.leeds.ac.uk/act/coexpanalyser.php; [16]) is used in the example presented here and *AtWAKL10* (At1g79680) is the GOI.
2. The co-expression analysis is performed by selecting Option 2: "Co-expression analysis over available array experiments."
3. A probe ID is required for the input and can be obtained from the AGI code, At1g79680, using the "ID exchanger."

The corresponding probe ID, from the ATH1-121501 (22k)-type array (*see* **Note 2**), in this case, 261394_at, is copied into the "Probe ID" input box.

4. The number of correlated genes to be returned can be selected by typing the number required in the "limited to the first" box. For the purposes of this analysis an ECGG of ≥50 is required; however it is a good idea to leave the limit blank to obtain the correlation values for all available genes.
5. The results are returned as a table in the browser (*see* **Note 3**) and can be copied and pasted as unicode text into an Excel spreadsheet for further analysis.
6. The top 50 genes with the highest Pearson correlation coefficient (*r*-value) are extracted from the list and along with *AtWAKL10* comprise the *AtWAKL10*-ECGG (total 51 genes). This gene list (Table 1) is used in subsequent analysis (*see* **Note 4**).

3.2 Gene Ontology and Functional Enrichment Analysis

1. A functional enrichment analysis is performed to identify functional annotations associated with the individual co-expressed genes and additionally determine if there are any enrichments in specific functional terms associated with the ECGG as a whole (*see* **Note 5**). The FatiGO tool (http://babelomics.bioinfo.cipf.es/functional.html) in the Babelomics version 4.3 suite is used to analyze the *AtWAKL10*-ECGG since this tool performs the above requirements and returns results in tabulated files.
2. A text file containing the ATG IDs for the genes to be analyzed (in this case the *AtWAKL10*-ECGG, 51 genes) is uploaded by selecting the "upload data" tab at the top left of the page.
3. The "Functional analysis" tab and the "FatiGO" option are selected next.
4. The "Id List vs. Rest of genome" comparison is selected and the previously uploaded ID file is selected for List 1 under "Browse server." "Never" remove duplicates is also selected.
5. Under databases, "*Arabidopsis thaliana*" is selected as the organism and all the available annotation databases are checked to maximize the scope of the analysis. Finally "Run" is selected.
6. When the analysis is complete, the results can be accessed by clicking on the name of the job. The "Abstract" section lists the number of ID annotations identified for each database analyzed for both the ECGG input genes and the complete genome. In the Significant Results section a summary of the databases found to contain significant enrichments and the number of significant terms associated with each database are listed under "Number of significant terms per DB." Following this the results for each individual database analyzed are listed and those

Table 1
The top 25 genes' expression correlated with *AtWAKL10*

Probe set	*r*-Value	*p*-Value	*e*-Value	Gene ID	Annotation
247145_at	0.913	0.0	0.0	AT5G65600	Legume lectin family protein
254408_at	0.907	0.0	0.0	AT4G21390	S-locus lectin protein kinase family
258037_at	0.895	0.0	0.0	AT3G21230	4-coumarate-CoA ligase
254905_at	0.894	0.0	0.0	AT4G11170	Disease resistance prot. (TIR-NBS-LRR class)
246988_at	0.893	0.0	0.0	AT5G67340	Armadillo/beta-catenin repeat family
261021_at	0.877	0.0	0.0	AT1G26380	FAD-binding domain-containing protein
246406_at	0.869	0.0	0.0	AT1G57650	Disease resistance protein (NBS-LRR class)
263948_at	0.868	0.0	0.0	AT2G35980	Harpin-induced family (YLS9)
259033_at	0.868	0.0	0.0	AT3G09410	Pectinacetylesterase family
261763_at	0.866	0.0	0.0	AT1G15520	ABC transporter family
265008_at	0.865	0.0	0.0	AT1G61560	Seven-transmembrane MLO family
258787_at	0.862	0.0	0.0	AT3G11840	U-box domain-containing protein
256017_at	0.862	0.0	0.0	AT1G19180	Expressed protein
261748_at	0.861	0.0	0.0	AT1G76070	Expressed protein
258277_at	0.859	0.0	0.0	AT3G26830	Cytochrome P450
252739_at	0.858	0.0	0.0	AT3G43250	Cell cycle control protein
246368_at	0.858	0.0	0.0	AT1G51890	Leucine-rich repeat protein kinase
254103_at	0.857	0.0	0.0	AT4G25030	Expressed protein
251176_at	0.856	0.0	0.0	AT3G63380	Calcium-transporting ATPase
245662_at	0.855	0.0	0.0	AT1G28190	Expressed protein
251950_at	0.854	0.0	0.0	AT3G53600	Zinc finger (C_2H_2 type) family
247026_at	0.852	0.0	0.0	AT5G67080	Protein kinase family
259734_at	0.850	0.0	0.0	AT1G77500	Expressed protein
248700_at	0.848	0.0	0.0	AT5G48400	Glutamate receptor family
245252_at	0.845	0.0	0.0	AT4G17500	Ethylene-responsive element-binding protein
246927_s_at	0.883	0.0	0.0	AT5G25250	Expressed protein
				AT5G25260	Expressed protein

The highlighted text is an example of when multiple genes hybridize to a single probe ID and should therefore not be included in the analysis

Note: Only the top 25 expression correlated genes are presented due to spatial limitations; however, the top 50 genes are used in downstream analysis

that contain significant enrichments can be downloaded in Excel format.

7. In the example presented here, the *AtWAKL10*-ECGG was found to contain enrichments in a number of gene ontology categories including a number that are associated with pathogen defense responses in the biological process category. These include defense response, immune response, and response to biotic stimulus which suggests that *AtWAKL10* may function in pathogen defense responses.

3.3 Stimulus-Specific Expression Profiling

1. The online Genevestigator tool (https://www.genevestigator.com/gv/; [17]) is next used to screen the transcriptional responses and identify specific conditions that induce differential expression of the genes (*see* **Note 6**).
2. To begin, the "Plant Biology" option is selected, then "Start" the analysis tool, and login. An overview of the available tools is presented.
3. To start an analysis, "New" is selected under Sample Selection. The organism is next selected, in this case "*Arabidopsis thaliana*" and the "ATH1: 22k" near full genome array which has the greatest number of experiments. There is an option to filter for specific experiments or conditions; however, for this analysis no filtering is selected so as to examine the maximum number of experiments (>7,000 arrays).
4. The IDs of the specific genes to be analyzed, in this case *AtWAKL10* and a sample of genes from the ECGG (top 5), are next added by clicking "New" under Gene Selection. The ATG IDs (e.g., At1g79680) or the probe IDs (e.g., 261394_at) can be used as input (*see* **Note 7**).
5. The expression of the genes can be viewed in a number of different formats (*see* **Note 8**). In the "Samples" tool, the normalized signal intensity for the selected genes is displayed on each individual array chip. The "Anatomy" tool displays the signal intensity of the selected genes in a range of different tissue types (*see* **Note 9**). The "Development" tool displays the signal intensity of the genes throughout the developmental stages of the organism. In the "Perturbation" tool the fold-change of the genes, in comparison to their appropriate controls, is displayed in response to a range of different conditions and genotypes. The list of experiments displayed can be filtered according to fold changes and *p*-value in order to identify conditions that induce the largest changes (up and down) in the expression of the genes. The results can be displayed in a "Tree" or a "List" format with heat map and scatterplot options available for both. In the study presented here, the "List"-heat map and -scatterplot displays clearly indicate that *AtWAKL10* and the five most correlated genes are extremely induced in response to a range of

biotic stress-related conditions including bacterial and fungal pathogens as well as their associated elicitors.

6. There are also other tools available in Genevestigator including a number of "Similarity Search Tools" that perform clustering and co-expression analysis (*see* **Note 10**).

3.4 Promoter Analyses

1. The visualization tool in the *Arabidopsis thaliana* Expression Network Analysis (ATHENA; http://www.bioinformatics2.wsu.edu/cgi-bin/Athena/cgi/visualize_select.pl; [18]) Web site is used to detect the presence of enrichments in known plant *cis*-elements in the promoters of the *AtWAKL10*-ECGG that may mediate their co-expression.
2. To run the analysis, the ATG IDs for the *AtWAKL10*-ECGG are copied into the "Accessions" box. From the drop-down boxes "Compact" is selected for "Visualization Type," for "Upstream Range" select "1000," and the option to "Cutoff at adjacent genes" is not selected. All other options are left as default and the analysis is started by selecting display.
3. In the results page, a visualization of the individual genes and their promoters is displayed, with zooming options, which illustrates the name and location of individual TFBS identified in each promoter. Lower on the page, the "Transcription Factors" table contains an overall summary of the TFBS identified in the ECGG, with significantly enriched TFBS displayed at the top. This table contains values for each TFBS identified which include the *p*-value, the number of promoters that contain the site (#P), and the total number of predicted TFBS (#S) present in all promoters of the ECGG. The details of the identified TFBS, including the consensus sequence, can be obtained by clicking on the name and following links. All the figures, tables, and data can be downloaded and saved for further analysis.
4. For the *AtWAKL10*-ECGG example, a significant enrichment (*p*-value 10^{-4}) in the W-box motif/element (TTGACY) was identified with the promoters of 44/51 genes (86.3 %) containing the motif a total of 99 times. The W-boxes are known to bind WRKY TFs which have been shown to regulate defense-related gene transcription [19].

4 Notes

1. There is a range of co-expression tools available for analysis on different organisms including ATTED-II that covers Arabidopsis and rice orthologs (http://atted.jp; [20]), the Rice Oligonucleotide Array Database (ROAD) (http://www.ricearray.org/coexpression/coexpression.shtml; [21]), the Coexpressed Gene Database (COXPRESdb; http://coxpresdb.hgc.jp) that covers for a range

of animal species [22], and the Comprehensive Systems-Biology Database (CSB.DB; http://csbdb.mpimp-golm.mpg.de/) that covers prokaryotic, eukaryotic, and plant model organisms [23]. In addition, some tools are available that can perform specialized co-expression analysis such as ATTED-II that can determine condition-specific co-expression and the Arabidopsis CressExpress tool (http://www.cressexpress.org; [24]) that measures co-expression in specific tissue types.

2. There are two types of Affymetrix array platforms available in the ACT, the AtGenome1 (8k) and the ATH1 (22k) chip. Since the ATH1 has near full genome coverage, it is the chip of preference.
3. The output from ACT is arranged in decreasing order of correlation. The corresponding gene ID is provided for each probe as well as a description of the annotated function. The *r*-value is the calculated Pearson correlation coefficient which is a measure of the strength and direction of the linear relationship between two variables; the GOI and all other genes represented on the microarray chip (ranging from −1 to +1). The *p*-value represents the statistical significance of the *r*-value while the e-value represents the expected frequency of the observed value occurring by chance and accounts for multiple hypothesis testing in whole genome studies.
4. When extracting the ECGG, it is important to only include genes that are represented by a single unique probe. In some cases, a single probe is known to hybridize with multiple genes and thus the corresponding *r*-value is not representative of a single gene (Table 1). In addition, some probes are not assigned to ATG ID and thus should not be included in the ECGG (i.e., the gene ID is blank). The ECGG should include a minimum of 50 genes so as to provide an adequate representative sample size for downstream analysis.
5. There are a number of tools available to perform functional enrichment analysis, including BINGO (http://www.psb.ugent.be/cbd/papers/BiNGO/; [25]), EasyGO (http://bioinformatics.cau.edu.cn/easygo/; [26]), AmiGO (http://amigo.geneontology.org; [27]), and FatiGO (http://babelomics.bioinfo.cipf.es/functional.html; [28]). For the purposes of this type of analysis it is essential that the tool selected returns functional annotation information and calculates enrichment frequencies for the ECGG in comparison to the rest of the genome.
6. Genevestigator is an extremely useful tool that provides gene expression values for a wide range of experimental conditions across numerous organisms including *Arabidopsis thaliana*, *Drosophila melanogaster*, *Escherichia coli*, *Homo sapiens*, *Mus*

musculus, *Oryza sativa*, and *Physcomitrella patens* among others. This tool has limited access for free public usage with only the "Sample" and "Anatomy" tools of those described having open access.

7. The ability to have multiple entries in "Gene Selection" option differs according to the account being used. The anonymous account will only allow one gene per analysis; the academic account allows up to 50 entries per analysis while the full version can handle up to 400 entries per analysis.
8. There are three main types of analysis that can be performed in Genevestigator, with each containing a number of options. As described above, the Condition search can be used to identify conditions that alter the transcription of the genes in different tissue types, stages of development, and in response to various perturbations. Gene search identifies genes that are specifically expressed in one or a subset of conditions in different databases. The Similarity search includes a group of tools designed to identify groups of genes that have similar expression profiles.
9. All the options are described in the overview on the main screen. Details about specific experiments can be obtained by placing the cursor over the experiment name which brings up a pop-up box that can be frozen by pressing "F2" on the keyboard. To obtain information on precise/specific transcriptional responses for specific treatments or time points, it is necessary to download the relevant data sets from repositories.
10. There is a help button within each tool that provides a detailed explanation of the specific display.

References

1. Newton RP, Smith CJ (2004) Cyclic nucleotides. Phytochemistry 65:2423–2437
2. Salmi ML, Morris KE, Roux SJ, Porterfield DM (2007) Nitric oxide and cGMP signaling in calcium-dependent development of cell polarity in *Ceratopteris richardii*. Plant Physiol 144:94–104
3. Stöhr C, Stremlau S (2006) Formation and possible roles of nitric oxide in plant roots. J Exp Bot 57:463–470
4. Durner J, Wendehenne D, Klessig DF (1998) Defense gene induction in tobacco by nitric oxide, cyclic GMP, and cyclic ADP-ribose. Proc Natl Acad Sci USA 95:10328–10333
5. Pasqualini S, Meier S, Gehring C, Madeo L, Fornaciari M et al (2009) Ozone and nitric oxide induce cGMP-dependent and -independent transcription of defence genes in tobacco. New Phytol 181:860–870
6. Ludidi N, Gehring C (2003) Identification of a novel protein with guanylyl cyclase activity in *Arabidopsis thaliana*. J Biol Chem 278: 6490–6494
7. Kwezi L, Meier S, Mungur L, Ruzvidzo O, Irving H et al (2007) The *Arabidopsis thaliana* brassinosteroid receptor (AtBRI1) contains a domain that functions as a guanylyl cyclase in vitro. PLoS One 2:e449
8. Meier S, Ruzvidzo O, Morse M, Donaldson L, Kwezi L et al (2010) The arabidopsis *wall associated kinase-like 10* gene encodes a functional guanylyl cyclase and is co-expressed with pathogen defense related genes. PLoS One 5:e8904
9. Mulaudzi T, Ludidi N, Ruzvidzo O, Morse M, Hendricks N et al (2011) Identification of a novel *Arabidopsis thaliana* nitric oxide-binding molecule with guanylate cyclase activity in vitro. FEBS Lett 585:2693–2697

10. Qi Z, Verma R, Gehring C, Yamaguchi Y, Zhao Y et al (2010) Ca^{2+} signaling by plant *Arabidopsis thaliana* pep peptides depends on AtPepR1, a receptor with guanylyl cyclase activity, and cGMP-activated Ca^{2+} channels. Proc Natl Acad Sci USA 49:21193–21198
11. Kwezi L, Ruzvidzo O, Wheeler JI, Govender K, Iacuone S et al (2011) The phytosulfokine (PSK) receptor is capable of guanylate cyclase activity and enabling cyclic GMP-dependent signaling in plants. J Biol Chem 286: 22580–22588
12. Allocco D, Kohane I, Butte A (2004) Quantifying the relationship between co-expression, co-regulation and gene function. BMC Bioinformatics 5:18
13. Jansen R, Greenbaum D, Gerstein M (2002) Relating whole-genome expression data with protein–protein interactions. Genome Res 12:37–46
14. Stuart JM, Segal E, Koller D, Kim SK (2003) A gene-coexpression network for global discovery of conserved genetic modules. Science 302:249–255
15. Pandey SP, Somssich IE (2009) The role of WRKY transcription factors in plant immunity. Plant Physiol 150:1648–1655
16. Manfield IW, Jen C-H, Pinney JW, Michalopoulos I, Bradford JR et al (2006) Arabidopsis co-expression tool (ACT): web server tools for microarray-based gene expression analysis. Nucleic Acids Res 34:W504–W509
17. Hruz T, Laule O, Szabo G, Wessendorp F, Bleuler S et al (2008) Genevestigator v3: a reference expression database for the meta-analysis of transcriptomes. Adv Bioinformatics 2008:420747
18. O'Connor TR, Dyreson C, Wyrick JJ (2005) Athena: a resource for rapid visualization and systematic analysis of arabidopsis promoter sequences. Bioinformatics 21:4411–4413
19. Maleck K, Levine A, Eulgem T, Morgan A, Schmid J et al (2000) The transcriptome of *Arabidopsis thaliana* during systemic acquired resistance. Nat Genet 26:403–410
20. Obayashi T, Nishida K, Kasahara K, Kinoshita K (2011) ATTED-II updates: condition-specific gene coexpression to extend coexpression analyses and applications to a broad range of flowering plants. Plant Cell Physiol 52:213–219
21. Jung K-H, Dardick C, Bartley LE, Cao P, Phetsom J et al (2008) Refinement of light-responsive transcript lists using rice oligonucleotide arrays: evaluation of gene-redundancy. PLoS One 3:e3337
22. Obayashi T, Kinoshita K (2011) COXPRESdb: a database to compare gene coexpression in seven model animals. Nucleic Acids Res 39:D1016–D1022
23. Steinhauser D, Usadel B, Luedemann A, Thimm O, Kopka J (2004) CSB.DB: a comprehensive systems-biology database. Bioinformatics 20:3647–3651
24. Srinivasasainagendra V, Page GP, Mehta T, Coulibaly I, Loraine AE (2008) CressExpress: a tool for large-scale mining of expression data from arabidopsis. Plant Physiol 147: 1004–1016
25. Maere S, Heymans K, Kuiper M (2005) BiNGO: a cytoscape plugin to assess overrepresentation of gene ontology categories in biological networks. Bioinformatics 21:3448–3449
26. Zhou X, Su Z (2007) EasyGO: gene ontology-based annotation and functional enrichment analysis tool for agronomical species. BMC Genomics 8:246
27. Carbon S, Ireland A, Mungall CJ, Shu S, Marshall B et al (2009) AmiGO: online access to ontology and annotation data. Bioinformatics 25:288–289
28. Al-Shahrour F, Minguez P, Tárraga J, Medina I, Alloza E et al (2007) FatiGO^{+}: a functional profiling tool for genomic data. Integration of functional annotation, regulatory motifs and interaction data with microarray experiments. Nucleic Acids Res 35:W91–W96

Chapter 16

Identification and Characterization of Cyclic Nucleotide Phosphodiesterases

Erin B. Purcell and Rita Tamayo

Abstract

Cyclic nucleotide phosphodiesterases regulate cellular levels of small molecule second messengers that control important biological processes in all kingdoms of life. Identifying and characterizing these enzymes is necessary for basic research and pharmaceutical applications. Here, we describe the use of thin layer chromatography to analyze cellular extracts or purified proteins for cyclic nucleotide phosphodiesterase activity.

Key words Phosphodiesterase, Cyclic nucleotide, cAMP, cGMP, Cyclic dinucleotide, Thin layer chromatography, Phosphorimagery

1 Introduction

Cyclic nucleotide monophosphates (cNMPs) are second messenger molecules regulating animal, plant, and microbial signaling pathways. They are synthesized from purine nucleotide triphosphates by enzymes known as cyclases and degraded into mononucleotides by phosphodiesterases (PDEs). The expression of PDE genes, and subcellular distribution and regulation of PDE enzymes, determines the duration and localization of cNMP-mediated signaling events.

When cNMPs were identified, the transient nature of cyclic nucleotide signals and the discovery of chemical inhibitors that prolong these signals by preventing cNMP turnover indicated the necessity of hydrolyzing enzymes as well as synthases, but it was decades before individual PDEs were identified [1–4]. Early attempts to identify PDEs usually centered on screening known PDE inhibitors against whole tissue extracts and often produced negative or conflicting results. The era of genomic sequencing has revealed that most eukaryotic PDEs have multiple, often cell-type specific, isoforms, which exhibit different sensitivities to activators and inhibitors [5, 6]. In addition, because of multiple layers of

Chris Gehring (ed.), *Cyclic Nucleotide Signaling in Plants: Methods and Protocols*, Methods in Molecular Biology, vol. 1016, DOI 10.1007/978-1-62703-441-8_16, © Springer Science+Business Media New York 2013

transcriptional and posttranscriptional regulation [6], the presence of a PDE-encoding gene in a genome is no guarantee of an active PDE protein in a cell at a given moment, or under a particular condition. Quick, reproducible screening of tissue extracts, cell fractions, or purified proteins for PDE activity is crucial for a detailed understanding of the signaling pathways and biological readouts controlled by these molecules.

Thin layer chromatography (TLC) easily and quantifiably resolves cyclic nucleotide substrates from their mononucleotide catabolites. The protocol described herein allows the separation of mononucleotides based on their defined migration rates, which are influenced by the attraction of individual nucleotides to the stationary phase (the TLC plate surface) and their solubility in the mobile phase (the solvent). High pressure liquid chromatography (HPLC) can be used to analyze PDE reaction products [7], but the low expense, ease of use, and ability to analyze multiple parameters simultaneously makes TLC an appealing strategy. It should be noted that these methods are also applicable for analysis of PDE enzymes that hydrolyze other cyclic nucleotides, such as the cyclic dinucleotides c-di-GMP and c-di-AMP used by bacteria [8–10]. With TLC, the use of radioactive cyclic nucleotide substrates allows testing of diverse biological samples for PDE activity, as well as kinetic characterization and inhibitor screening of purified PDEs.

The following describes a standard procedure using TLC to detect PDE activity and analyze the reaction products. This protocol involves the use of radiolabeled [^{32}P]-cNMP substrate due to the ease of detection of ^{32}P. The use of ^{32}P labeled substrates requires in vitro synthesis using commercially available cyclic nucleotide cyclase enzymes, as detailed in Subheading 3.1. Alternatively, ^{3}H labeled cNMP substrates are available commercially and can be substituted; we describe in Subheading 4 the modified detection methods recommended for detecting ^{3}H-cNMPs. In addition, the use of phosphorimagery to detect radiolabeled nucleotides is described here, though autoradiographic detection is acceptable as well.

2 Materials

Prepare all solutions using sterile double distilled water and store at room temperature unless otherwise specified.

2.1 [^{32}P]-cNMP Synthesis

1. [α^{32}P]-ATP or [α^{32}P]-GTP.
2. Adenyl cyclase (Sigma Aldrich) or guanyl cyclase (Enzo LifeSciences).
3. Supplier-recommended cyclase reaction buffer.

4. Calf intestinal phosphatase (CIP).
5. Spin filters (Millipore Amicon Ultra 3000 MWCO, or equivalent).
6. Microcentrifuge capable of 14,000 × *g*.
7. To analyze cyclase reaction products, TLC materials are required (*see* Subheading 2.3).

2.2 PDE Assay

1. Sample to be analyzed (cell lysate, cell fraction, purified protein, etc.).
2. 10× reaction buffer: 0.5 M Tris–HCl (pH 7.5), 50 mM $MgCl_2$ (*see* **Note 1**).
3. Radiolabelled cAMP or cGMP.
4. To analyze cyclase reaction products, TLC materials are required (*see* Subheading 2.3).
5. Optional: broad-spectrum PDE inhibitor such as dipyridamole or 3-isobutyl-1-methylxanthine (IBMX) (*see* **Note 2**).

2.3 Thin Layer Chromatography

1. PEI-cellulose plates, 20 cm × 20 cm (SelectoScientific).
2. Glass chamber (Wheaton Science Products, No. 276860, or equivalent) or glass beaker with an inner diameter >20 cm. A glass cover is required for use with both the chamber and beaker.
3. Solvent: 0.5 M LiCl (*see* **Note 3**).
4. Phosphor storage screen.

   Optional: heat lamp (*see* **Note 4**).

2.4 Analysis

1. Phosphorimager.
2. Quantification software such as ImageQuant.

3 Methods

3.1 Optional: Wash PEI Cellulose Plates (See Note 5)

1. Soak plates in deionized water by submerging in a tray of water for 15 min, then air-dry completely (*see* **Note 4**).
2. Fill the TLC chamber with water approximately 0.5 cm deep and stand the PEI cellulose plate on edge in the chamber. Place the lid on chamber. Allow the water to run to the top of the plate (Fig. 1), then air-dry the plate completely.
3. Stand plate in the chamber on a second edge orthogonal to the first one, allow the water to run to top of plate, and then air-dry completely.
4. Wrap washed, dried plates in plastic wrap and store them at 4 °C. The plates may be cut into narrower strips for TLC experiments involving fewer samples.

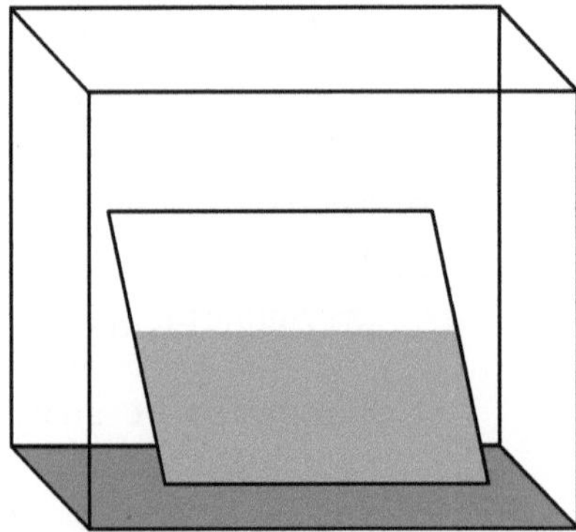

Fig. 1 PEI cellulose plate in solvent (*gray*). Aqueous solvents will reach the top of a 20 mm plate in approximately 3 h

3.2 [^{32}P]-cNMP Synthesis (See Note 6)

1. Prior to beginning an experiment, prepare a TLC plate by using a soft pencil and a ruler to mark the appropriate number of drop spots. The spots should be spaced at least 10 mm apart, and 15 mm from one edge (this will be the "bottom" of the plate, i.e., the edge that is submerged in solvent in the TLC chamber). The following cNMP synthesis experiment requires three drop spots. A 20 cm × 20 cm plate can be trimmed into a strip to accommodate this number of samples.
2. Combine cyclase, radioactive substrate, and cyclase reaction buffer according to supplier specifications.
3. Immediately spot 1 μL "$t = 0$" aliquot of the reaction onto a PEI cellulose plate on the first drop spot.
4. Incubate reaction 16 h at 37 °C, or as directed by the supplier.
5. Spot 1 μL onto the next drop spot on the PEI cellulose plate.
6. To remove phosphate radiolabel from unreacted [α^{32}P]-NTP, add 0.5 μL CIP and incubate 30 min at 37 °C (*see* **Note 7**).
7. Spot 1 μL onto the third drop spot on the PEI cellulose plate.
8. Perform TLC (Subheading 3.4) and expose the dried PEI cellulose plate to phosphor storage screen.
9. View the autoradiograph using a phosphorimager. If synthesis and CIP treatment were successful, the final spot on the PEI cellulose plate should contain only [^{32}P]-cAMP or [^{32}P]-cGMP approximately halfway up the plate. Any contaminating materials that may be present in the [α^{32}P]-NTP stock, such as unincorporated ^{32}Pi, will run near the bottom of the plate (Fig. 2, Table 1).
10. Load the reaction into a centrifugal filter and spin for 30 min at 14,000 × *g*. Save the filtrate as substrate for PDE assays. The material retained on the column contains CIP and can be discarded as appropriate.

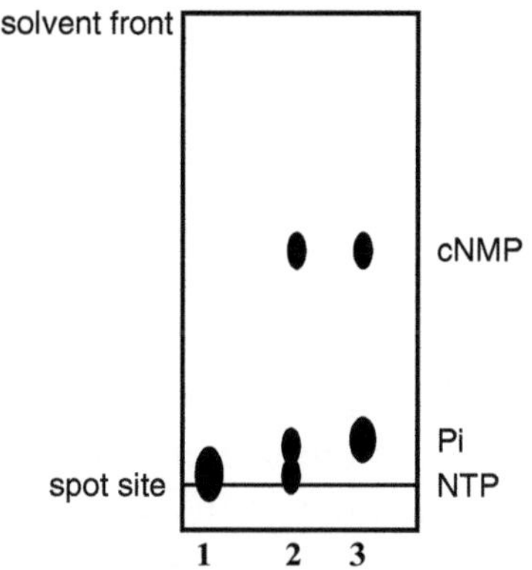

$$Rf = \frac{\text{(distance traveled from spot site)}}{\text{(distance traveled by solvent front)}}$$

Fig. 2 Sample autoradiograph of a cyclase reaction run on a TLC plate in 0.5 M LiCl. *Lane 1*: "$t=0$" aliquot, containing radioactive NTP. *Lane 2*: overnight reaction, containing unreacted NTP substrate, unincorporated phosphate, and the desired cNMP product. *Lane 3*: CIP-treated reaction, containing cNMP and unincorporated inorganic phosphate (P_i). Retardation factor (R_f) values for each species are calculated with the formula above

Table 1
Expected relative mobilities (R_f) of ^{32}P species in LiCl solvent

	0.3–0.5 M LiCl	1 M LiCl	References
NTP	0–0.05	0	[17, 18]
NDP	0–0.10	0.17–0.26	[17, 18]
P_i	0.05–0.10	0.45	[17]
NMP	0.10–0.27	0.40–0.52	[17–19]
cNMP	0.48–0.55	0.7	[19, 20]

3.3 PDE Assay

1. Prior to beginning the PDE experiment, prepare a TLC plate by using a soft pencil and a ruler to mark the appropriate number of drop spots 15 mm from one edge (this will be the bottom of the plate, and the edge that is submerged in solvent in the TLC chamber). Space the spots at least 10 mm apart. The following PDE experiment requires 15 drop spots.
2. For an initial characterization, assemble three 20 μL reactions containing 0, 4, and 16 μL of PDE sample (e.g., cell lysate, cell fraction, or purified protein). If desired, assemble an additional control reaction including PDE inhibitor (*see* **Note 2**). Bring the volume to 17 μL by adding deionized water. Add 2 μL of 10× buffer. Mix thoroughly by pipetting.

3. To initiate the reaction, add 1 μL of the appropriate radioactive substrate, [^{32}P]-cAMP or [^{32}P]-cGMP.
4. As the substrate is added to a reaction, immediately spot 1 μL "t = 0" aliquots of each reaction onto the prepared PEI cellulose plate, prior to initiating the next reaction.
5. Incubate the reactions at ambient temperature (25 °C). At designated time points, remove 1 μL aliquots from reactions and spot immediately onto plates. The spots dry quickly, terminating the enzymatic reaction. For an initial characterization, remove aliquots at 0, 1, 5, 30, and 90 min.
6. Perform TLC (Subheading 3.4) and expose the dried PEI cellulose plate to phosphor storage screen.
7. View the autoradiograph using a phosphorimager (*see* **Note 6**). If the analyzed sample contains PDE(s), the [^{32}P]-cAMP or [^{32}P]-cGMP spots should disappear or become fainter with increased reaction time and sample volume. Concomitantly, the respective reaction product, [^{32}P]-AMP or [^{32}P]-GMP should appear in increasing amounts. If the sample contains secondary phosphodiesterase activity, [^{32}P]-Pi may also appear; this is easily distinguished from cyclic nucleotides by TLC (Fig. 2, Table 1).

3.4 Thin Layer Chromatography

1. Fill the glass chamber with solvent (here, 0.5 M LiCl) to a depth of approximately 5 mm.
2. Place the bottom edge of spotted PEI cellulose plate (i.e., the edge along which samples are spotted) in solvent and allow the solvent front to run completely to top of plate. Remove the plate from the chamber and allow it to air-dry (*see* **Note 4**).
3. Wrap the plate in plastic wrap and expose it to a phosphorscreen overnight. Exposure times allowing detection of reaction products vary with intensity of the radioactive signal, so shorter exposures are possible.
4. View the autoradiograph using a phosphorimager.

3.5 Analysis

1. To quantify PDE activity, use ImageQuant or comparable software to measure the intensity of the radioactive spots corresponding to the substrate and product (*see* **Note 8**).
2. At each time point, calculate the extent of hydrolysis. Percent hydrolysis = 100 × (product/(substrate + product)) (Fig. 3).

3.6 Optimization

Optimize the reaction conditions by varying the pH, ionic strength, ion content, amount of PDE sample, PDE–substrate ratio, and reaction time (*see* **Note 1**). Once the reaction parameters have been optimized, studies for the characterization of PDE activity can be performed.

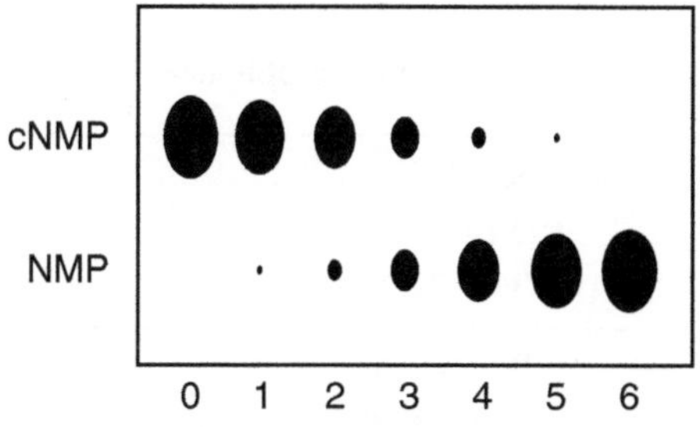

Time	cGMP	NMP	% Hydrolysis
0	100	0	0
1	90	10	10
2	75	25	25
3	50	50	50
4	25	75	75
5	10	90	90
6	0	100	100

Fig. 3 Sample PDE time course. At each time point, cNMP and NMP can be quantified

3.7 PDE Characterization Using TLC

1. PDEs in vivo act on timescales varying from milliseconds to hours [6]. An initial, small reaction of several hours (or overnight) can confirm that a protein is active.
2. The time intervals for more detailed kinetic characterization must be determined experimentally. Once appropriate time intervals for a given sample have been established, TLC is well suited for quickly measuring activity across a range of substrate concentrations in order to establish a Km for pure protein or specific activity for impure protein samples.
3. It should be noted that PDEs display varying levels of substrate specificity; thus, it may be necessary to determine substrate specificity by measuring the relative hydrolysis of cAMP and cGMP [6].
4. The methods described may be used to test potential activators or inhibitors of PDE activity [11, 12]. To do this, reaction parameters empirically determined to detect PDE activity should be repeated in the presence and absence of the putative activator or inhibitor, at a range of concentrations.

4 Notes

1. Several variables can affect PDE activity, so reaction buffer conditions will need to be experimentally optimized for individual PDEs. For example, some PDEs use Mg^{2+}, while others use Mn^{2+} in their metal coordination site. Table 2 provides guidance in starting values [13–16].
2. If working with heterogenous samples such as tissue extracts, assemble a control reaction containing 100 μM of a broad-spectrum PDE inhibitor such as dipyridamole or IBMX to confirm that loss of substrate is due to specific PDE activity [11].
3. In LiCl solvent, less charged species migrate faster. All phosphate species migrate faster in 1.0 or 1.5 M LiCl than in 0.5 M. If contaminating monophosphate, diphosphate, or triphosphate

Table 2
Recommended reaction buffer components for optimization of PDE reactions

Buffer	Tris–HCl HEPES Acetic acid–NaOH	5–50 mM
pH	6.0–7.5	
Divalent cation	$MgCl_2$ $MnCl_2$ $MgSO_4$	0.8–5.0 mM
Salt	KCl	0–100 mM

nucleotides are a concern, such as when cell lysates are assayed, these species can be resolved from unincorporated Pi using a more concentrated solvent.

4. Use of a heat lamp can substantially speed the TLC drying process and will not affect the quality of the autoradiograph. Wet TLC plates in a standard laboratory forced air hood will dry faster than on a bench top.
5. Impurities in some commercially available plates can affect the migration of molecules through the plate. Washing the plates prior to use, while not strictly necessary, can improve how molecules run.
6. Synthesis of [^{32}P]-cNMP substrate is not required. Tritiated cNMPs are commercially available (GE Healthcare) and can serve the same purpose; these may be exposed to ^{3}H screens (GE Healthcare) (*see also* **Note 8**).
7. CIP treatment of the cyclase reaction has two purposes: to remove radioactive signal from unused substrate from subsequent assays, and to provide an indication of the percentage of substrate that was incorporated into cNMP. Removal of the NTP signal by CIP treatment is critical for clearly discerning hydrolysis of cNMP to NMP. Contaminating [^{32}P]-NTP, as well as degraded [^{32}P]-NDP or [^{32}P]-NMP, may be removed by more rigorous treatment with CIP, which should then be removed via spin filtration.
8. If a phosphorimager is not available, the TLC plate may be exposed to X-ray film for a qualitative assessment of activity. This is most easily applicable to experiments performed with [^{32}P]-labeled substrate, since standard X-ray film detects ^{3}H poorly. Chemical enhancers of beta particle emitters, such as EN^{3}HANCE™ Spray Surface Autoradiography Enhancer (Perkin Elmer) which is compatible with TLC methods, may enhance detection of tritium-labeled nucleotides using film.

Alternatively, use of radioactivity may be avoided altogether by using high performance liquid chromatography to measure PDE activity upon cyclic nucleotide substrates, as described in [13].

References

1. Butcher RW, Sutherland EW (1962) Adenosine 3′,5′-phosphate in biological materials. I. purification and properties of cyclic 3′,5′-nucleotide phosphodiesterase and use of this enzyme to characterize adenosine 3′,5′-phosphate in human urine. J Biol Chem 237:1244–1250
2. Charbonneau H, Beier N, Walsh KA, Beavo JA (1986) Identification of a conserved domain among cyclic nucleotide phosphodiesterases from diverse species. Proc Natl Acad Sci USA 83:9308–9312
3. Chen CN, Denome S, Davis RL (1986) Molecular analysis of cDNA clones and the corresponding genomic coding sequences of the *Drosophila* Dunce+ gene, the structural gene for cAMP phosphodiesterase. Proc Natl Acad Sci USA 83:9313–9317
4. Sass P, Field J, Nikawa J, Toda T, Wigler M (1986) Cloning and characterization of the high-affinity cAMP phosphodiesterase of *Saccharomyces cerevisiae*. Proc Natl Acad Sci USA 83:9303–9307
5. Beavo JA, Reifsnyder DH (1990) Primary sequence of cyclic nucleotide phosphodiesterase isozymes and the design of selective inhibitors. Trends Pharmacol Sci 11:150–155
6. Conti M, Beavo J (2007) Biochemistry and physiology of cyclic nucleotide phosphodiesterases: essential components in cyclic nucleotide signaling. Annu Rev Biochem 76: 481–511
7. Simm R, Morr M, Kader A, Nimtz M, Romling U (2004) GGDEF and EAL domains inversely regulate cyclic di-GMP levels and transition from sessility to motility. Mol Microbiol 53:1123–34
8. Christen M, Christen B, Folcher M, Schauerte A, Jenal U (2005) Identification and characterization of a cyclic di-GMP specific phosphodiesterase and its allosteric control by GTP. J Biol Chem 280:30829–37
9. Rao F, See RY, Zhang D, Toh DC, Ji Q, Liang ZX (2010) YybT is a signaling protein that contains a cyclic dinucleotide phosphodiesterase domain and a GGDEF domain with ATPase activity. J Biol Chem 285:473–482
10. Tamayo R, Tischler AD, Camilli A (2005) The EAL domain protein VieA is a cyclic diguanylate phosphodiesterase. J Biol Chem 280:33324–30
11. Szmidt-Jaworska A, Jaworska K, Kopcewicz J (2008) Involvement of cyclic GMP in phytochrome-controlled flowering of *pharbitis nil*. J Plant Physiol 165:858–867
12. Teng Y, Xu W, Ma M (2010) cGMP is required for seed germination in *Arabidopsis thaliana*. J Plant Physiol 167:885–889
13. Abel S, Nurnberger T, Ahnert V, Krauss G, Glund K (2000) Induction of an extracellular cyclic nucleotide phosphodiesterase as an accessory ribonucleolytic activity during phosphate starvation of cultured tomato cells. Plant Physiol 122:543–552
14. Chen J, Yoshida T, Bitensky MW (2008) Light-induced translocation of cyclic-GMP phosphodiesterase on rod disc membranes in rat retina. Mol Vis 14:2509–2517
15. Schenk T, Breel GJ, Koevoets P, van den Berg S, Hogenboom AC et al (2003) Screening of natural products extracts for the presence of phosphodiesterase inhibitors using liquid chromatography coupled online to parallel biochemical detection and chemical characterization. J Biomol Screen 8:421–429
16. Temkitthawon P, Viyoch J, Limpeanchob N, Pongamornkul W, Sirikul C, Kumpila A, Suwanborirux K, Ingkaninan K (2008) Screening for phosphodiesterase inhibitory activity of Thai medicinal plants. J Ethnopharmacol 119:214–217
17. Bar HP, Hechter O (1969) Adenyl cyclase assay in fat cell ghosts. Anal Biochem 29:476–489
18. Randerath K, Randerath E (1964) Ion-exchange chromatography of nucleotides on poly-(ethyleneimine)-cellulose thin layers. J Chromatogr 16:111–125
19. Riegel JA, Maddrell SH, Farndale RW, Caldwell FM (1998) Stimulation of fluid secretion of malpighian tubules of *Drosophila melanogaster* meig. by cyclic nucleotides of inosine, cytidine, thymidine and uridine. J Exp Biol 201:3411–3418
20. Bressan RA, Ross CW (1976) Attempts to detect cyclic adenosine 3′:5′-monophosphate in higher plants by three assay methods. Plant Physiol 57:29–37

Chapter 17

Detection of Reactive Oxygen Species Downstream of Cyclic Nucleotide Signals in Plants

Robin K. Walker and Gerald A. Berkowitz

Abstract

Cyclic nucleotides act in plant cell signal transduction cascades by activating cyclic nucleotide gated cation-conducting ion channels (CNGCs). Activation of CNGCs results in inward cation (including Ca^{2+}) conductance across the plasma membrane. Elevation of cytosolic Ca^{2+} is an early step in numerous plant cell signal transduction cascades, including plant immune responses to pathogens. CNGC involvement, along with cyclic nucleotides cAMP and cGMP, in pathogen defense programs is one relatively well-studied area of cyclic nucleotide signaling in plants. During plant immune responses, CNGC-dependent Ca^{2+} elevations lead to a signaling cascade that results in the generation of defense molecules such as hydrogen peroxide and nitric oxide, and induction of defense gene expression. This pathogen defense response is discussed, and methods to detect some of the downstream signaling steps in the pathway are presented.

Key words Calcium signaling, cAMP, cGMP, Cyclic nucleotide gated channel, Diaminobenzidine, Nitric oxide, Pathogen associated molecular pattern, Reactive oxygen species

1 Introduction

Cyclic nucleotide (cAMP or cGMP) activated (nonselective) cation channels (CNGCs) comprise a large family of cation conducting channels in plants. Of the 56 coding sequences identified at present as cation conducting channels in the *Arabidopsis thaliana* genome, 20 are members of the CNGC family [1]. CNGCs have been functionally characterized by expression in heterologous systems, or by analysis of cation-related phenotypes of mutant plants (typically Arabidopsis) that have specific CNGC genes silenced [2, 3]. Functional analyses of members of this channel family have associated many of them with inward K^+ and Ca^{2+} currents, and at least in several cases, Na^+ conductance [2]. CNGC function in plant biology may be more related to their ability to conduct Ca^{2+} rather than monovalent cations into plant cells; this Ca^{2+} conductance has been shown to be activated by elevation of cytosolic levels of cyclic

Chris Gehring (ed.), *Cyclic Nucleotide Signaling in Plants: Methods and Protocols*, Methods in Molecular Biology, vol. 1016, DOI 10.1007/978-1-62703-441-8_17, © Springer Science+Business Media New York 2013

nucleotides [4]. In contrast to animals, plants lack genes encoding cyclic nucleotide-activated protein kinases [5]; therefore signaling downstream from the generation of the secondary messenger molecules cAMP or cGMP is thought to be predominantly (or exclusively) facilitated by CNGCs in plants. Ca^{2+} signaling in plants through CNGCs has been implicated in plant immune responses to pathogens [6]. The involvement of CNGCs in immune signaling cascades is one of the most well-studied areas of cyclic nucleotide signaling in plants.

Plant cells face assault by many potentially pathogenic microorganisms during their life cycle. It should be noted that in contrast to animal cells, plants are typically sessile, they cannot run or hide in response to an external assault and they do not have the mobile macrophage sentry system of animals. Therefore, plants have developed their own complex multilayered immune system that allows for each individual cell to recognize pathogens as "nonself" and transduce such perception to downstream defense responses. One way in which plant (and animal) cells perceive the presence of a pathogen is by recognition of evolutionarily conserved components of pathogenic microbes that are required for some essential function of the microbe; a pathogen associated molecular pattern (PAMP).

Similar to animal cells, plant cells also utilize Ca^{2+} as a pivotal secondary messenger in numerous signaling cascades. Ca^{2+} signaling has been shown to play an essential physiological role in plant cell response to abiotic and biotic stresses. A Ca^{2+} elevation in the cytosol has been known to be an important early event in plant cell perception of pathogen infection and leads to the activation of downstream immune responses [6]. However, much is still unknown about the molecular events that link pathogen signal perception to the cytosolic Ca^{2+} rise and what gene products are responsible for Ca^{2+} entry into the cytosol as an early event in this signal transduction cascade.

Prior studies from this lab [4, 7, 8] provide the basis for a model involving cyclic nucleotide elevation occurring upon pathogen perception by plant cells, and the change in the level of this cytosolic secondary messenger initiates a pathogen response signaling cascade that has, as an early step, an elevation of cytosolic Ca^{2+}. Previous studies showing a mutation of Arabidopsis CNGC2 (in the *dnd1* mutant) results in impaired cAMP- and cGMP-dependent cytosolic Ca^{2+} elevation [4, 7, 8]. Generation of the signaling molecule nitric oxide (NO) in response to the PAMP lipopolysaccharide (LPS) is impaired in *dnd1* cells as well. Moreover, LPS-induced NO synthesis in plant cells can be quenched by the Ca^{2+} channel blocker Gd^{3+} [4]. NO is an important downstream component of pathogen defense programs evoked upon pathogen recognition [9]. Pathogen induced cAMP elevation in plant cells can be blocked by the application of an adenylyl cyclase (AC) inhibitor. The addition of an AC inhibitor can also lead to abolishment of pathogen induced Ca^{2+} elevation in the cytosol and also block steps of the pathogen response signaling cascade downstream from the Ca^{2+} elevation, such as generation of NO as well as other antimicrobial and signaling molecules such as H_2O_2 (and other reactive oxygen species (ROS)). It is thought that ROS and NO may work synergistically in antimicrobial action and in pathogen defense signaling cascades leading to increased expression of pathogen defense genes [10] and the balance between the two may be required for pathogen-induced programmed cell death, i.e., the hypersensitive response (HR) [11].

ROS has been well characterized as a signaling molecule associated with HR occurring in response to pathogen infection [12]. Respiratory burst NADPH oxidases have been linked to the production of ROS [13, 14], and enhanced HR. In all biotic stress responses, ROS is thought to occur in two phases, first as a rapid, transient, low amplitude oxidative burst generating within minutes of pathogen perception; the second burst occurs as a sustained phase with a higher amplitude [15]. ROS has various roles in biotic stress responses. During pathogen attack, ROS strengthens the cell wall, triggers the onset of systemic signaling, fosters vesicle trafficking, activates the MAP kinase pathway, as well as indirectly kills bacteria/pathogens through pH changes and the activation of proteases, along with many other responses [16].

Generation of a Ca^{2+} signal in the cytosol of a plant cell under assault by a pathogen is required for the production of the antimicrobial and immune signaling molecules NO and ROS [4, 16]. Plant immune signal cascades downstream from the Ca^{2+} signal also lead to transcriptional reprogramming [8]. As discussed above, cyclic nucleotide activation of CNGCs, leading to the generation of a Ca^{2+} signal is one way in which plant cell immune responses in the cytosol are initiated. Thus, cyclic nucleotide signaling can be studied by monitoring ROS or NO generation, or changes in the expression of specific pathogen-responsive genes.

2 Materials

2.1 Chemiluminescent Assay for H_2O_2 Production in Leaf Disks

1. Stock Solution: 25 mM $FeSO_4$, 25 mM $(NH_4)_2SO_4$, 2.5 M H_2SO_4 (50 mL of 10 N H_2SO_4 150 mL) diH_2O in a final volume of 200 mL.
2. Xylenol Orange Solution: 1 mL of Stock Solution in 100 mL of 125 μM Xylenol Orange and 100 mM sorbitol made up to a final volume of 101 mL.
3. Modulators/Inhibitors: *N*-[6-aminohexenyl]-5-chloronaphthalenesulfonamide (W7, used at 50 μM; W7 is a CaM antagonist that affects CNGC activation-W7 prevents CaM inactivation

of CNGCs activated by cyclic nucleotides), dibutryl (db)-cAMP (1 mM, lipophylic analog of cAMP that activates CNGCs), forskolin (50 μM, leads to cAMP generation in both animal and plant cells), dedioxyadenosine ((DDA, 200 μM), AC inhibitor), *N*-(6-aminohexyl)-1-naphthalenesulfonamide (W5, 50 μM), inactive analog of W7 used as a negative control).

4. PAMP elicitor of pathogen defenses: Lipopolysaccharide (LPS; 100 μg/mL) dissolved in Arabidopsis cell culture medium.
5. Additional Materials: *Arabidopsis thaliana* cell culture medium, Erlenmayer flasks, rotating platform and a microcentrifuge.

2.2 Diaminobenzidine Leaf Staining

1. 3,3′-Diaminobenzidine (DAB) (Sigma-D-8001) solution: 1 mg DAB/1 mL diH_2O). Cover the tube with foil since the solution is light-sensitive and add 20 μL of hydrochloric acid to lower pH and dissolve the DAB.
2. Place the tube in a 42 °C water bath for 6–8 h (or overnight at room temperature and then a couple of hours at 42 °C prior to use), mix solution during the incubation process to solubilize DAB.

2.3 Real-Time Quantitative PCR (qRT-PCR) Analysis of Changes in Expression of Genes

1. Qiagen RNeasy Plant Mini kit
2. NanoDrop or standard spectrophotometer
3. Bio-Rad Real-Time PCR plates
4. Bio-Rad iScript Select cDNA synthesis kit
5. Bio-Rad IQ SYBR green Supermix
6. Bio-Rad IQ5 Real-time PCR detection system

3 Methods

3.1 In Vivo *ROS Measurement in (Arabidopsis) Leaf Guard Cells*

Intact guard cells can be prepared using leaf epidermal peels; the guard cell pairs forming each stomatal complex are left intact along with epidermal cells, while virtually all of the mesophyll cells of the leaf interior are destroyed. Floating the peels in solutions containing various compounds is a relatively straightforward individual cell-based assay that can be used to evaluate treatment effects on ROS generation. In this assay, ROS generation is monitored using a fluorescence dye reagent. Generation of ROS is monitored in intact guard cells by monitoring fluorescence occurring in guard cells using an epifluorescence microscope, using the filter typically supplied for monitoring fluorescence of cells expressing recombinant green fluorescent protein (GFP).

1. As mentioned above, NO is downstream from cyclic nucleotide signaling in pathways evoked during pathogen immune

responses in plants. There is also evidence that NO generation is involved in other cyclic nucleotide signaling in plants, such as senescence programming [17]. The method described here is suitable for monitoring ROS in guard cells and can be adapted for a cell-based assay of NO (*see* **Note 1**). Make 2 mm epidermal peels from detached leaves from the whole plant.

2. Proceed to peel the lower epidermal layer away from the upper epidermis (a sharp pair of forceps can be used) while maintaining intact guard cells (forming the stomatal pore) and epidermal cells.
3. Place the epidermal peels in ROS loading buffer (10 mM KCl, 25 mM MES-KOH, and 0.1 mM $CaCl_2$, at pH 6.15) for 1–2 h to stabilize tissue from wounding effects.
4. Fresh dye stock solution of 25 mM 2′,7′-dichlorodihydrofluorescein diacetate (H_2DCF-DA*, D399, Invitrogen, Inc.) should be made and stored at −20 °C (*see* **Note 2**). The dye should be added at a working concentration of 50 μM to the ROS loading buffer containing the peels and incubated for 30 min (in petite petri dishes).
5. After dye loading, the peels should be washed three times (3 petite petri dishes with 3 mL of ROS loading buffer only).
6. The peels are placed in their respective treatments for 5–10 min.
7. Mount the peels on a glass slide with 40 μL of ROS loading buffer containing additions for a specific treatment. Place coverslip on slide and view under the microscope at 600× magnification (bright field and GFP wavelength-recommended for the specific ROS dye).

3.2 Chemiluminescent Assay for H_2O_2 Production in Leaf Disks

This assay allows for measurement of ROS production in leaf tissue (including mesophyll cells of the leaf interior) responding to treatment compounds added to the solution in which leaf disks are incubated.

1. Generate leaf disks (from ~1–2 month old leaves if using Arabidopsis) approximately 1 mm wide and float overnight in 200 μL diH_2O in 96 black-well assay plates.
2. Remove the water and replace it with fresh Luminol solution (0.15 mL supplied with 20 μM Luminol (Sigma Aldrich) and 1 μg horseradish peroxidase (Fluka).
3. Add your desired treatment into the wells. Immediately place your 96-well plate into a microplate reader with a luminometer function (e.g., FLUOstar Optima microplate reader, BMG Labtech; Cary, NC) (*see* **Note 3**).
4. Collect luminescence signals for 30 min.

3.3 Xylenol Orange-Dependent H_2O_2 Production Assay in Arabidopsis Cells

This colorometirc assay has been used to monitor ROS production from cultured leaf cells. Cultured Arabidopsis cells are exposed to various compounds that affect CNGCs. ROS is measured as a change in solution absorbance at a specific wavelength using a spectrophotometer.

1. Take 20 mL of cell suspension and place into a 125 mL Erlenmeyer flask. Place flask on a rotating platform at 180 rpm.
2. Begin adding modulators/inhibitors to flask and incubate for a specified length of time depending on modulator utilized (*see* **Note 2**). After incubation periods with modulators/inhibitors add 20 μL of LPS to flask. Generate a time course curve to determine optimal time to measure H_2O_2 production after elicitation.
3. Take 125 mL flask with cells, modulators, and elicitor and take a 300 μL aliquot (4–5 replications for each treatment) and place into a microcentrifuge tube. Centrifuge the 300 μL aliquot for 15 s at maximum speed.
4. Take 100 μL of the supernatant and mix with 1 mL of xylenol orange solution (*see* **Note 3**). Incubate for 30 min. Then read absorbance, or optical density (OD) at a wavelength of 560 nm (OD_{560}).
5. Spectrophotometric measurement assay use tubes and the blank is 100 μL Arabidopsis cell culture media and 1 mL of Xylenol Orange solution and the sample volume for the measurements is 1.1 mL.

3.4 Diamino-benzidine Staining in Leaves

This method allows for qualitative evaluation of ROS generation in intact leaves by visual observation of a ROS-dependent precipitate. This assay can monitor, for example, ROS generation in leaves inoculated with pathogens, and is adapted from Torres et al. [15].

1. After administering treatments to leaves, vacuum-infiltrate the DAB solution (*see* **Note 4**) into treated leaves.
2. Place 50 mL of DAB solution (*see* **Note 5**) in a 200 mL beaker, add treated leaves, place the container holding the leaf in the DAB solution under a vacuum for 1 min, and then dry leaves with soft tissue paper.
3. Place the vacuumed, dried leaves in a sealable plastic box for 5–6 h.
4. Transfer the leaves to a 50 mL tube containing a fixative and clearing solution (3:1:1 (v/v/v) of ethanol: lactic acid: glycerol).
5. When chlorophyll is completely removed transfer leaves to a 60 % glycerol tube to keep for an extended period of time if needed.

6. Mount leaves on slides to visualize the brown precipitate formed inside of leaf tissue as a measurement of ROS production under the microscope. Photographs can capture the treatment-induced differences in precipitate formation in regions of leaves subjected to treatments.

3.5 Real-Time Quantitative PCR (qRT-PCR) Analysis of Changes in Expression of Genes

1. Extract total RNA from plant tissue using, e.g., a Qiagen RNeasy Plant Mini Kit.
2. Quantify the RNA in your tissue extracts with a "NanoDrop" or conventional spectrophotometer, measuring absorbance at 260 nm. An OD_{260} of 1 corresponds to an RNA concentration of 40 μg/mL. The quality of the RNA preparation can be estimated by also measuring OD_{280}. An OD_{260}/OD_{280} ration of 1.8–2.0 indicates optimal quality.
3. Generate cDNA from your RNA (use 1 μg of your RNA) with the Bio-Rad iScript Select cDNA synthesis kit. This will yield 1 μg of cDNA. Follow the protocol provided by the supplier.
4. Set-up your standards using your control cDNA. Make a 20 ng/μL dilution of your control cDNA and serial dilutions of 10, 5, and 2.5 ng/μL. Use diH_2O as a 0 ng/μL standard.
5. Use 10 ng/μL of your cDNA for all samples in a 96-well qRT-PCR plate.
6. Then use Bio-Rad IQ SYBR green Supermix to make master mixes of desired primers, the supermix, and diH_2O. Follow the instructions provided by the supplier.
7. Real-time PCR plates are loaded into a Bio-Rad IQ5 Real-time PCR detection system.

4 Notes

1. This assay of NO uses the fluorescence dye diaminofluorescein diacetate. This guard cell-based assay of NO is similar to that described above in Subheading 3.1 for ROS and details of the NO assay method can be found in Ali et al. [4].
2. In the in vivo ROS measurement in leaf guard cells the *H_2DCF-DA stock solution only lasts 1 month. If inhibitors are used then one should incubate the epidermal peels with them 15 min prior to, and then along with treatments.
3. Chemiluminescent assay for H_2O_2 production in leaf disks need to be performed at least in triplicate. For W7 and forskolin incubate 5 min, for db-cAMP incubate for 30 min (*see* Subheading 2.1, **items 3** and **4** for description of activators/inhibitors of cyclic nucleotide signaling).

4. In the Xylenol Orange-dependent H_2O_2 production assay in Arabidopsis cells, the solution will turn a brown-yellowish color indicating it is ready for use. Do not use solution once it turns red.
5. The DAB staining in leaves requires that the solutions must be made fresh daily because they only last 6–8 h.

References

1. Ward JM, Mäser P, Schroeder JI (2009) Plant ion channels: gene families, physiology, and functional genomics analyses. Annu Rev Physiol 71:59–82
2. Ma W, Yoshioka K, Gehring CA, Berkowitz GA (2010) The function of cyclic nucleotide gated channels in biotic stress. In: Demidchik V, Maathuis FJM (eds) Ion channels and plant stress responses. Springer, Berlin, pp 159–174
3. Talke IN, Blaudez D, Maathuis FJ, Sanders D (2003) CNGCs: prime targets of plant cyclic nucleotide signalling? Trends Plant Sci 8:286–293
4. Ali R, Ma W, Lemtiri-Chlieh F, Tsaltas D, Leng Q, von Bodman S, Berkowitz GA (2007) Death don't have no mercy and neither does calcium: Arabidopsis cyclic nucleotide gated channel2 and innate immunity. Plant Cell 19:1081–1095
5. Kaplan B, Sherman T, Fromm H (2007) Cyclic nucleotide-gated channels in plants. FEBS Lett 581:2237–2246
6. Dangl JL, Dietrich RA, Richberg MH (1996) Death don't have no mercy: cell death programs in plant-microbe interactions. Plant Cell 8:1793–1807
7. Ma W, Qi Z, Smigel A, Walker RK, Verma R, Berkowitz GA (2009) Leaf senescence signaling: the Ca^{2+}-conducting Arabidopsis cyclic nucleotide gated channel2 acts through nitric oxide to repress senescence programming. Proc Natl Acad Sci U S A 106:20995–21000
8. Qi Z, Verma R, Gehring C, Yamaguchi Y, Zhao Y, Ryan CA, Berkowitz GA (2010) Ca^{2+} signaling by plant Arabidopsis thaliana Pep peptides depends on AtPepR1, a receptor with guanylyl cyclase activity, and cGMP-activated Ca^{2+} channels. Proc Natl Acad Sci U S A 107:21193–21198
9. Dangl J (1998) Innate immunity. Plants just say NO to pathogens. Nature 394:525–527
10. Neil S, Barros R, Bright J, Desikan R, Hancock J, Harrison J et al (2008) Nitric oxide, stomatal closure, and abiotic stress. J Exp Bot 59:165–176
11. Delledonne M, Zeier J, Marocco A, Lamb C (2001) Signal interactions between nitric oxide and reactive oxygen intermediates in the plant hypersensitive disease resistance response. Proc Natl Acad Sci U S A 98:13454–13459
12. Bolwell GP (1999) Role of active oxygen species and NO in plant defence responses. Curr Opin Plant Biol 2:287–294
13. Jabs T (1999) Reactive oxygen intermediates as mediators of programmed cell death in plants and animals. Biochem Pharmacol 57:231–245
14. Torres MA, Dangl JL, Jones JD (2002) Arabidopsis gp91phox homologues AtrbohD and AtrbohF are required for accumulation of reactive oxygen intermediates in the plant defense response. Proc Natl Acad Sci U S A 99:517–522
15. Torres MA, Jones JD, Dangl JL (2006) Reactive oxygen species signaling in response to pathogens. Plant Physiol 141:373–378
16. Torres MA (2010) ROS in biotic interactions. Physiol Plant 138:414–429
17. Ma W, Smigel A, Walker RK, Moeder W, Yoshioka K, Berkowitz GA (2010) Leaf senescence signaling: Ca^{2+} accumulation mediated by Arabidopsis cyclic nucleotide gated channel2 acts through nitric oxide to repress senescence programming. Plant Physiol 154:733–743

Chapter 18

Measurement of Nitric Oxide in Plant Tissue Using Difluorofluorescein and Oxyhemoglobin

Ndiko Ludidi

Abstract

Nitric oxide (NO) is now well established as a signalling molecule in plants, regulating various physiological processes ranging from development to responses to pathogens and changes in the physical environment. Various methods for the detection of NO in plant tissue have been described, and all of these methods have serious limitations that impact their utility for accurate detection of NO in plant tissues. Despite such limitations, both difluorofluorescein diacetate and oxyhemoglobin present convenient and relatively easy approaches for measuring NO in plant tissue and their utility can be enhanced by including appropriate controls to address some of the limitations that these two methods have. This chapter provides methods for measuring or detecting NO production in plant tissue using either difluorofluorescein diacetate or oxyhemoglobin.

Key words 4-Amino-5-methylamino-2,7-difluorofluorescein diacetate, 2-(4-Carboxyphenyl)-4,4,5,5-tetramethylimidazoline-L-oxyl-3-oxide, Cyclic guanosine monophosphate, Nitric oxide, Oxyhemoglobin, *N*-ω-nitro-L-arginine (L-NNA)

1 Introduction

Endogenously synthesized NO serves as a vital signalling molecule that regulates various plant physiological processes [1–3]. Some of the signalling functions of NO are transduced via NO-dependent induction of cyclic guanosine monophosphate (cGMP) biosynthesis by NO-responsive guanylate cyclases [4–6]. Because of its importance in the regulation of plant physiology, various methods have been developed for the detection and quantification of NO in plant tissue [7]. Diaminofluorescein fluorescent dyes have been used extensively for detecting NO in plant tissues, whereas an oxyhemoglobin-based assay, which presents a fairly sensitive and inexpensive method for NO detection, remains rather underutilized in plant science research.

The relative ease, low cost, and sensitivity of the two methods, together with their reproducibility and requirement of only a few

Chris Gehring (ed.), *Cyclic Nucleotide Signaling in Plants: Methods and Protocols*, Methods in Molecular Biology, vol. 1016, DOI 10.1007/978-1-62703-441-8_18,

specialized pieces of equipment that can be used for various other purposes, make 4-amino-5-methylamino-2,7-difluorofluorescein diacetate (DAF-FM DA) (which we refer to simply as difluorofluorescein diacetate) and oxyhemoglobin the preferred reagents in methods for measuring plant NO content in many laboratories. Furthermore, DAF-FM DA is more photo-stable and less sensitive to pH variations than other diaminofluorescein fluorescent dyes used for NO detection in plants. We have used the two methods successfully for various plant tissue types ranging from roots [8] to root nodules [9].

2 Materials

DAF-FM DA must be stored desiccated at −20 °C, at which the solution is stable for up to a year. Solutions of catalase and superoxide dismutase must be stored at 4 °C where they are stable for at least 6 months. Hemoglobin must be stored at 4 °C and freshly prepared just before use.

2.1 Reagents and Equipment for the DAF-FM DA Method

1. DAF-FM DA loading buffer: 10 mM 3-(*N*-Morpholino)propanesulfonic acid sodium salt (MOPS sodium salt), pH 7.0, 10 μM 4-amino-5-methylamino-2,7-difluorofluorescein diacetate (DAF-FM DA). This must be prepared fresh every day of use and kept in the dark.
2. DAF-FM DA wash buffer: 10 mM 3-(*N*-Morpholino)propanesulfonic acid sodium salt (MOPS sodium salt), pH 7.0.
3. Standard microscope slides and coverslips.
4. Confocal laser scanning microscope (e.g., we use the Zeiss Axiovert 200 M LSM 510 META confocal microscope).
5. Vibratome (e.g., we use the Leica VT1200 S vibrating blade microtome).
6. Software for analyzing pixel intensity (e.g., AlphaEase FC imaging software, Alpha Innotech Corporation).

2.2 Reagents and Equipment for the Oxyhemoglobin Method

1. Hemoglobin (we use lyophilized hemoglobin from bovine blood).
2. Catalase (we use catalase from *Corynebacterium glutamicum*, supplied as a suspension at ~500,000 Units/mL.
3. Superoxide dismutase (e.g., superoxide dismutase from bovine liver supplied as a suspension at 2,000–6,000 Units/mg protein).
4. Sodium hyposulfite ($Na_2S_2O_4$).
5. Sephadex G-25 beads.
6. Potassium phosphate buffer, pH 7.4 (50 mM): Mix 40.1 mL of K_2HPO_4 with 9.9 mL of KH_2PO_4 and make up the volume

to 1 L with distilled H_2O. Check that the pH is 7.4, otherwise adjust it to 7.4 using 0.5 M KOH or concentrated HCl.

7. Extraction buffer: 0.1 M sodium acetate, 1 M NaCl, 1 % (w/v) ascorbate, pH 6.0. Adjust pH with acetic acid and NaOH.
8. Spectrophotometer capable of kinetic readings at wavelengths between 401 and 421 nm.

3 Methods

All procedures are carried out at room temperature unless specified otherwise.

3.1 Detection of Nitric Oxide in Plant Tissue Using Difluorofluorescein Diacetate

1. Obtain fresh tissue (e.g., leaves, roots, etc.) using tweezers and scalpel blades to make sections of the plant tissue.
2. Use quick-setting superglue to secure the plant tissue onto the sectioning platform of the vibratome (Note that thin leaves do not need to be sectioned on a vibratome and would in fact not be suitable to attachment to the vibratome's sectioning platform with superglue, hence these leaves can be stained directly with DAF-FM diacetate without sectioning).
3. Once the superglue is dry, apply a few drops of DAF-FM DA wash buffer onto the plant tissue to prevent it from drying.
4. Section the plant tissue to produce 100–200 μm sections using the vibratome (normally, standard razor blades perform well for the sectioning on the vibratome).
5. Transfer the tissue sections to DAF-FM DA loading buffer (1 mL loading buffer in 1.5 mL microcentrifuge tubes for small sections or 10 mL loading buffer in Petri dishes for large sections that would not fit into 1.5 mL microcentrifuge tubes) (*see* **Note 1**).
6. Incubate these samples at room temperature in the dark for 30 min.
7. Remove the DAF-FM DA loading buffer from the tissue and add an equal volume of DAF-FM DA wash buffer and incubate the samples in the dark for 10 min to wash off unbound DAF-FM DA.
8. Repeat the washing step once more, and then transfer the tissue section onto a microscope slide.
9. Add a few drops of DAF-FM DA wash buffer to the sample on the slide just sufficient to cover the sample and to prevent bubbles when a coverslip is placed onto the specimen, and then place a coverslip on the specimen.
10. Set-up the confocal laser scanning microscope such that the excitation wavelength from the argon laser is 480–490 nm and emission is 520–530 nm.

11. Collect green fluorescence (as a result of reaction of DAF-FM-2T with NO) images and analyze green pixel intensities using appropriate image analysis software (Fig. 1).

3.2 Measurement of Plant Nitric Oxide Content Using the Oxyhemoglobin Method

1. Add 5 g of Sephadex G-25 beads to 25 mL of potassium phosphate buffer in a 100 mL beaker and allow the matrix to swell overnight at 4 °C.
2. Carefully transfer the matrix into a chromatography column and let the liquid flow through until only just enough liquid is left in the column to just cover the matrix.
3. Wash the matrix twice with 4 mL of potassium phosphate buffer.
4. Gently dissolve 25 mg of hemoglobin in 1 mL of 50 mM potassium phosphate buffer in a 100 mL bottle.
5. Add 1.5 mg of sodium hyposulfite and mix the solutions. Continue adding the sodium hyposulfite until the solution turns dark purple (maximum 2.5 mg of sodium hyposulfite).
6. Aerate the solution by blowing air into the solution using a 2–5 mL pipette for 5 min, and then incubate the solution at room temperature for 30 min (at this stage the solution should turn from dark purple to light red, indicating conversion from methemoglobin to oxyhemoglobin).
7. Transfer the oxyhemoglobin into the Sephadex column to desalt and purify the oxyhemoglobin.
8. Collect 4–5 fractions (200–250 μL each) of the oxyhemoglobin onto Eppendorf/microcentrifuge tubes by passing the solution through the column.
9. Dilute the fractions 20-fold and determine the absorbance of each fraction at 415 nm. The fraction containing oxyhemoglobin will be the one with the highest absorbance at 415 nm.
10. Calculate the concentration of the oxyhemoglobin fraction by using the extinction coefficient of 131 mM^{-1} cm^{-1} for oxyhemoglobin at 415 nm.
11. Obtain tissue extracts by grinding 100 mg of plant tissue into a fine powder in liquid nitrogen and transferring the powder into 300 μL of Extraction Buffer in a microcentrifuge tube.
12. After vortexing the mixture for 30 s, spin the sample for 20 min at 10,000 × *g*.
13. Transfer the supernatant to a clean tube and add 200 Units of catalase and 200 Units of superoxide dismutase (*see* **Note 2**).
14. Mix the sample by mild vortexing for 10 s and incubate the reaction for 10 min.
15. Add freshly prepared oxyhemoglobin to a final concentration of 10 μM. Mix by gentle vortexing for 10 s.

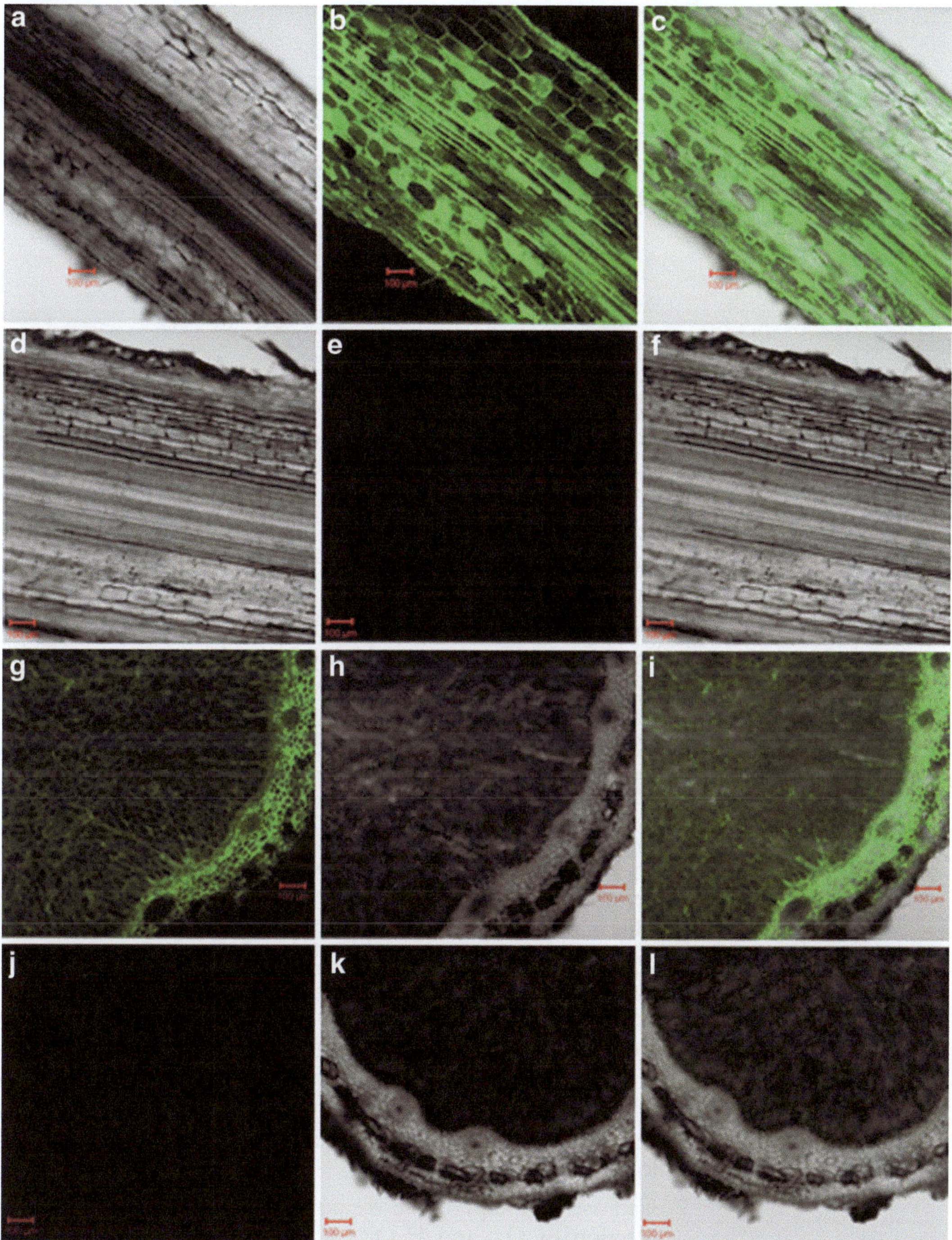

Fig. 1 Typical results obtained for NO detection using DAF-FM DA in maize roots [(**a**)–(**e**), *see* ref. 8] and soybean root nodules [(**g**)–(**l**), *see* ref. 9)]. (**a**), (**d**), (**h**), and (**k**) are bright-field images; (**b**), (**e**), (**g**), and (**j**) are fluorescence images, whereas (**c**), (**f**), (**i**), and (**l**) are overlay images of the bright-field and fluorescence images. (**d**), (**e**), and (**f**) represent the control in which no DAF-FM DA was added (DAF-FM DA loading buffer lacking DAF-FM DA), whereas (**j**), (**k**), and (**l**) represent controls where 1 mM L-NNA was added before addition of the DAF-FM DA loading buffer

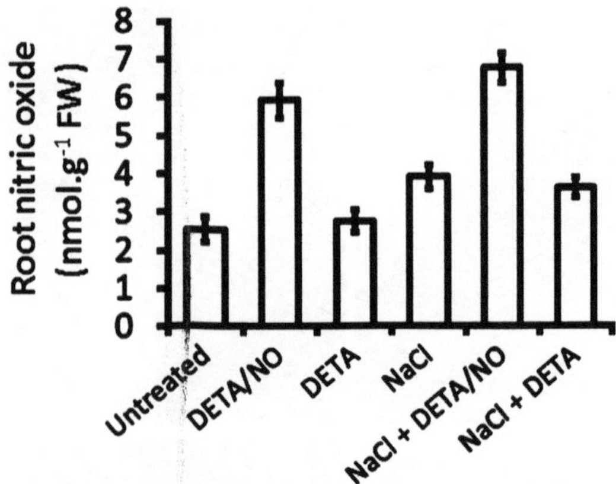

Fig. 2 Typical results obtained for NO detection using the oxyhemoglobin methods in maize roots (*see* ref. 8). DETA/NO is a nitric oxide donor, whereas DETA is a control for DETA/NO as it is chemically analogous to DETA/NO but lacks the NO group, and thus DETA does not release NO

16. Incubate the reaction at room temperature for 10 min.
17. Measure absorbance of the reaction mixture at 401 and 421 nm, and then calculate NO concentration by subtracting the absorbance at 421 nm from the absorbance at 401 nm, and then dividing the difference by 77 mM^{-1} cm^{-1} (Fig. 2) (*see* **Note 3**).

4 Notes

1. It is important that appropriate controls for DAF-FM DA staining are included in the experiment to ensure that the fluorescence detected is a result of reaction of NO with the fluorophore. This is most effectively done by pretreating the tissue sections (for 15 min) with the NO scavenger 2-(4-carboxyphenyl)-4,4,5,5-tetramethylimidazoline-L-oxyl-3-oxide (carboxyPTIO) at a final concentration of 400 μM or, in cases where nitric oxide synthase-like activity is thought to be the source of the NO, pretreating the tissue sections (for 30 min) with the nitric oxide synthase inhibitor *N*-ω-nitro-L-arginine (L-NNA) at a final concentration of 1 mM before adding the DAF-FM DA loading buffer.
2. Catalase and superoxide dismutase are added to prevent the reaction of O_2^- and H_2O_2 with NO, which would interfere with accurate measurement of NO content.
3. The calculation yields NO content as nM, which can be translated to nmol/g^{-1} of tissue fresh weight (calculated back to the 100 mg used in the assay) or nmol/mg^{-1} of protein content in the tissue (predetermined from the amount of tissue used, based on a Bradford assay performed on the tissue).

References

1. Baudouin E (2011) The language of nitric oxide signalling. Plant Biol 13:233–242
2. Besson-Bard A, Pugin A, Wendehenne D (2008) New insights into nitric oxide signalling in plants. Ann Rev Plant Biol 59:21–39
3. Corpas FJ, Barroso JB, Carreras A, Quiros M, Leon AM et al (2004) Cellular and subcellular localization of endogenous nitric oxide in young and senescent pea plants. Plant Physiol 136:2722–2733
4. Bai X, Todd CD, Desikan R, Yang Y, Hu X (2012) N-3-oxo-decanoyl-L-homoserine-lactone activates auxin-induced adventitious root formation via hydrogen peroxide- and nitric oxide-dependent cyclic GMP signaling in mung bean. Plant Physiol 158:725–736
5. Wilson ID, Neill SJ, Hancock JT (2008) Nitric oxide synthesis and signalling in plants. Plant Cell Environ 31:622–631
6. Ederli L, Meier S, Borgogni A, Reale L, Ferranti F, Gehring C, Pasqualini S (2008) cGMP in ozone and NO dependent responses. Plant Signal Behav 3:36–37
7. Mur LA, Mandon J, Cristescu SM, Harren FJ, Prats C (2011) Methods of nitric oxide detection in plants: a commentary. Plant Sci 181:509–519
8. Keyster M, Klein A, Ludidi N (2012) Caspase-like enzymatic activity and the ascorbate-glutathione cycle participate in salt tolerance of maize conferred by exogenously applied nitric oxide. Plant Signal Behav 6:349–360
9. Keyster M, Klein A, Ludidi N (2010) Endogenous NO levels regulate nodule functioning: potential role of cGMP in nodule functioning? Plant Signal Behav 5:1679–1681

Chapter 19

Infrared Gas Analysis Technique for the Study of the Regulation of Photosynthetic Responses

Alex Valentine, Oziniel Ruzvidzo, Aleysia Kleinert, Yun Kang, and Vagner Bennedito

Abstract

Homeostatic maintenance of physiological and biochemical processes is a key requirement for survival and adaptive responses of multicellular organisms such as plants. These important processes are in part mediated by various plant enzymes and hormones, many of which are in part, controlled by cyclic nucleotides and/or other signalling molecules. Infrared gas analysis (IRGA) technique is one of the modern methods which allows for rapid and accurate measurements of cyclic nucleotide mediated photosynthetic responses to plant hormones, and thus makes it a powerful and useful tool to study aspects of downstream cell signalling events in plants. In this chapter the basic protocols enabling the use of the IRGA technique to study signalling molecules, such as cyclic nucleotides on photosynthetic responses, are outlined.

Key words Cyclic nucleotides, Cyclic 3′, 5′-guanosine monophosphate, Plant natriuretic peptide, Infrared gas analysis, Photosynthesis, Leaf gas exchange

1 Introduction

Maintenance of growth, development, and physiological processes in general is a key requirement for the survival of multicellular organisms such as plants and animals [1]. Enzymes and hormones produced by these organisms play important roles in such processes [1]. In plants, the action of a hormone involves its perception and the initiation of a specific response pathway that may be independent of transcriptional or translational control, and/or a more sustained response that requires the hormone-dependent regulation of transcription [2, 3]. In many cases, the early responses evoked within a few minutes are at the level of ion fluxes and generation of second messengers such as calcium ions [4], cyclic nucleotides [5], cytosolic pH [6], reactive oxygen species [3], nitric oxide [7], and lipids [8]. Besides mediating the responses, these second messengers may also amplify the signal and sustain the generated response [1, 2].

Chris Gehring (ed.), *Cyclic Nucleotide Signaling in Plants: Methods and Protocols*, Methods in Molecular Biology, vol. 1016, DOI 10.1007/978-1-62703-441-8_19,

Plant natriuretic peptides (PNPs) are one class of plant hormonal molecules that elicit a number of physiological responses important for homeostasis and growth [9]. Several of these molecules have since been immunoaffinity purified from a number of different species [10–12] and an *Arabidopsis thaliana* PNP (AtPNP-A; At2g18660) has been cloned and partially characterized [13]. Exogenous application of recombinant PNPs or their native forms to various plant tissues promotes a number of physiological responses, which include stomatal guard cell movements [14], net water uptake [11], tissue specific ion movements [15], photosynthesis, transpiration [16], and leaf dark respiration [17]. The induction of some of these responses by PNPs has been shown to be closely associated with cytosolic transients in cGMP levels [11, 13, 14, 18], thus implicating a cyclic nucleotide-dependent signalling system for this class of plant hormones.

Infrared gas analysis (IRGA) is a modern, powerful technique used for the rapid and accurate detection of physiological responses in whole plant leaves to various factors such as hormones, enzymes, chemical and physical environmental factors [19]. The technique utilizes infrared analysis to measure parts-per-million fluxes in CO_2 between leaves and the external atmosphere, under controlled environmental conditions [19]. The method is based on the principle that hetero-atomic molecules such as CO_2 can absorb infrared radiation in specific infrared wavelengths [19]. This method has been extensively used in exploring the various biophysical and biochemical processes of leaf photosynthetic gas exchange [17, 19] whose responses are mediated by various second messengers including the cyclic nucleotides. For this reason, the IRGA can be applied to undertake quantitative measurements of various cyclic nucleotide-dependent leaf responses such as photosynthesis, respiration, transpiration as well as the stomatal physiology. To date, the use of the IRGA in cyclic nucleotide research has enabled scientists to gain significant insight into leaf physiology and function. In this chapter, we detail specific protocols for the application of this technique for the rapid and convenient detection and quantification of physiological parameters in response to various signalling elicitors, e.g., cyclic nucleotides.

2 Materials

1. Portable infrared gas analyzer (LC-pro+, ADC Bioscientific Ltd, Herts, England; LI-6400, LI-COR Biosciences, Lincoln, Nebraska, USA), with the capabilities to conduct photosynthetic light response assays and photosynthetic CO_2 response assays.
2. Cultivated plant species (for example, *Plectranthus eklonii* Benth.), with sufficient broad leaves to fit into and fill the leaf

chamber of the chosen infrared gas analyzer model. Ensure that the youngest, fully expanded leaves are used for these measurements.

3. 100 μg/ml solution of the recombinant PNP fusion protein (*see* refs. 13, 17).
4. 100 μM cyclic 3′,5′-guanosine monophosphate (cGMP) solution. Weigh 0.03 mg of cGMP sodium salt into an Eppendorf and dissolve completely in deionized water. Store at −20 °C (*see* **Note 1**).
5. Sterile deionized water.

3 Methods

1. Germinate plant seeds (for example, *Plectranthus ecklonii* Benth. seeds) (*see* **Note 2**) in potting soil inside an atmospherically controlled greenhouse with average daylight conditions of 570–650 μmol/s/m^2, average day/night temperatures of 25/16 °C, average day/night humidity of 30/70 %, and a CO_2 concentration which is similar to current atmospheric levels (approximately 400 ppm). Propagate the seedlings under similar conditions for 8 weeks.
2. Collect three plants during day time (09:00 to 16:00 h) and place them in a laboratory that has shady (2.0–4.0 μmol photons/m^2 s) and room temperature (24 °C) conditions.
3. On each plant, select three pairs of the youngest, fully expanded leaves and label them X, Y, and Z.
4. For all subsequent assaying procedures, always treat leaf X with the sterile deionized water (negative control leaf), leaf Y with the recombinant PNP solution (experimental leaf), and leaf Z with the cyclic nucleotide solution (positive control leaf) (*see* **Note 1**).
5. Do the respective leaf treatments individually by applying 25 μl of the solution to the abaxial surface and another 25 μl of same solution to the adaxial surface and evenly distributing the liquids across both surfaces through a gentle spreading with the long flat side of a sterile pipette tip (Fig. 1).
6. Immediately (within 30 s), enclose the treated leaf into the leaf chamber of a portable infrared gas analyzer (*see* Fig. 1), which can be set to keep all other environmental factors at ambient states, except for light intensity or CO_2 concentration that should be varied for conducting either the photosynthetic light response assays or photosynthetic CO_2 response assays, respectively.
7. For the light response assays, vary the light intensity between 0 and 1,600 μmol/photons/m^2 s 100 μmol photons/m^2 s intervals and at each interval, record values for the photosynthetic rate,

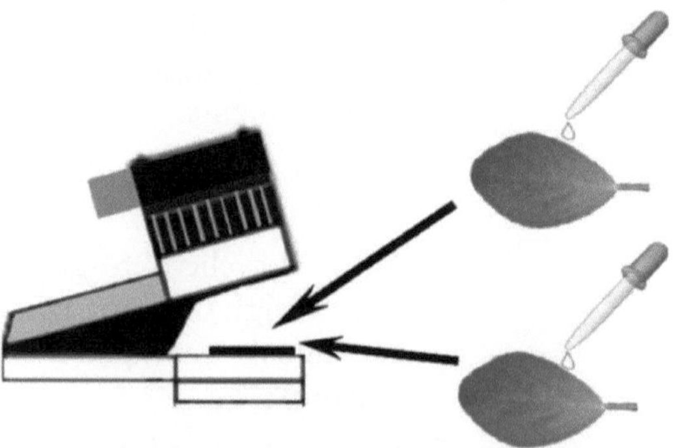

Fig. 1 Application of H_2O, PNP or cyclic nucleotide solution to the abaxial and adaxial surfaces of an attached leaf. Prior to leaf photosynthesis (light or CO_2) measurements, the treatment solution should be applied very carefully on the underside (abaxial) of the leaf where the highest concentration of stomata can be found, and then on the topside (adaxial) of the leaf surface. Ensure that the application time and exposure to the solution remain constant between the three leaves. Once inside the leaf chamber of the infrared gas analyzer, the treated leaf can be used for photosynthetic light response or CO_2 response curves

transpiration rate as well as the stomatal conductance. Use the photosynthetic values to construct a light response curve, with photosynthetic rate on the *y*-axis and photosynthetically active radiation on the *x*-axis (Fig. 2) (*see* **Note 3**).

8. For the CO_2 response assays, use a fresh set of leaves similarly treated with the three respective solutions and as outlined in **steps 4** and **5** above. For each leaf, use the light level at which maximum saturating photosynthetic rate was obtained, and vary the CO_2 concentrations between 50 and 2,000 ppm, at 100–150 ppm intervals, ensuring that the ambient CO_2 concentrations (approximately 400 ppm) are covered in this response assay. To construct a photosynthetic CO_2 response curve, use the photosynthetic values with photosynthetic rate on the *y*-axis and sub-stomatal (internal leaf) CO_2 concentration on the *x*-axis (Fig. 3) (*see* **Note 4**).
9. From the photosynthetic light response curve, calculate and derive the following associated parameters: light saturating photosynthetic rate (P_{max}), light compensation point (LCP), apparent quantum yield (θ), dark respiration (Dr), stomatal conductance (Gs), transpiration (E), and photosynthetic water use efficiency (PWUE) (*see* **Notes 5–11**).
10. From the CO_2 response curve, calculate and derive the following associated parameters, whose responses are also cyclic nucleotide dependent: rubisco carboxylation (Vc_{max}), percentage stomatal limitation of photosynthesis (Ls), and electron transport rate (J_{max}) (*see* **Notes 12–14**).

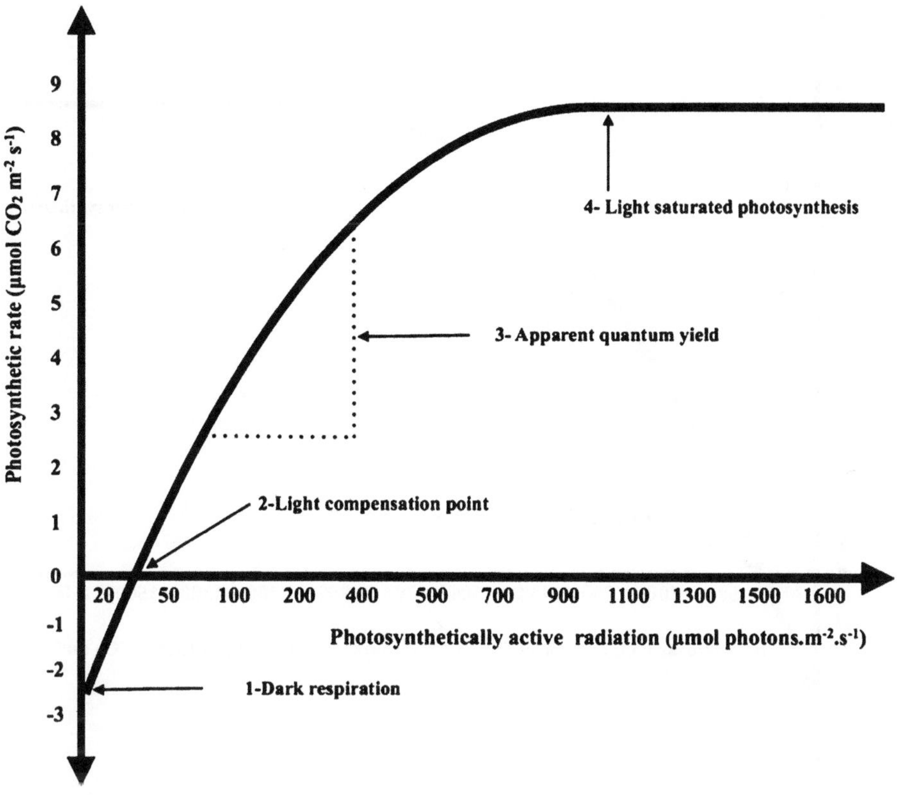

Fig. 2 Photosynthetic light response curve. The response of photosynthetic rate to increasing irradiances of photosynthetically active radiation (PAR), measured at a constant CO_2 concentration. The light level should be varied with 11–13 increments in the range of 0–1,600 μmol photons/m^2 s on the infrared gas analyzer. At each light level, the photosynthetic rate should be recorded once the values have stabilized (usually between 6 and 10 min, but may take longer depending on the plant species)

4 Notes

1. Experiments can be equally performed using other related cyclic nucleotide signalling molecules such as 100 μM cyclic 3′,5′-adenosine monophosphate (cAMP), 100 μM 8-bromo-3′,5′-cyclic guanosine monophosphate (8-Br-cGMP), 100 μM 8-bromo-3′,5′-cyclic adenosine monophosphate (8-Br-cAMP). All solutions should be prepared by dissolving the cyclic nucleotide salts into sterile deionized water and keeping them at −20 °C.
2. The *Plectranthus ecklonii* plant was used for the undertaking of the physiological assays in this study because it is a species that has previously been used to test responses to native and recombinant PNPs from different species [16, 17]. However, any other higher plant species with sufficient broad leaves to fit into and fill the infrared gas analyzer leaf chamber can still be similarly used.

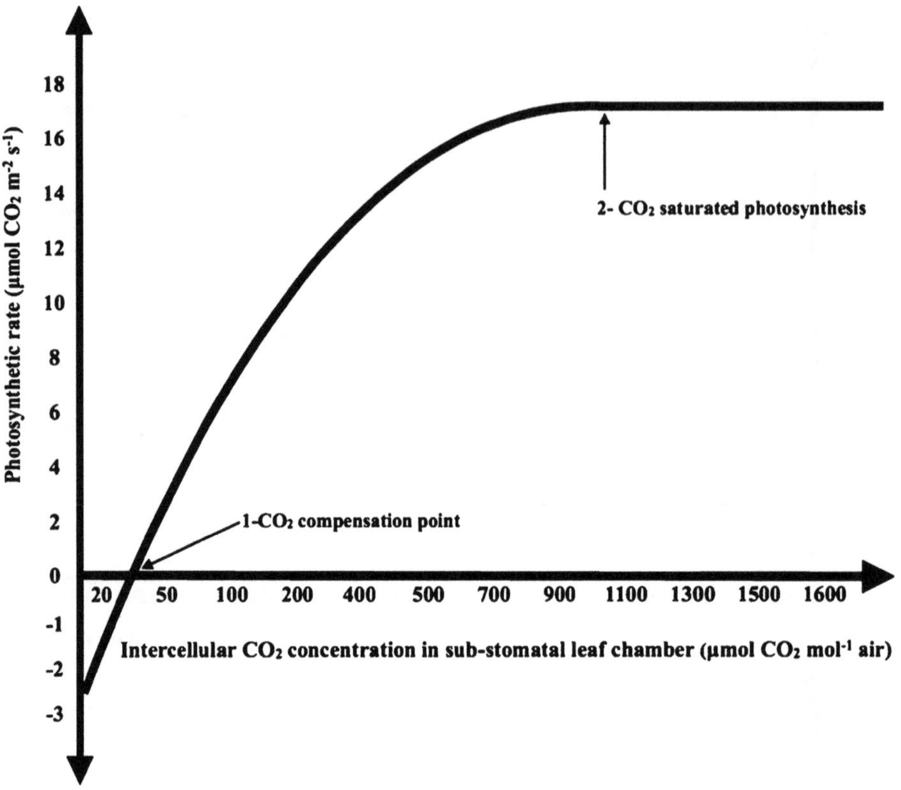

Fig. 3 Photosynthetic CO_2 response curve. The response of photosynthetic rate to increasing CO_2 concentrations in the sub-stomatal leaf chamber, measured at a constant light saturation level of the PAR (from the light response curve Fig. 2). The external CO_2 level should be varied with 11–13 increments in the range of 0–2,000 μmol CO_2/m^2 s on the infrared gas analyzer. At each CO_2 level, the photosynthetic rate should be recorded once the values have stabilized (usually between 6 and 10 min, but may take longer depending on the plant species)

3. In this work, we noted that treatment of plant leaves with either the recombinant PNP or cyclic nucleotides resulted in rapid and significant increases in stomatal conductance and hence stomatal opening occurred [16, 17]. Further, the effects were light intensity dependent and highly pronounced at higher irradiances. Apparently, these increases in stomatal conductance caused by either the recombinant PNP or the cyclic nucleotides also concurred with the observed higher leaf transpiration rates [16, 17]. Once again, these increases also strongly correspond to the PNP and/or cyclic nucleotide dependent increases in leaf photosynthetic rates [16, 17]. In addition, the overall efficiency of light utilization during the photosynthetic CO_2 fixation was apparently enhanced, as is evidenced by the higher water use efficiencies and apparent photon yields, while leaf dark respiration rates were somewhat four- to eightfold higher after the application of PNP and/or cyclic nucleotides [16, 17].

4. The CO_2 response curve can be used to calculate the contribution of rubisco carboxylation and electron transport capacity in the chloroplasts to the underlying photosynthetic processes [20]. Further, the CO_2 response curves can also be used to calculate the carboxylation-limited rubisco activity (Vc_{max}) and electron transport capacity (J_{max}), using the equations of Von Caemmerer and Farquhar [20]. The percentage stomatal limitation of photosynthesis can be obtained from the CO_2 response curve using the equations of Farquhar and Sharkey [21].
5. Light saturating photosynthetic rate (P_{max})—is the photosynthetic rate when CO_2 assimilation starts to level off at increasing photon flux densities until it reaches a saturation point, where after further increases in photon flux densities do not lead to an increase in the photosynthetic rate. At this point, light is no longer the rate limiting step but rather other factors like rubisco activity or triose phosphate metabolism.
6. Light compensation point (LCP)—is the photon flux density level at which the amount of CO_2 released by mitochondrial respiration is balanced by the CO_2 fixed via photosynthesis, i.e., the net CO_2 exchange is zero.
7. Apparent quantum yield (θ)—is the efficiency of CO_2 fixation per unit of light reaction product. This is the slope of the initial limiting portion of the photosynthetic light response curve.
8. Dark respiration (Dr)—is the respiration occurring in plants irrespective of the presence of light to differentiate it from photorespiration that occurs in the presence of light.
9. Stomatal conductance (Gs)—is a measurement of the flux of water and carbon dioxide through the stomata, in and out of the leaf. This is a measurement of the opening/aperture of the stomatal pores that facilitate the free diffusion of gases from and into the leaf.
10. Transpiration (E)—is evaporation of water from the surface of leaves and stems. The ratio of transpiration is defined as the ratio of water lost to carbon gained during photosynthesis. Transpiration ratio gives an indication of the effectiveness of plants in managing water loss through the stomata while allowing sufficient uptake of CO_2 for photosynthesis.
11. Photosynthetic water use efficiency (PWUE)—is the ratio of photosynthesis to transpiration and is a reflection of the moles of CO_2 fixed by photosynthesis per mole of H_2O lost via transpiration. Higher values indicate increased water-use efficiencies.
12. Rubisco carboxylation (Vc_{max})—the velocity of rubisco carboxylation is a reflection of the first committed step of the dark reaction of photosynthesis. The carboxylation of the CO_2 acceptor molecule, ribulose 1,5-bisphosphate, is the first committed enzymatic step of the Calvin-Benson Cycle.

13. Electron transport rate (J_{max})—is a reflection of the maximum rate of photosynthetic electron transport under light saturation during the light reaction of photosynthesis. The products of the light reaction are ATP and NADPH.
14. Percentage stomatal limitation of photosynthesis (Ls)—the calculated estimation of the degree by which the stomata are limiting the photosynthetic rate. This may be of particular importance in the application of PNP and/or cyclic nucleotides, and their subsequent effects on photosynthesis and its other related biochemical responses.

Acknowledgment

This material is based upon work supported financially by the National Research Foundation, South Africa.

References

1. Johri MM (2008) Hormonal regulation in green plant lineage families. Physiol Mol Biol Plants 14:23–38
2. Vogler H, Kuhlemeier C (2003) Simple hormones but complex signalling. Curr Opin Plant Biol 6:51–56
3. Apel K, Hirt H (2004) Reactive oxygen species: metabolism, oxidative stress, and signal transduction. Ann Rev Plant Biol 55:373–399
4. Sanders D, Pelloux J, Brownlee C, Harper JF (2002) Calcium at the crossroads of signaling. Plant Cell 14(Suppl):S401–S417
5. Newton RP, Smith CJ (2004) Cyclic nucleotides. Phytochemistry 65:2423–2437
6. Felle HH (2001) pH: signal and messenger in plant cells. Plant Biol 3:577–591
7. Besson-Bard A, Pugin A, Wendehenne D (2008) New insights into nitric oxide signaling in plants. Ann Rev Plant Biol 59:21–39
8. Meijer HJG, Munnik T (2003) Phospholipid-based signaling in plants. Ann Rev Plant Biol 54:265–306
9. Gehring C, Irving H (2003) Natriuretic peptides—a class of heterologous molecules in plants. Int J Biochem Cell Biol 35:1318–1322
10. Billington T, Pharmawati M, Gehring CA (1997) Isolation and immunoaffinity purification of biologically active plant natriuretic peptide. Biochem Biophys Res Commun 235:722–725
11. Maryani MM, Bradley G, Cahill D, Gehring C (2001) Natriuretic peptides and immunoreactants modify osmoticum-dependent volume changes in *Solanum tuberosum* L. mesophyll cell protoplasts. Plant Sci 161:443–452
12. Rafudeen S, Gxaba G, Makgoke G, Bradley G, Pironcheva G et al (2003) A role for plant natriuretic peptide immuno-analogues in NaCl- and drought-stress responses. Physiol Plant 119:554–562
13. Morse M, Pironcheva G, Gehring C (2004) AtPNP-A is a systemically mobile natriuretic peptide immunoanalogue with a role in *Arabidopsis thaliana* cell volume regulation. FEBS Lett 556:99–103
14. Pharmawati M, Maryani MM, Nikolakopoulos T, Gehring CA, Irving HR (2001) Cyclic GMP modulates stomatal opening induced by natriuretic peptides and immunoreactive analogues. Plant Physiol Biochem 39:385–394
15. Ludidi N, Morse M, Sayed M, Wherrett T, Shabala S, Gehring C (2004) A recombinant plant natriuretic peptide causes rapid and spatially differentiated K^+, Na^+ and H^+ flux changes in *Arabidopsis thaliana* roots. Plant Cell Physiol 45:1093–1098
16. Gottig N, Garavaglia BS, Daurelio LD, Valentine A, Gehring C, Orellano EG et al (2008) *Xanthomonas axonopodis* pv. *citri* uses a plant natriuretic peptide-like protein to modify host homeostasis. Proc Natl Acad Sci U S A 105:18631–18636
17. Ruzvidzo O, Donaldson L, Valentine A, Gehring C (2011) The *Arabidopsis thaliana* natriuretic peptide AtPNP-A is a systemic regulator of leaf dark respiration and signals via the phloem. J Plant Physiol 168:1710–1714

18. Wang YH, Gehring C, Cahill DM, Irving HR (2007) Plant natriuretic peptide active site determination and effects on cGMP and cell volume regulation. Funct Plant Biol 34:645–653
19. Hunt S (2003) Measurements of photosynthesis and respiration in plants. Physiol Plant 117: 314–325
20. Caemmerer S, Farquhar GD (1981) Some relationships between the biochemistry of photosynthesis and the gas exchange of leaves. Planta 153:376–387
21. Farquhar GD, Sharkey TD (1982) Stomatal conductance and photosynthesis. Ann Rev Plant Physiol 33:317–345

INDEX

Chris Gehring (ed.), *Cyclic Nucleotide Signaling in Plants: Methods and Protocols*, Methods in Molecular Biology, vol. 1016, DOI 10.1007/978-1-62703-441-8, © Springer Science+Business Media New York 2013

R

S

T

V

W

X

Y

MIX
Papier aus verantwortungsvollen Quellen
Paper from responsible sources
FSC® C105338

If you have any concerns about our products,
you can contact us on
ProductSafety@springernature.com

In case Publisher is established outside the EU,
the EU authorized representative is:
Springer Nature Customer Service Center GmbH
Europaplatz 3, 69115 Heidelberg, Germany

Printed by Libri Plureos GmbH
in Hamburg, Germany